TREE DISEASE CONCEPTS

PAUL D. MANION

State University of New York,
College of Environmental Science and Forestry
Syracuse, New York

Prentice-Hall, Inc., Englewood Cliffs, New Jersey 07632

Library of Congress Cataloging in Publication Data

Manion, Paul D.
Tree disease concepts.

Includes bibliographies and index.
1. Trees—Diseases and pests. 2. Trees—Wounds and injuries. I. Title.
SB761.M22. 634.9'63 81-2477
ISBN 0-13-930701-X AACR2

TO MY WIFE, NANCY,
AND MY SONS, WILL AND ED

Printed in the United States of America

10 9 8 7

Editorial/production supervision
by Zita de Schauensee
Cover design by Edsal Enterprises
Manufacturing buyer: John Hall

Prentice-Hall International, Inc., *London*
Prentice-Hall of Australia Pty. Limited, *Sydney*
Prentice-Hall of Canada, Ltd., *Toronto*
Prentice-Hall of India Private Limited, *New Delhi*
Prentice-Hall of Japan, Inc., *Tokyo*
Prentice-Hall of Southeast Asia Pte. Ltd., *Singapore*
Whitehall Books Limited, Wellington, *New Zealand*

CONTENTS

PREFACE

The forest pathology profession began about a century ago when Robert Hartig published his observations on the interrelationship of fruit bodies or conks on the outside of trees and fungus hyphae causing decay in the wood of the tree. Up until the late 1940s, there were a very limited number of forest pathologists in the world, but they contributed the major share of our present understanding of the causes of various diseases of trees. The forest pathology text written by John Shaw Boyce in 1938, and updated in 1948 and 1961, is a testimonial to the industriousness of early forest pathologists.

The tone for forest pathology set by the early workers centered on the cause and control of specific disease problems. Teaching of forest pathology and texts in forest pathology have emphasized these aspects. The student is generally exposed to a massive array of diseases. He or she is expected to identify the disease, causal agent, and be able to remember the control measure. If one looks at the list of control measures, it becomes obvious that we do not have control measures for diseased plants but rather recommendations on how to prevent the problem in the first place. Therefore, the student of forest pathology has been traditionally trained to write proper epitaphs for diseased trees with little emphasis on the basic biological understanding of representative disease systems, the ecological role of disease, and the economic interaction of disease in forest and urban management systems.

Rather than write an updated treatise cataloging diseases of trees for the specialist, I have attempted to write an introductory book which emphasizes the biological understanding, ecological considerations, and interactions of diseases of forest and urban trees with management practices for a limited number of selected diseases. These are intended to represent the types of tree disease problems that one might encounter.

The disease "types" will be emphasized by presenting a reasonable approximation of the mode of action and disease cycle of selected diseases. The reasonable approximation is necessary because for many diseases we do not have available from research results the total picture on mode of action and disease cycles.

My primary purpose in writing this book was to provide an introductory treatment of concepts of tree diseases for students in the forest and shade tree pathology course in the College of Environmental Science and Forestry of the State University of New York. The course, which is suitable for students of varying background, is auto-tutorial, self-paced, and primarily self-directed. Because lectures are neither the primary source of information nor the pacesetter for the course, it is necessary to have a comprehensive source of basic information for the student to draw upon as needed.

A minimum of previous exposure to basic biology is sufficient to understand disease-related phenomena. Therefore, laypersons will also find these concepts meaningful in understanding the forces at work on urban and forest trees.

The thrust of the book is to present tree diseases in such a way that beginners can understand and categorize things they see around them. If the majority of beginners can be stimulated to be more critically observant of the role of disease in management practices as they relate to both urban and forest trees, my task will have been successfully completed. To the few who will eventually become professionals in forest pathology or related areas, I hope the initial stimulus will provide direction for the vast array of fascinating and potentially extremely important interactions among trees, microorganisms, and environments.

Although specific references are not cited in the text, the reference lists should make it readily evident where many of my concepts originate. These lists are by no means complete but will provide the interested reader with a place to begin a more comprehensive search of the literature.

The synthesis of a book of this type requires more than can be acquired through reading the literature. I am very much in debt to many of my associates for ideas, concepts, criticism, suggestions, and encouragement. If I were to properly acknowledge the many people who have contributed ideas that I have freely drawn upon, the list would begin with the professors and graduate students at the University of Minnesota, where I began my training in forestry and phytopathology. The list would continue through many of the participants of the national and northeastern regional meetings of the

American Phytopathological Society and Northeastern Forest Pathology Workshops, and further through many colleagues in various parts of the United States and Canada who have graciously hosted me as a visitor to their region.

I especially wish to acknowledge Dr. Harrison Morton, Dr. Robert Zabel, Dr. Edson Setliff, Dr. Wayne Sinclair, Dr. William MacDonald, and Dr. Gary Lahey for their assistance in reviewing drafts of this manuscript. I am also grateful for specific comments and criticism from my graduate students, in particular Dr. Robert Bruck, Dr. Patricia Gowen, Miss Anne Mycek, and Miss Barbara Schultz, who on many occasions have had to interpret what I have written for the undergraduate students in my courses. I would also like to gratefully acknowledge the typing assistance of Mrs. Julia Thomas, Mrs. Regina Carlin, and Miss Penny Weiman and the photographic assistance of Mr. George Snyder.

Syracuse, New York Paul D. Manion

1

INTRODUCTION TO TREE DISEASE CONCEPTS

- Historical perspective of plant diseases
- Forest pathology in relation to plant pathology
- Role of tree diseases in natural ecosystems
- Importance of tree diseases
- Disease in relation to other disorders of plants
- Symptoms of tree diseases as a reflection of disturbed physiological function
- Proof of pathogenicity
- Categorizing types of tree diseases
- Recognition of biotic, abiotic, and decline diseases
- Brief outline of the book

Plants interact with their environment and other organisms in a wide range of ways. The plants most fit to survive are in balance with their environment. In the short run, imbalance caused by the presence of disease agents may produce serious economic and ecologic effects. These are the concerns of the plant pathologist.

HISTORICAL PERSPECTIVE OF PLANT DISEASES

To properly understand plant pathology today, it is helpful to look back and see how plant pathology developed. The perspective of the past should improve our capacity to comprehend the present and may sharpen our ability to predict and influence the future.

Your immediate reaction to the idea of "predicting and influencing the future may be that all you want is a little understanding of the topic of tree diseases. But a little knowledge is dangerous. You will find or may already have found that people expect you to be conversant on a wide array of topics related to trees and plants. It is difficult to separate the casual conversation from real questions, but misinformation in either case may be costly. You will be asked to make judgments on problems and make suggestions on how to "control" the problem, which is really influencing the future.

Instead of starting this historical account at the beginning, let us start at the present and work backward.

1978–1900: Modern plant pathology In the United States today, there are approximately 3000 professional plant pathologists. What role do they play? They are engaged in a cooperative effort with other professionals in the agricultural and forestry fields to provide a stable supply of food and fiber for our modern industrialized society. Agricultural technology in the United States allows an average farmer to supply food for 80 people rather than just himself and his immediate family. One can appropriately question specific environment, societal, and economical aspects of the agricultural technology, but there should be no doubt in anyone's mind that the majority of the people in this country and a large number of people in other countries of the world are highly dependent upon the success of each year's food crop.

What role do plant pathologists play in the success of each year's crop? It is difficult to single out the role of plant pathologists when, in reality, an array of disciplines are integrated into a successful system. Plant pathologists are involved in the development of disease resistance, fungicidal controls, monitoring disease buildups, making recommendations on when and what to spray, predicting yields and losses, and a host of other aspects of agricultural production.

Rather than continue to elaborate in general terms, let us look back just a few years to 1970, when a major disease epidemic reduced corn production 15% nationwide. Losses due to the corn leaf blight caused by *Helminthosporium maydis* were as high as 50% in some states. Even more severe losses were predicted for 1971, but because of dryer weather conditions, the fungus disease epidemic never materialized. By 1972, new resistant varieties were available, thereby reducing the threat of major losses, at least for the time being.

Throughout most of this century, plant pathologists have been highly effective in reducing losses due to major epidemics. Today we rarely have a major disease epidemic. Twentieth-century plant pathologists can be credited with eliminating from agricultural production the merciless ravages of disease epidemics that have plagued humankind throughout recorded and unrecorded history.

Over the past 50 years the number of professional plant patholigists has increased from about 300 to 3000. During this period, viruses and mycoplasmas have become recognized as agents of plant disease. A wealth of information on diagnoses and control of plant diseases is available today to anyone who is interested, through extension specialists associated with state agencies and agricultural colleges.

1900–1850: Beginning of plant pathology During the latter half of the nineteenth century, modern plant pathology began. It was spawned out of the activities of three great scientists: a botanist, Anton deBary; a chemist, Justis Freiherr von Liebig; and a chemist-bacteriologist, Louis Pasteur.

A major scientific figure of the first half of the nineteenth century, Franz Unger, firmly established in scientific thinking the doctrine of spontaneous generation. This doctrine asserted that organisms spontaneously emerged from diseased or decomposing matter. Obviously, plant pathology, medical pathology, and microbiology could not emerge until it was recognized that fungi and bacteria were the cause rather than the product of the disease. It was a major uphill battle to disprove spontaneous generation and explain in chemical and biological terms what was really taking place during fermentation and decomposition of organic matter. Research conducted independently by deBary, von Liebig, Pasteur, and others eventually set the foundation for our present concepts of these processes.

In 1853, deBary demonstrated the causal nature of rust and smut fungi in diseases of cereals, and later identified *Phytophthora infestans* as the cause of potato late blight. Based on his contributions, deBary is credited with being the father of modern plant pathology.

Julius Kuhn, in 1858, published a text on plant pathology that incorporated the concepts of deBary on the causal nature of fungi and mycological concepts of early nineteenth-century mycologists.

1847–1845: Late blight epidemic Moving back just a bit further, to 1845–1847, we recognize the potato famine of Europe as a major stimulus for botanists and mycologists to apply themselves to problems associated with the economic welfare of mankind. The death by malnutrition of a million people, and the emigration of another million and a half, reduced the population of Ireland by one-half in a period of 5 years. The need for plant pathologists was clear. It took almost a century to develop the concepts and trained professionals to fill the need.

1800–1700: Classification period A botanist, Carolus Linnaeus, published a two-volume work, *Species Plantarum,* in 1853. These volumes were characteristic of the eighteenth century, a period of classification and taxonomy for botanical sciences. A number of other authors attempted disease classification. Diseases were named and classified more for the sake of classification than for the value to agriculture.

During the eighteenth century, occult influences that presumed diseases to be the wrath of angry gods still persisted, although there was some recognition of the effects of external environmental factors. During this period, the first pruning wound dressing was developed for fruit trees. This appears to be the origin of the idea that wound dressings do something for trees. This occult-fostered idea persists to this day.

1700–1600: Renaissance of earlier philosophical writings For plant pathology, the seventeenth century was a period of renaissance or revival of interest in the writings of early philosophers; there were few additional contributions or interpretations. This followed a period, A.D. 500 to 1600, the dark or middle ages, during which science and learning appeared to slumber.

A.D. 500–300 B.C.: Philosophers The writings of such philosophers as Theophrastus and Pliny during the period 300 B.C. to A.D. 500 were the source of concepts for the renaissance plant pathologists. These ancient philosophers observed, described, and speculated on the nature of diseases. It is interesting to note that Theophrastus (about 300 B.C.) recognized that wild trees were not liable to the ravages of disease, whereas cultivated plants were subject to an array of devastating diseases.

Beginning of recorded history Moving back to the beginning of recorded history, we find references to plant diseases in Greek and Hebrew writings. Various crop maladies, such as blighting, blasting, rust, mildews, and smuts, were assumed to result from the wrath of gods. Biblical references can be found in Gen. 41:23; 1 Kings 8:37; Deut. 28:22; Amos 4:9; Hag. 2:16, 17; and 2 Chron. 6:28.

There is every reason to assume that plant diseases have plagued human beings from the very beginnings of plant cultivation.

To summarize the development of plant pathology, we see a series of transitions from philosophical and occult interpretation to descriptive and taxonomic classification, to the application of scientific investigation to plant-related economic and social problems. The latter half of the nineteenth century saw the expansion of plant pathologists to most of the developed countries of Europe. The turn of the twentieth century saw a shift to North America and an explosion of plant pathological study and application to problems of agriculture.

FOREST PATHOLOGY IN RELATION TO PLANT PATHOLOGY

Forest pathology is the branch of plant pathology that deals with diseases of woody plants growing in natural forests, in plantations, and in urban environments. For historical and other reasons, the deterioration of forest products is often included in forest pathology. Some would exclude from forest pathology the woody plant diseases of commercial fruit and nut crops. The individual importance of these crops has resulted in a series of subdisciplines associated with the individual crops. In my opinion, it is important not to exclude aspects of fruit crops from the topic of forest pathology.

Few people are satisfied with the name "forest pathology," but none can suggest a better or more acceptable title for this profession, which embraces numerous disciplines.

Although there are about 3000 plant pathologists in the United States, there are just a few hundred forest pathologists in the world. Obviously, by number alone, their impact on societal needs is somewhat less apparent than the impact of plant pathologists on agriculture. In many ways, the development of forest pathology as a profession is about 75 to 100 years behind that of agricultural pathology. The major disease epidemics of this century such as chestnut blight, white pine blister rust, and Dutch elm disease have had a stimulating effect on the profession much as the late blight epidemic of potatoes had on agriculture over a century ago. Just like the agricultural pathologists of the nineteenth century, forest pathologists have a well-developed taxonomic classification of diseases but are just beginning to understand how to manipulate conditions to reduce the impact of disease.

The foregoing comments are in no way intended to belittle the forest pathology profession but rather to point out some historical similarities in the development of professional identities. In many ways the forest pathology profession is a contributing partner with agricultural pathology in advancing our present understanding of disease.

The forest pathology profession is still plagued with wound dressings as a disease-preventive measure. Some persons continue to utilize "blood-letting" activities such as the selective removal of diseased individuals for canker and decay problems, and "sugar-pill" therapy, such as the use of fertilizer and water for an array of diseases. The value of such occult-based activities is difficult to disprove; indeed, such methods may be warranted in certain cases. But a major task of present-day forest pathologists is to separate the witchcraft from the proven, scientifically based therapy.

ROLE OF TREE DISEASES IN NATURAL ECOSYSTEMS

The ancient philosopher Theophrastus recognized the difference between natural ecosystems and cultivated plants. Natural ecosystems were not subject to the destructive effects of diseases as were cultivated crops. I would suspect that something may have been lost in the translation and interpretation of the works of this observant philosopher, because if one critically observes natural ecosystems, one sees the destructive effects of disease. The same disease-causing organisms are present in the natural ecosystem, culling out weakened and less-fit individuals. The spread rate and overall visual impact of these organisms is tempered by the buffering effect of genetic and/or species diversity.

In the natural ecosystem, the pathogen and host populations have evolved a balanced relationship. The natural population of plants may have genetic diversity, age diversity, and species diversity. There is little or no selection pressure on pathogens to increase rapidly. Under this type of system, diseases play a role in eliminating less vigorous plants and in facilitating succession. Widespread lethal diseases are unknown because there is no way for a large population of susceptible plants to develop. The pathogen population would quickly shift to eliminate the development of such a population.

Therefore, a superficial look or the misinterpretation of a concept might cause an observer to assume that diseases are not as significant to a natural forest as to they are to a cultivated crop. But in actual practice, disease-causing factors are a dominant part of the balance.

IMPORTANCE OF TREE DISEASES

Evaluation of the importance of tree diseases is often based on evaluation of the lethal effects of diseases. A number of major diseases have produced extensive losses of this type. Syracuse, New York, which in 1950 had 53,000 elms along its streets, now has fewer than 300 as a result of the ravages of Dutch elm disease. The chestnut, once the major hardwood timber species in the eastern United States, has been reduced to a useless brush species by

TABLE 1-1 Annual Percent Growth Loss, Mortality, and Total Growth Impact Caused by Various Agents on the Supply of Saw Timber in the United States. (Data are for 1952 from "Timber Resources for America's Future," U.S. Forest Service, 1958)

	Percent of total loss (43.8 billion board feet)		
	Growth loss	Mortality	Growth impact
Disease	40	5	45
Insects	8	11	19
Fire	15	2	17
Weather	1	8	9
Other	7	3	10
Total	71	29	100

chestnut blight. Western white pine, one of the most valuable timber species of the northwestern United States, is so adversely affected by white pine blister rust that management objectives suggest discrimination against white pine in favor of any other species.

Although dramatic, these few examples of the importance of diseases do not begin to represent the total picture. In 1958, the U.S. Forest Service published "Timber Resources for America's Future," in which they attempted, in the forest protection section, to evaluate the roles of disease, insects, fire, animals, and weather on losses. They separated losses into mortality and growth loss, which added together were called growth impact. Table 1-1 shows that diseases rank as the most destructive agent. Approximately 45% of all losses in saw timber are attributable to diseases. The actual loss due to diseases is equivalent to about one-half of the annual amount cut.

One may question the specifics and applicability of these figures 28 years later. Times have changed. We no longer harvest vast quantities of old-growth western timber, so it may be heart rots have been reduced. Possibly other factors, such as root rots, have increased. Management of southern pines has changed the picture there also.

There is no way to get an overall picture of the problem today but I see no reason to assume that the overall effect of disease has lessened. In fact, with increased management we probably have more wood but also more losses due to disease.

DISEASE IN RELATION TO OTHER DISORDERS OF PLANTS

Disease is a disturbance in the normal physiologic functioning of a plant, has many causes, and exhibits an array of appearances and results. Plant pathologists do not agree on a precise definition of plant disease. A useful concept of disease should distinguish it from plant injury, from disease symptoms, and from disease incitants (pathogens). There is a major difference between the pathogen-host interaction of a rust fungus and a white pine tree compared to the interaction of a camper-wielded hatchet and the same tree. One is considered disease, the other injury. These are two extremes in a continuum of plant-incitant interactions that are damaging to the plant.

Disease will be defined here as any deviation in the normal functioning of a plant caused by some type of persistent agent.

How long must an agent persist in its interaction with a plant to cause disease? This is where the continuum comes in. The hatchet blow is a very short interaction. An air pollutant such as fluoride released suddenly in large amounts as a result of an industrial accident also causes injury. The same pollutant continuously released in small quantities as the result of an ongoing industrial process causes disease. The boundary line between when something causes injury only or results in disease is not particularly important, since the problems that fall into this area can be handled as specific cases. The more important point is to recognize that disease is generally caused by a persistent biotic or abiotic agent.

Any agent that causes disease is called a pathogen. As we shall see, pathogens may be either biotic agents such as fungi or abiotic agents such as air pollution. Some pathogens are parasites, but not all parasites are pathogens. Any organism that lives on and derives nutrients from another organism is a parasite. Only those parasites that cause a disruption in the normal physiological function of the host are called pathogens.

SYMPTOMS OF TREE DISEASES AS A REFLECTION OF DISTURBED PHYSIOLOGICAL FUNCTION

Disease symptoms resulting from the interaction of specific pathogens and hosts are characteristic signatures of the pathogen and host. The plant pathologist can often readily recognize the presence of a specific pathogen based on symptoms alone. Why are symptoms so characteristic?

The symptoms of diseases are expressions of disturbed or abnormal physiology of the host plant. The woody plant has evolved a complex structure to separate and yet tie together various functions necessary for competitive survival. In Fig. 1-1, an elementary understanding of the structure and function of the woody plant is superimposed on the diagrammatic tree. There is a division of function, and therefore a limit to the range of expression, which various parts of trees can produce in response to invasion by pathogens.

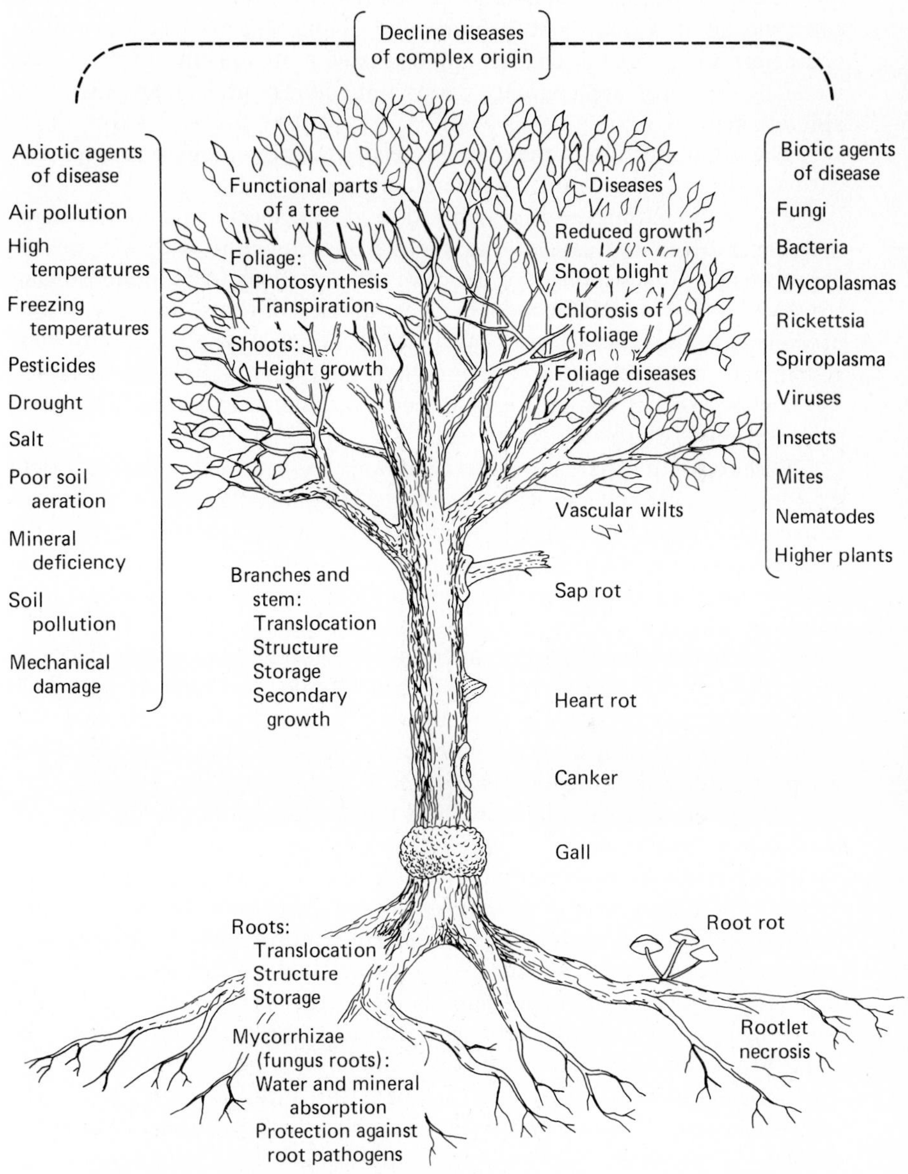

FIGURE 1-1 Summation of abiotic and biotic agents involved in diseases of trees, types of diseases, and functional parts of the tree. Decline diseases are caused by a combination of biotic and abiotic agents.

Along the right and left columns of the diagram are listed the various abiotic and biotic agents of disease. At the center top a third category of diseases, declines, is tied to both biotic and abiotic agents. Declines will be characterized later in this chapter and discussed more fully in Chapter 18. As we shall see, they are complex diseases involving interacting biotic and abiotic agents.

The center portion of the diagram relates the physiological functions of various parts of trees with the general categories of disease.

The biotic pathogens have evolved to fit into specific niches. A fungus that has evolved the capacity to survive by competing with soil microorganisms and the responses of tree roots is most likely to be found causing a disease of the roots. If we recognize the function of roots, we see why root problems produce characteristic symptoms of decay and necrosis of the root system and an overall appearance of mineral deficiency in the rest of the tree.

Pathogens that disrupt DNA-directed meristematic cell division result in cancerous-like growths called galls.

Pathogens that parasitize the cambium, phloem, and sapwood xylem cells for available sugars and other nutrients result in the death of the invaded area. Death of a localized stem area prevents secondary growth in the affected area. The bark may change color. A depressed area on the stem results from the lack of stem enlargement in the diseased area. As the stem is being completely girdled by the invasion of an aggressive canker fungus, the roots are the first remote part of the tree to deteriorate. The canker interrupts the production and maintenance of functional phloem. Therefore, transport of photosynthetic products of the leaves down to the roots is restricted, and eventually eliminated, as the canker enlarges. The roots deteriorate and water and mineral absorption is reduced.

Heart rot pathogens have evolved the capacity to utilize the cell-wall materials of woody plants, (i.e., cellulose and lignin). They differ from sap rot fungi in that the heart rot fungus is able to tolerate the dynamic chemical and morphological defense mechanisms of the living stem. The sap rot fungi do not compete well against the chemical and morphological defenses of a vigorous living stem. Therefore, they successfully invade dead or dying branches, large wounds, and eventually the dead or dying tree. The effects of decay or rot fungi are to weaken the structural integrity of the stem, roots, or branches.

Vascular wilt pathogens are adapted for survival in vessels of the sapwood xylem. Disruption of xylem vessels by wilt pathogens reduces the capacity of the vessels to translocate water from the roots to the top of the transpiring tree. During hot, dry periods, insufficient water is translocated to the leaves, causing them to wilt and die.

Foliage diseases affect the photosynthetic activity of trees. Viruses induce subtle color changes such as mottling and chlorosis, as well as other morphological and metabolic abnormalities. Obligate parasites such as rust and mildew fungi disrupt photosynthetic activity without causing serious mortality of leaves. Other fungi and bacteria cause necrosis of invaded portions, thereby reducing the effective area of the leaf. Abiotic toxicants, including salt and heavy metals from the soil, pesticides, and air pollutants, accumulate in leaves, disrupting or reducing photosynthetic activity.

Chlorosis of foliage may result from the direct effects of biotic and abiotic factors on leaves or the indirect effects of biotic and abiotic factors on roots. The most common symptom of mineral deficiency in plants is chlorosis.

Shoot blight is caused by microorganisms that aggressively parasitize succulent, rapidly growing shoots. These fungi and bacteria are also foliage and canker pathogens. They may gain access to shoots through infection of foliage, and may persist as stem cankers at the base of the infected shoot. The effect of shoot blight on young seedlings is more pronounced because killing the terminal shoot may destroy a great deal of the aboveground portion of the plant. As trees get larger, the killing of a shoot or shoots induces lateral buds to take over and compete for dominance as the new leader. A bushy-crowned tree may result from the inability of one lateral to gain dominance, or from the successive deaths of new leaders.

Reduced growth may occur as a consequence of the effects of any one or a combination of the problems discussed above. Reduced growth is a characteristic symptom of declines and may result from the interaction of at least three factors. Reduced growth may be the only aboveground symptom of some destruction of the root system. But reduced growth is a very subjective symptom, which may not be caused by disease agents. One must keep in mind that the capacity to grow is a combination of the genetic makeup of the tree, environmental effects on those genes, and possible pathogens.

Although the profile of a tree has been emphasized in developing this introduction to disease symptoms as a reflection of disturbed physiological functions, it is appropriate also to think of the functions of a tree in cross section. The tree stem is a complex structure consisting of (1) inner xylem (heartwood), functioning basically for structure; (2) outer xylem (sapwood), for storage and translocation of water; (3) cambium, as the meristematic layer of cells which, by mitotic division, produces xylem cells on one side and phloem cells on the other; (4) phloem, as a region where photosynthetic products, produced in the leaves, are translocated down to the stem and roots; and (5) bark, as a protective envelope of dead cells surrounding the living cells and providing a physical as well as a chemical barrier to invasion by microorganisms. This is a rather simplified characterization of the stem cross section, but it gives a framework on which one can impose the activity and effects of various diseases of the stem and branches.

This quick survey of disease as a reflection of disturbed physiologic function is meant simply as an overview. Subsequent chapters will provide details regarding the interaction of pathogens and hosts.

PROOF OF PATHOGENICITY

Proof of the pathogenicity of specific biotic agents has generally been accomplished by the following set of procedures originally proposed by Robert Koch (1843–1901). The modified procedures are as folows:

1. There must be constant association of the suspected causal agent and the disease.
2. The suspected causal agent must be isolated and grown in a pure culture.
3. When inoculated into healthy plants, the agent that has been isolated must induce the disease.
4. Re-isolation from the disease-induced plants must yield the same causal agent.

Certain modifications of the procedures are necessary for specific types of disease agents that cannot be cultured—viruses, nematodes, mycoplasms, and some fungi. These are covered later in the appropriate sections.

CATEGORIZING TYPES OF TREE DISEASES

Biotic Plant Diseases

Biotic plant disease is the product of three interacting factors: the plant, the pathogen, and the environment (Fig. 1-2). It is important to recognize that all three factors interact to produce diseases and that we may therefore prevent or control diseases by manipulating any one of the three.

By tradition, plant pathology is concerned with all diseases of plants except those caused by insects. This is a very artificial separation of an important group of pathogens. In actual practice, a useful plant pathologist must also recognize and understand insect-plant interactions.

Abiotic Plant Diseases

Diseases can also be caused by abiotic agents such as high or low temperature, phytotoxic gases, nutritional imbalance, soil-oxygen deficiency, moisture stress, and other abiotic factors. Abiotic diseases are sometimes very similar to injury, so that separation of disease from injury is often more academic than practical.

Decline Plant Diseases

Major emphasis in plant pathology has been directed toward single biotic or abiotic primary-causal-agent diseases. There is a third category of diseases,

FIGURE 1-2 Biotic plant disease is the product of three interacting factors.

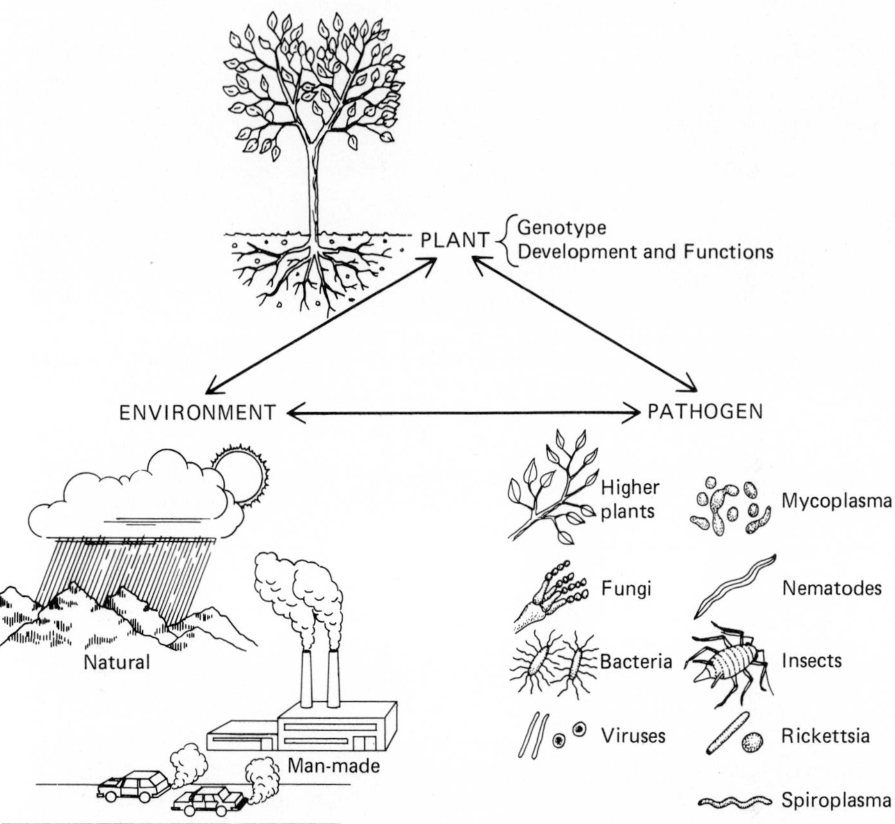

called decline, which results not from a single causal agent but from an interacting set of factors (Fig. 1-3). Terms that denote the symptom syndrome, such as dieback and blight, are commonly used to identify these diseases. One must be cautious, though, because these are not terms used exclusively for declines.

RECOGNITION OF BIOTIC, ABIOTIC, AND DECLINE DISEASES

The reasons for separating disease into three groups will become more evident once an understanding of each of the three types has been developed, but a few generalizations may be helpful at this point. If we shift from the theoretical concept that three types of disease exist to the practical question of how a person involved in the management of trees recognizes the three types, the concept should be more understandable.

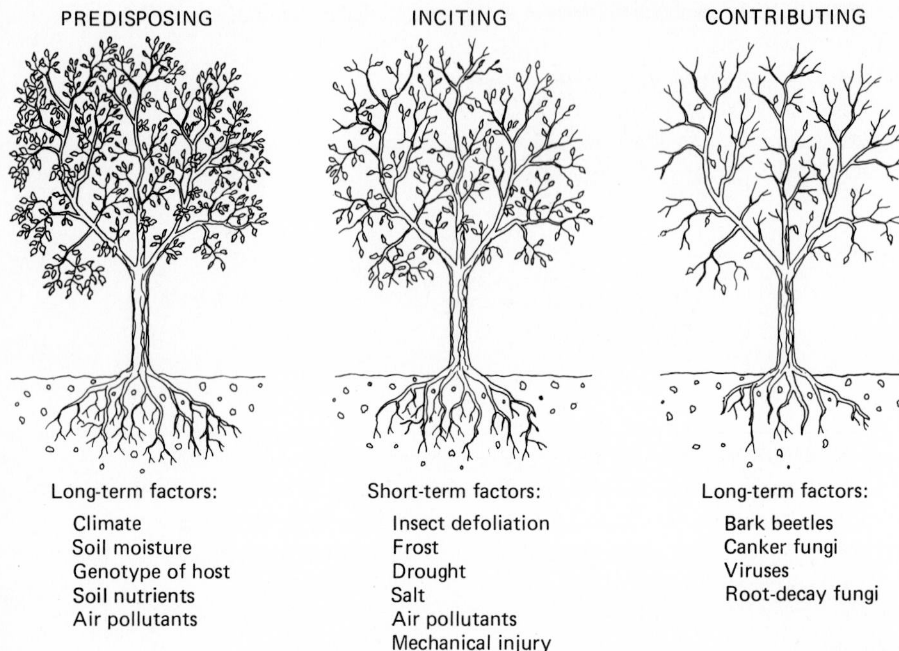

FIGURE 1-3 Factors influencing declines.

One must look at both the diseases of individual trees and the diseased trees in a population to recognize differences among the three categories of disease. The biotic, abiotic, and decline diseases will be compared using four criteria: symptoms, signs, host specificity of disease, and spatial distribution.

Symptoms

Biotic agents of disease produce symptoms on specific plant parts. The affected parts are usually not randomly distributed on the plant; more often, only a portion of a plant is affected. Environmental or inoculum dispersal factors may account for the unevenness of disease symptoms. A progressive invasion of tissues is also a good symptom of biotic-induced disease.

Abiotic disease may or may not be plant-part-specific but is usually uniform in its symptom expression on the plant. There may be exceptions to a general distribution throughout the plant, caused by portions of the plant not being exposed. For example, the lower branches of a tree may be covered with snow and therefore not exposed to air pollution, salt spray, or the desiccating effects of winter winds. Abiotic disease does not generally occur as a progressive invasion like biotic infection, so that evidence of callous ridges on stem infection, or necrosis of leaves surrounded by chlorotic or other colored tissue, is usually not seen.

The most characteristic symptom of decline is the progression of symptom expression on individual plants and between plants. Another characteristic symptom is a reduction in growth. Some trees show very slight symptoms, others are dead, and others are intermediate in condition. A range of symptom expression may also occur with biotic diseases as a result of genetic variation and the spread patterns of the pathogen.

Signs

Signs are fruiting or other structures of biotic causal agents of diseases. Signs are most useful with fungal-induced diseases. The fungus can be identified and recognized as a causal agent of disease by the presence of specific fruiting structures.

With abiotic diseases, a confusing array of fungal structures may be seen. These can be identified and recognized as known saprophytic organisms and are therefore not signs of tree pathogens.

Signs associated with decline are not uncommon either. Some of these may be the confusing saprophytes such as those often found on trees suffering from abiotic diseases, and others are facultative or weak parasites that contribute to the decline. Identification of a specific organism as a known contributor to declines is a good indication of decline.

Host Specificity

Biotic diseases are usually host-specific or occur on limited numbers of related or unrelated hosts. This concept fits best for fungal and bacterial pathogens. As you will see later, other pathogens may be more general in their hosts.

The most characteristic feature of abiotic disease is the occurrence of similar symptoms on two or more totally unrelated hosts.

Declines are host-specific problems, but more than one tree species in a region may have its own specific decline syndrome.

Spatial Distribution

Biotic diseases, because they are caused by infectious agents, usually show a clumping distribution pattern of diseased individuals. Inoculum produced by diseased individuals is most concentrated around the diseased individuals, thereby contributing to a higher incidence of disease in localized areas. Only with initial infection caused by inoculum dispersed from a distance does the distribution of disease approach randomness. Topographic features that produce moisture or temperature conditions favorable for inoculum production, dispersal, and infection may contribute to clumped disease distribution patterns typical of biotic disease.

Abiotic disease is usually random in a population except when the agent is distributed in a nonrandom fashion. For example, a point source of pollution will produce a progressive intensification of symptoms as one nears the source. Over distance, the distribution of symptoms is progressive, but at a given distance the individuals affected will be randomly distributed.

Decline diseases have a random symptom distribution pattern within a given location.

All three types of diseases occur nonrandomly if one looks at a region as a whole. Thus, differences from one stand to another can be caused by many factors—site, environmental, and genetic factors of the host, to name just a few.

BRIEF OUTLINE OF THE BOOK

Each of the three types of diseases will be discussed at length in subsequent chapters. The abiotic diseases will be discussed first. The biotic agents of disease will then be developed, and finally the declines will be presented as a complex interaction of at least three factors from the biotic and abiotic groups.

A series of overview chapters discussing plant disease epidemics, genetic control of resistance, diseases of seedlings in the nursery, pathological considerations of urban tree management, and pathological considerations of intensively managed forest plantations will draw upon earlier chapters to develop some of the concepts applied to forest management systems.

REFERENCES

ANONYMOUS. 1972. Genetic vulnerability of major crops. National Academy of Sciences, Washington, D.C. 307 pp.

HEPTING, G. H., and G. M. JEMISON. Forest protection. *Section of* Timber resources for America's future. USDA For. Serv. For. Resource Rep. 14, pp. 184–220.

HORSFALL, J. G., and E. B. COWLING, eds. 1977. Plant disease: an advanced treatise. Vol I. How disease is managed. Academic Press, Inc., New York. 465 pp.

STAKMAN, E. C., and J. G. HARRAR. 1957. Principles of plant pathology. The Ronald Press Company, New York. 581 pp.

WHETZEL, H. H. 1918. An outline of the history of phytopathology. W. B. Saunders Company, Philadelphia.

PART ONE

ABIOTIC AGENTS OF TREE DISEASES

The number of abiotic factors that may be involved in disease and injury to trees is vast. To limit the subject to a reasonable number of major factors automatically eliminates many that others may consider important. Chapters 2, 3, and 4 will address plant diseases associated with soil factors, winter damage, and air pollution.

2

SOIL CONDITIONS AFFECTING TREE HEALTH

- Mineral nutrition
- Moisture
- Salt
- Soil aeration

The interaction of various physical factors of the soil, such as moisture, oxygen, mineral content, structure, and profile, with tree health is complex enough to make separation of single factors very difficult. For example, a heavy clay topsoil with impeded drainage will have a serious oxygen-deficiency effect, particularly during seasons of excessive rainfall. Anaerobic bacteria tie up nitrogen and sulfur under these conditions (see Chapter 7). Tree root development is impeded in this type of soil, resulting in mineral-deficiency symptoms in the tree crown. Application of fertilizer may temporarily alleviate the crown symptoms but will not solve the long-run imbalance between root regeneration and crown demands on the root system. Another complicating factor of soil environment on tree health involves the indirect effects of physical factors on the mycorrhizae (Chapter 9), pathogens (Chapter 16), and many interacting species of soil microorganisms.

Plant species differ in their tolerance of deficient soil environments. Those able to tolerate low oxygen resulting from high moisture are able to avoid competition from other, less-tolerant species in bog or wet sites. Those able to tolerate excessive vapor-pressure deficits in droughty sites avoid competition by growing in arid or highly drained sandy soils. A point to consider is that species that tolerate deficiencies do not necessarily have an obligate requirement for the deficient environment and may actually do better in a better soil if competition is removed.

MINERAL NUTRITION

An ideal environment should supply a balance of all necessary major nutrients, such as nitrogen (N), phosphorus (P), potassium (K), calcium (Ca), magnesium (Mg), and sulfur (S). Micronutrients, such as iron (Fe), manganese (Mn), zinc (Zn), boron (B), copper (Cu) and molybdenum (Mo) are needed in smaller quantities. The tree roots compete with microorganisms and the chemical attraction of soil structure bonds for these elements, so that availability is dependent upon more than just total concentration.

Serious deficiencies in chemical nutrients are uncommon in the "natural" forest because the species have evolved and occupy the sites that supply the required nutrients. Exotic trees planted on sites not normally occupied by the species may show deficiencies. Short chlorotic needles typical of potassium-deficiency symptoms in plantations of white spruce in central New York provide an example of this type of problem (Fig. 2-1).

Trees grown in urban environments or on highly disturbed sites also will often develop mineral-deficiency symptoms. Regular fertilization of ornamental plantings may improve their development on such imbalanced sites.

FIGURE 2-1 The white spruce trees in the center of the photograph have short chlorotic needles typical of a mineral deficiency. The problem is most evident in the poorly drained sections of the plantation. Scotch pines on the left do not show mineral-deficiency symptoms.

MOISTURE

Soil moisture is divided into water held in the soil because of impeded drainage to lower layers, water held in the macrocapillary structure, water held in the microcapillary structure of small soil particles, and water chemically bound (Fig. 2-2). Macrocapillary water is utilized by tree roots and soil microorganisms. Microcapillary water is generally not available to plants, nor is chemically bound water. The ideal soil has a large expanse of macrocapillary structure to hold water for extended periods. Humus or decomposing organic matter is the major source of macrocapillary water holding capacity; therefore, the amount of organic matter in the soil is very important to tree health. Excess water held in the soil, filling air voids, is detrimental to most tree roots.

The availability of moisture to plants is measured as soil vapor pressure. A well-watered, yet aerated soil may have a vapor pressure of -1 bar. At -15 bars most plants begin to permanently wilt.

The physical structure of the soil, as described above, is the major factor affecting vapor pressure in most soils. A second factor affecting vapor pressure results from the effects of ions dissolved in water. Dissolved ions decrease vapor pressure to a greater or lesser extent depending upon the concentration and chemical characteristics of the ions.

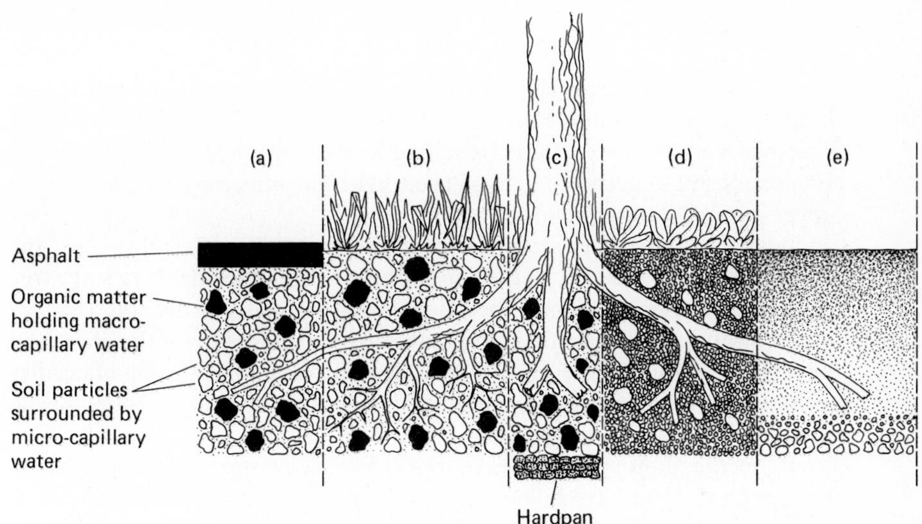

FIGURE 2-2 Interactions between tree roots and soil water with aeration, drainage, and soil structure. (a) Degenerate roots in a good soil that is poorly aerated because of asphalt or concrete covering the surface. (b) Healthy fibrous roots in a well-drained soil with extensive macrocapillary water. (c) Degenerate roots in a poorly aerated soil because a hardpan impedes free water drainage. (d) Degenerate roots in a soil with little macrocapillary water but much microcapillary water. (e) Degenerate roots in a highly compacted soil with little macrocapillary water and aeration.

SALT

Salt applied to de-ice highways generally does not accumulate in soil because it is very water-soluble and is washed out with spring runoff. In low areas where water accumulates or near intersections where salt application is excessive, salt will accumulate and be taken up by plants. Distribution of salt from high-speed highways to the foliage of conifers via small droplets of salt spray also occurs. This salt can be taken up by the foliage and accumulate in plants.

Along Interstate 95 (Connecticut Turnpike) during the winter of 1969–1970, sodium chloride was applied at approximately 35 tons per two-lane mile (about 20 metric tons/km) and calcium chloride was applied at 1.5 tons per two-lane mile (about 1 metric ton/km). Under these conditions, samples of foliage of white pine had chloride concentrations greater than 1% (dry-weight basis). A typical highway in New England receives 20 tons of salt per mile annually (about 11 metric tons/km).

The effects of salt on plants are threefold. One is a direct internal effect resulting from sodium and chloride uptake and accumulation. Chloride translocation to points of active growth concentrates the toxic chloride ion in leaves and growing shoots. Sodium displaces the uptake of potassium by the root system.

The second effect of salt on plants results from changes in soil structure. Accumulation of sodium cations in the soil causes a displacement of other ions absorbed to clay particles, resulting in a change in interparticle forces. Air spaces are reduced, and displaced essential mineral elements are no longer available to plant roots.

The third effect of salt is on moisture availability. Salt decreases the vapor pressure of the soil. As the vapor pressure of the soil decreases, the tree roots have more and more difficulty drawing water from the soil.

Symptoms of salt damage to plants vary depending upon the plant and the method of salt uptake. In conifers, tip chlorosis and browning of needles occur with low concentrations. With higher concentrations the entire needle turns brown. Usually, the bud is not killed, so that new growth in the spring maintains the photosynthetic leaf surface. By late summer, brown needles of the previous winter have fallen off and the trees look reasonably healthy. A somewhat thinner crown, caused by the loss of needles, may be apparent. Growth reduction and predisposition to infection by root, shoot, and foliage pathogens may complicate the damage due to salt.

Usually, the damage to conifers is most apparent on the road side of trees. Salt-spray droplets, resulting from high-speed driving on the deiced surface, are wind-blown to the trees. Foliage on the road side of the tree intercepts most of the salt spray. Symptoms expressed on foliage are related to the concentration of chloride directly taken up by the foliage.

On hardwoods, the symptoms of salt damage are expressed as a marginal chlorosis and browning of foliage. Twig dieback is common. The crown of a hardwood tree damaged by salt usually has a concentration of foliage along large branches and the stem. A scattering of small dead twigs over the outer surface of the crown gives a pincushion appearance to the tree (Fig 2-3).

Hardwood trees may accumulate salt by root uptake from soil or from droplets deposited on small twigs. The chlorides translocated to growing points generate the typical twig dieback and marginal leaf chlorosis symptoms.

Plant species vary in their tolerance to salt. Limited evaluation of salt damage has shown white pine and sugar maple to be very sensitive to salt, whereas Austrian pine and Norway maple are somewhat more tolerant to salt.

Salt is a pollution problem involving more than just plant damage. Rusting of automobiles, deterioration of concrete, and increasing salt concentrations in freshwater lakes are some of the concerns caused by salt in our environment. Damage to plants is probably well down the list in order of concern. Therefore, the best present methods for avoiding damage is to place trees well back from high-speed highways (at least 35 m was suggested for the Connecticut study), and to avoid planting trees in places where water runoff from roads accumulates salt in the soil.

FIGURE 2-3 Salt-spray damage on red maples along Interstate 81 in Syracuse, New York.

SOIL AERATION

Oxygen is as essential to the root system of plants as it is to the rest of the plant. Soil oxygen and carbon dioxide in the ideal soil should be at about the same level as in the atmosphere (approximately 20% 0_2, $<1\%$ $C0_2$). Decreased levels of 0_2 are more significant than increased levels of CO_2 to the health of plant roots. Factors such as heavy clay soils, soil compaction, filling over root systems with asphalt or concrete, or saturation of soil because of excessive rainfall or impeded drainage reduce the exchange of gases in the soil with the atmosphere and thereby reduce 0_2. Increased metabolism of soil microorganisms increases $C0_2$ and consumes 0_2, which needs to be exchanged with the atmosphere. Anything that interrupts air exchange is a potential problem.

In soil deficient in oxygen, root growth is retarded, amino acid leakage from roots to soil presumably increases, mycorrhizal development is reduced, and water and mineral absorption is consequently reduced. Avoiding conditions that reduce 0_2 in the soil, as well as planting species that tolerate low 0_2 levels, are ways to handle the oxygen problem.

Species that tolerate low oxygen also tolerate excesses of water, so as a general rule, bog species such as American elm, sycamore, hackberry, and honeylocust are preferred hardwood species for low-0_2 environments. You may note that these are species commonly used for urban plantings. Soil-oxygen deficiency is very common in urban areas. It is evident why these species have proven to be the best for use in the urban environment.

REFERENCES

HACSKAYLO, J., R. F. FINN, and J. P. VIMMERSTEDT. 1969. Deficiency symptoms of some forest trees. Ohio Agr. Res. Dev. Cent. Res. Bull. 1015. 68 pp.

HOFSTRA, G., and R. HALL. 1971. Injury on roadside trees: leaf injury in relation to foliar levels of sodium and chloride. Can. J. Bot. *49*:613–622.

SMITH, W. H. 1970. Salt contamination of white pine planted adjacent to an interstate highway. Plant Dis. Rep. *54*:1021–1025.

WESTING, A. H. 1969. Plants and salt in the roadside environment. Phytopathology *59*:1174–1181.

YELENOSKY, G. 1963. Soil aeration and tree growth. Int. Shade Tree Conf. *39*:16–25.

YELENOSKY, G. 1964. Tolerance of trees to deficiencies of soil aeration. Int. Shade Tree Conf. *40*:127-147.

ZAK, B. 1961. Aeration and other soil factors affecting southern pines as related to little leaf disease. USDA For. Serv. Tech. Bull. 1248. 30 pp.

3

WINTER DAMAGE TO TREES

- Physiological chlorosis
- Desiccation
- Rapid temperature changes
- Low temperature
- Late spring and early fall frosts

Winter damage to trees is a topic of concern in climates where subfreezing temperatures occur. Native vegetation has evolved methods for advanced recognition of seasons when freezing conditions will occur. The triggering or recognition signals are probably decreasing temperatures and shortening of the photoperiod. Regional adaptation of vegetation for recognition of specific dates when freezing temperatures occur and the photoperiod shortens generally avoids damage to natural vegetation.

When people move plants about, regional adaptation for one site may not fit the new site, and under these conditions, winter damage will often occur. Therefore, as a general rule, we may say that winter damage is relatively unimportant in locally adapted vegetation but is often a problem with any introduced plant, even plants of the same species as native vegatation. Movement of plants more than 100 miles north or south of the seed origin can potentially result in winter damage. Nurseries supplying planting stock should attempt to coordinate seed sources and planting locations. With exotics, one can only hope that some of the population of plants will by chance have the proper timing to the environment in which the plants are to be grown. Those that do not will not survive or will be injured year after year. Tulip poplars planted in central New York sometimes survive, but often the plants are severely injured by winter conditions. There are pecan trees growing in Syracuse, New York, but these are just chance selections from a species normally adapted for warmer climates.

Winter damage takes various forms, such as physiological chlorosis, desiccation, rapid-temperature-change freezing, low-temperature damage, and late spring or early fall frosts.

PHYSIOLOGICAL CHLOROSIS

Although conifers and other evergreen plants do not go through the major physiological change of deciduous plants, they do produce changes in chlorophyll content, structure, and function during the dormant season. These changes may appear as chlorosis, as in some varieties of Scotch pine, or may appear as red to bronze color changes, as in some junipers.

DESICCATION

Many southern plants grown in cooler climates suffer damage due to desiccation. Water uptake capacity is decreased between 5 and 0° C, and therefore a southern plant such as rhododendron suffers desiccation even without having the root or stem frozen. In contrast, freeze resistant trees require extended continuous freezing conditions of roots and/or stem to prevent water uptake.

Plant species differ in their resistance to dehydration. Scotch, jack, and red pines are examples of dehydration-resistant plants. Injury is enhanced by wind speed, sunshine, and low humidity.

Symptoms of desiccation are yellow-brown to red-brown foliage and stems. In rhododendron, desiccation produces darkened brown or black zones that fade to brown or tan near leaf margins. Freezing damage in rhododendron produces a contrasting blackening of leaves, with no tan or brown outer edges. Symptom differences between desiccation and freezing damage in other plants are not well documented.

It should be possible to reduce desiccation damage to plants with antidesiccant sprays, but a better way to avoid the problem is by selection of tolerant plants.

RAPID TEMPERATURE CHANGES

Winter damage from rapid temperature changes is very common even among native trees. Foliage of conifers or bark of young sugar maples exposed to heating by the sun can be damaged by the rapid temperature drop at sunset or when intermittently shaded by clouds. A drop from 2°C to −8°C in 1 minute and to −12°C in 6 minutes was measured for foliage of arborvitae (northern white cedar) at sunset with an air temperature of −12°C. Intracellular ice forms upon rapid freezing of plant tissue. The ice formation inside the living cells results in disruption of cell membranes, denaturation of some proteins, and other less-understood cellular dysfunctions.

Winter damage due to rapid temperature change is recognized as dead foliage or dead patches of bark, always on the south-facing side of the tree (Fig. 3-1).

One can avoid the damage by wrapping the stems of young maples with paper or by shading the southern exposure of arborvitae. One can also avoid damage by recognizing that physical factors of the site, such as buildings, produce shadows that abruptly block out the sun. Planting in the zone of intermittent shadow of buildings should be avoided. Recognition of the potential problem of southern exposures and grouping different types of plants to produce a diffusion of sunlight rather than exposing individual plants to the direct effects of the sun provide a long-run solution to winter injury of this type.

LOW TEMPERATURE

Low temperatures generally do not produce damage to plants that can survive freezing conditions. Pines exposed to −195°C were not injured if they were hardened for 12 days at subfreezing temperatures. If the drop in

FIGURE 3-1 Winter sun scald on sugar maple 1 year later.

temperature is slow enough, cells accommodate to the lower temperatures by concentrating solutes in the cytoplasm to prevent death due to intracellular freezing.

Low temperatures properly applied produce relatively harmless extracellular freezing to freeze-resistant plants, but tropical plants may be severely damaged by any type of freezing.

Extracellular freezing is associated with at least six physiological alterations of the cells: (1) reduction of cell water content because of lower vapor pressure of ice and movement of water from the cells to an area of lower vapor pressure; (2) increase in solute concentration within the cell due to reduction in water; (3) precipitation of solutes within the cell because of increased concentrations; (4) changes in pH caused by differential precipitation and changes in solute concentration; (5) reduction in spatial separation of macromolecules; and (6) cell shrinkage or plasmolysis as a physical expression of changes in water content.

The most pronounced effect of low temperatures on trees of northern climates that have evolved mechanisms of tolerance for the extracellular freezing is a frost cracking or physical splitting of the woody stem. Dehydration of outer xylem cells by extracellular ice produces tangential contraction without radial contraction. The forces are such that when the outer xylem splits, a sound like a rifle shot can be heard. Additional splitting or cracking is enhanced by water which gets into the crack and speeds up the extracellular ice formation of cells on the margin of the crack as well as physical expansion of ice forming within the crack.

Frost cracks may result because of low temperatures and/or may interact with wood decay and wound defects to produce a nonhealing vertical seam (Fig. 3-2).

Prevention of low-temperature damage involves recognition of sites where extremely low temperatures are expected to occur. Within cold regions, valleys or basins in the topography will generally have measurably lower temperatures than those in the surrounding upland country. Planting should be avoided in such areas. When planting exotics it is important to recognize the tolerance or lack of tolerance for freezing conditions of the planting stock. Match the plant to the climate of the site.

LATE SPRING AND EARLY FALL FROSTS

Damage to plants by freezing temperatures during periods when plants are not dormant is related to hardiness. Hardiness is a nebulous, variable, physiologically based, state of plant development. During active growth, hardiness, the ability to withstand freezing conditions, is nonexistent. The

FIGURE 3-2 Seam in beech.

effects of extracellular freezing are disruptive to actively growing plants. Those plants adapted to colder climates recognize cooler temperatures and reduced photoperiods as indicators of forthcoming freezing conditions and produce physiological changes to accommodate the conditions. Increases in storage sugars, particularly raffinose, a nonreducing trisaccharide composed of glucose, fructose, and galactose, is a common phenomenon in hardiness. The functions of sugars are not fully resolved, but they are assumed to (1) retard the growth of ice crystals, (2) protect proteins by replacing some water of hydration or holding the water of hydration more firmly, and (3) prevent bud break and growth during warm weather during the winter. Other physiological changes in protein, lipids, amino acids, and nucleic acids can be correlated with hardiness, so the actual chemical basis of hardiness is not fully understood.

Any factor that reduces or slows growth seems to induce hardiness. Decreased temperature and shortening of the photoperiod are examples. Fertilization increases growth and therefore should be avoided in late summer.

Chemically induced protection from spring frosts in apples has been accomplished experimentally with application of 2-chloroethyl trimethylammonium chloride (CCC), N-dimethylaminosuccinamic acid (Alar or B-nine), or malic hydrazide (Alar). How practical such chemicals may be for ornamentals is not yet known.

REFERENCES

ALDEN, J. and R. K. HERMANN. 1971. Aspects of cold-hardiness mechanism in plants. Bot. Rev. *37*:37–142.

BARNARD, J. E. and W. W. WARD. 1965. Low temperature and bole canker of sugar maple. For. Sci. *11*:59–65.

GERHOLD, H. D. 1959. Seasonal discoloration of Scotch pine in relation to microclimatic factors. For. Sci. *5*:33–343.

HERRINGTON, L., J. PARKER, and E. B. COWLING. 1964. The coefficient of expansion of wood in relation to frost cracks. Phytopathology *54*:128.

JONES, J. K. 1971. Seasonal recovery of chlorotic needles in Scotch pine. USDA For. Serv. Res. Pap. NE 184. 9 pp.

MAZUR, P. 1969. Freezing injury in plants. Annu. Rev. Plant Physiol. *20*:419–448.

SAKAI, A. 1970. Mechanism of desiccation damage of conifers wintering in soil-frozen areas. Ecology *51*:657–664.

SCHUBERT, G. H. 1975. Silviculturist's point of view on use of nonlocal trees. USDA For. Serv. Gen. Tech. Rep. RM-11. 12 pp.

ZALASKY, H. 1975. Low-temperature-induced cankers and burls in test conifers and hardwoods. Can. J. Bot. *53*:2526–2535.

4

TREE DISEASES CAUSED BY AIR POLLUTION

- Plant pathogenic air pollutants and sources
- Toxicity and symptoms of pollution on plants
- Weather conditions affecting air pollution buildup
- Real-world interaction of pollutants and weather systems
- Acid precipitation

Increased urbanization and industrialization, together with increased demands for mobility and the use of electricity, have placed excessive demands upon our environment to supply the needed raw materials and reabsorb the waste products. The problems produced fall into ecological, social, economic, and political areas of specialization which are interrelated into what seems like an unsolvable dilemma. I do not propose to develop or express the needed course of action out of this dilemma, because there is no single easy solution. My approach will be to describe the problems of air pollutants as they affect trees. You can develop the answers to our pollution problems. What this country needs is someone with the "right answer" who can convince others that he or she is right and who has enough money and political influence to instigate the solutions. In the meantime, those of us in biology who do not have the answer can strive to develop the background information that will help in the development of acceptable solutions.

Air pollutants, like many other abiotic factors, cause both disease and injury. If the pollutant is a persistent agent, it results in disease. If the agent interaction with the plant is of short duration, it produces injury. Let us not quibble over the separation of disease and injury, but rather consider phytotoxic gases of the atmosphere as a unified topic.

PLANT PATHOGENIC AIR POLLUTANTS AND SOURCES

In the past, pollution problems related to vegetation usually involved recognition of symptoms, proving that the injury was caused by an airborne gas, and locating the source of the phytotoxicant. Air pollution of this type

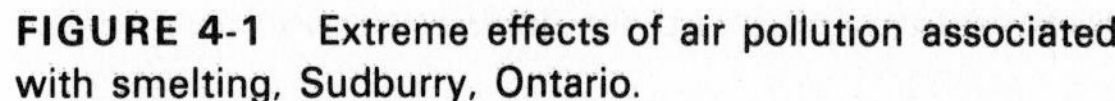

FIGURE 4-1 Extreme effects of air pollution associated with smelting, Sudburry, Ontario.

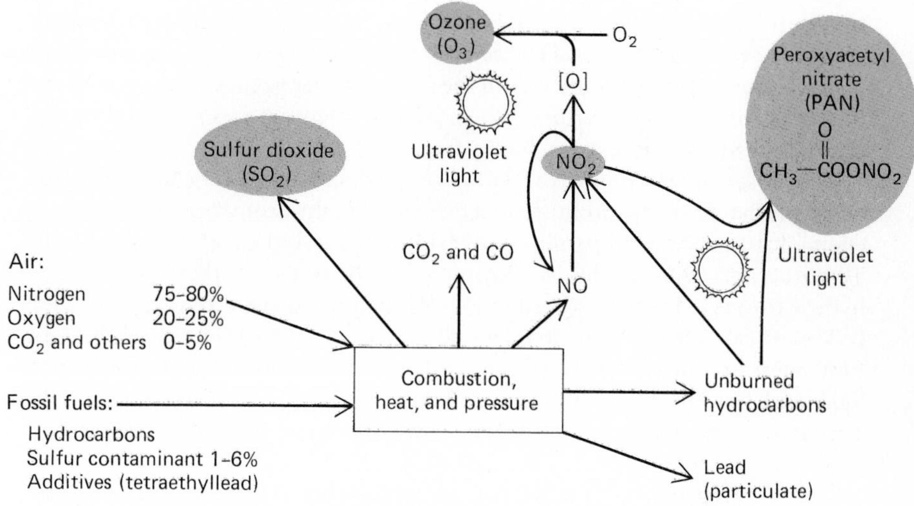

FIGURE 4-2 Plant-pathogenic air pollutants resulting from the combustion of fossil fuels.

centered around smelting (Fig. 4-1), chemical processing, and coal-burning electrical-generating installations. Present recognition of extensive damage over large areas with no point source of pollutant changes our perspective of pollution injury to vegetation.

Many of the conveniences associated with modern civilization emanate from the combustion of fossil fuels. Fossil fuels are a source of entrapped, life-giving energy of the sun but also a source of many life-destroying toxic gases when combusted.

Because combustion of fossil fuels is central to the pollution problem, it is important to understand the products of combustion. In Fig. 4-2, the combustion process is summarized. Ideally, hydrocarbons are oxidized to produce CO_2 and water. In actual practice the common contaminant sulfur is also oxidized to SO_2, and additives, particularly in gasoline, are also released to the environment. The hydrocarbons are seldom totally combusted, so some are released into the atmosphere as carbon monoxide and unburned hydrocarbons.

One can improve the combustion and reduce the amount of unburned hydrocarbons by increasing the temperature and pressure of combustion, but this is not a solution without problems. Another pollutant, nitrogen oxide, is produced under high temperature and pressure by chemical fusion of nitrogen and oxygen.

The array of primary pollutants from fossil fuels are carbon monoxide (CO), sulfur dioxide (SO_2), nitrogen oxides (NO) and (NO_2), lead (Pb), and hydrocarbons (HC). Of these, SO_2 and NO_2 are presently recognized as plant toxicants. Lead and other additives are being investigated and may represent additional plant problems. The picture regarding CO is not clear, but some plants appear to incorporate CO in the same manner as they do carbon dioxide. Most of this CO is broken down by soil microorganisms. No primary plant toxicant role is known for hydrocarbons.

Two secondary plant toxicants are produced by photochemical reactions of the primary pollutants. Ultraviolet light from the sun supplies energy for the chemical oxidation of NO to NO_2. Hydrocarbons play a role as reductants during this oxidation. The NO_2 can further react with hydrocarbons in the presence of ultraviolet light to form peroxyacetyl nitrate (PAN). It can, also with ultraviolet light, form ozone (O_3). The ozone formation again produces NO. Nitrogen oxide can be reoxidized with ultraviolet light and hydrocarbons to form NO_2. In this way, a limited quantity of nitrogen oxide, acting as a catalyst, can produce a very large amount of ozone.

Another group of plant-toxic air pollutants are by-products of industrial processes rather than fuel combustion. Of these, fluoride produced in the smelting of copper and other ores is the most significant plant toxicant.

The major sources of pollution and their relative contribution to the total plant pathogenic air pollution problem are presented in Table 4-1.

TABLE 4-1 Relative Amounts of Plant Pathogenic Air Pollutants Emitted into the Air by Various Sources (From Wood, 1968)

Pollutant	Transportation	Industry	Generation of electricity	Space heating	Refuse disposal
Sulfur oxides	1	9	12	3	<1
Hydrocarbons	12	4	<1	1	1
Nitrogen oxides	6	2	3	1	<1
Fluorides		<1			
Particulates	1	6	3	1	1
Miscellaneous others	<1	2	<1	<1	<1
Total	<21	<24	<20	<7	<5
Percent	28	30	26	9	7

TOXICITY AND SYMPTOMS OF POLLUTION ON PLANTS

Air pollutants vary in their toxicity to plants. Table 4-2 summarizes the concentrations and times necessary to produce injury as well as the most characteristic symptoms of injury; see also Figs. 4-3 to 4-9. One must also recognize that pollutants can produce growth reductions in the absence of obvious foliage symptoms.

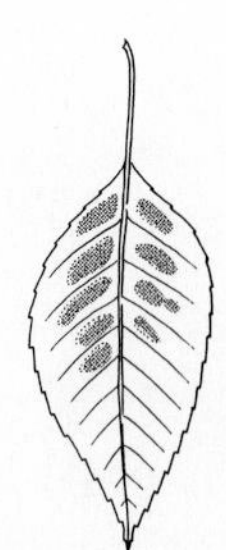

Sulfur dioxide: Interveinal brownings or necroses

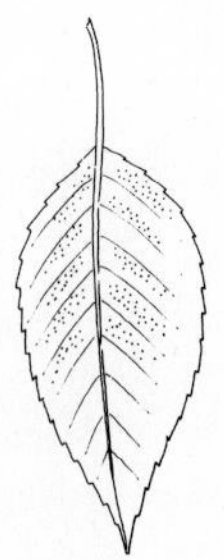

Ozone: Chlorotic or brown flecking on upper surface

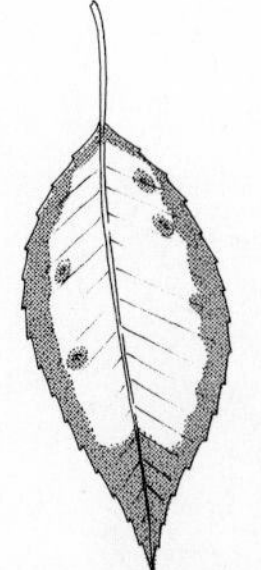

Fluoride: Marginal chloroses or necroses

Sulfur dioxide: Tip burn

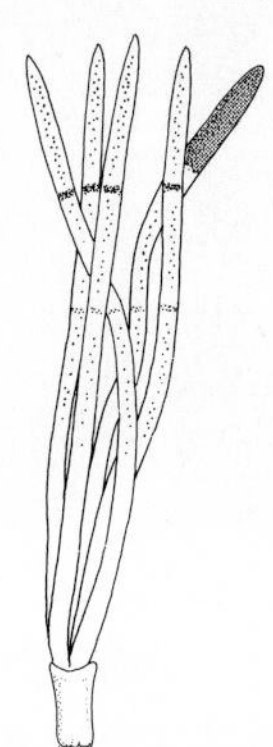

Ozone: Chlorotic flecking, banding, or terminal necroses (tip burn)

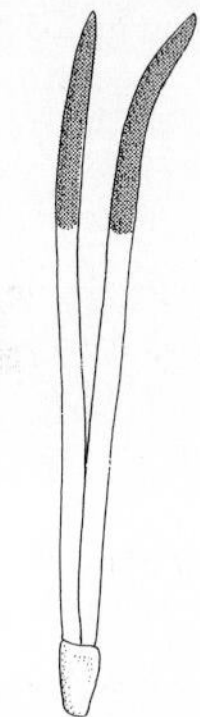

Fluoride: Tip burn

FIGURE 4-3 Typical air-pollutant injury of hardwood leaves and conifer needles associated with various air pollutants.

TABLE 4-2 **Concentrations Capable of Causing Injury and Typical Symptoms of Damage to Trees**

Concentrations of pollutants causing injury	Hardwoods	Conifers
SO_2, 50pphm [a] for 4 hr	Interveinal necrosis	Tip burning
NO_2, 200 pphm for 4 hr	Marginal or interveinal chlorosis	?
Fluorides [b]	Marginal chlorosis	Tip burning
Ozone, 7 pphm for 4 hr	Flecking on upper surface	Flecking and tip burning
PAN, 1 pphm for 6 hr	Silvering or glazing of lower surface	?

[a] pphm, parts per hundred million.
[b] Fluorides accumulate in the plant. When the level of 50–200 ppm is reached, symptoms appear.

(a) (b)

FIGURE 4-4 Ozone injury (50 pphm for 7.5 hours) on (a) honeylocust and (b) "Bloodgood" London plane. (Photographs compliments of Dr. David Karnosky.)

FIGURE 4-5 SO_2 injury (100 pphm for 7.5 hours) on (a) honeylocust and (b) "Bloodgood" London plane. (Photographs compliments of Dr. David Karnosky.)

(a) (b)

FIGURE 4-6 Fumigation chamber for testing air pollution sensitivity of ornamental plants. (Photograph compliments of Dr. David Karnosky.)

FIGURE 4-7 Chlorotic dwarf eastern white pine in Pennsylvania associated with air pollution. The small tree in the center of the picture is about 4 ft tall. Its nonsensitive neighbors of the same age are 20 ft tall.

FIGURE 4-8 Probably ozone injury on eastern white pine in Wisconsin. (Photograph compliments of Dr. David Karnosky.)

FIGURE 4-9 Leaf symptoms caused by (a) herbacide applied to the soil of pin oak or (b) flecking on sugar maple from leafhopper feeding may be confused with air pollution.

(a)

(b)

Fluoride is a serious pollutant even at low levels because it accumulates in the plant.

PAN and O_3 are toxic in concentrations of less than 10 parts per hundred million. It is hard to visualize how dilute "parts per hundred million" actually is, so I will provide an analogy. If one mixes a martini by adding one jigger of vermouth to a tank car of gin, one gets approximately one part per million and a very dry martini. One hundredth of a jigger in a tank car is in the range of parts per hundred million.

Sulfur dioxide is slightly less toxic than ozone or PAN.

Nitrogen dioxide is much less toxic and therefore is more important as a reactant in the photochemical production of O_3 and PAN than as a primary pollutant.

If one integrates Tables 4-1 and 4-2, it is obvious that transportation is the single most important source of air pollution. Because we all contribute to this type of pollution, change brought about in this area will be slow. It is much easier but less effective to condemn industry or electrical-power generators and force them to clean up than it is to reduce our own personal contribution to air pollution.

WEATHER CONDITIONS AFFECTING AIR POLLUTION BUILDUP

Air pollution buildup is usually directly related to weather inversions that concentrate, rather than allow dispersal of, pollutants. Therefore, it is important to understand this type of weather condition.

An inversion is a condition in which warmer air sits on top of a layer of cooler air (Fig. 4-10). In the absence of an inversion, the temperature of the atmosphere decreases with altitude at a rate of 1 °C per 100 m. This is called the adiabatic lapse rate.

The reason air temperature decreases with increasing altitude is because of a decrease in pressure. Recall the relationship between temperature, pressure, and volume of a gas from an introductory physics course. The general gas law is $PV = MRT$ (pressure $\times$ volume $=$ mass $\times$ a constant $\times$ temperature).

If a volume of polluted air emitted from a smokestack is warmer than the air around the top of the stack, it will rise because it is lighter than the surrounding air. It will lose temperature as it rises at a rate of 1°C per 100 m and will theoretically continue to rise indefinitely. Loss of some heat as it mixes with the surrounding air increases the heat loss with altitude, so that eventually the polluted air equilibrates in temperature with the surrounding air and is widely dispersed.

If an inversion occurs in the atmosphere, the parcel of polluted air from the smokestack may not be as warm as the air above the inversion and

FIGURE 4-10 Valley inversion over Syracuse, New York. The exhaust from the industrial complex in the background is visible just at the top of the inversion; the buildings are lost in the haze. The water vapor and pollutants from the stack in the foreground exhaust above the inversion and are therefore dispersed away from the city.

FIGURE 4-11 Geographic inversion.

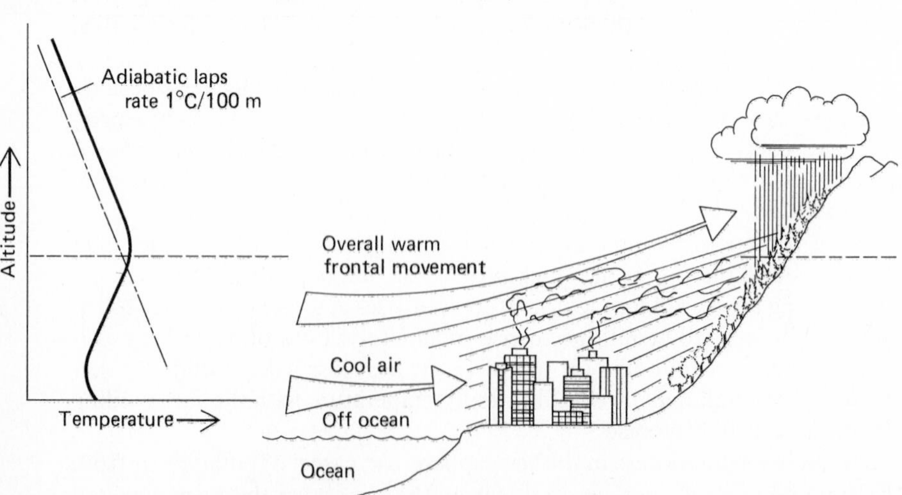

therefore will not rise through the warmer air. It is held below the inversion, which acts like a lid on a boiling kettle.

There are many types of inversions. Two are rather regional in occurrence, because specific land formations contribute to their prevalence. Others are more general in occurrence.

Geographic Inversions

The most notorious inversion problems in North America are associated with geographically induced inversions along the California coast (Fig. 4-11). A general westward movement of moisture-laden air is often trapped by the coastal range of mountains. To rise over the mountains, the air must condense the moisture it holds. Heat is released as moisture condenses, warming the air, so that eventually it will pass over the mountains. At the same time, a lower-level offshore wind brings inland cool air and fog underneath the warmer upper air that is attempting to get over the mountains. The final product is a mass of warm air on top of a mass of cool air.

Pollutants released into the cooler mass of air rise but are not warm enough to rise through the warmer upper layer and therefore are not dispersed from the coast. The photochemical reaction produces large amounts of O_3 and PAN. Damage results to vegetable, horticultural, and tree crops throughout the metropolitan coast and up into the lower elevations of the coastal range.

Valley Inversions

A second type of regional inversion is the valley-type inversion (Fig. 4-12). Radiation of heat from the surface of the earth on clear nights causes cooling

FIGURE 4-12 Valley inversion.

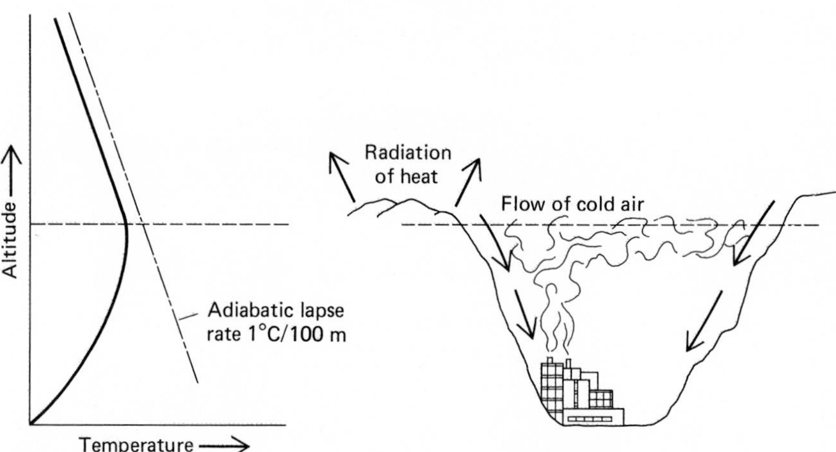

at the surface. The cool air, being heavier than the warm air, drains down into low-lying areas. If the topographic relief between hilltops and valleys is a few hundred meters, a mass of cool air may be trapped below a warm mass. A significant pollution problem occurs if there is a source of pollutant in the valley.

A town in a valley that expands its industrial or electrical-generating capacity must seriously consider the dispersion of pollutants out through the valley inversion. The engineering question of how to disperse the pollutants in this situation is easily handled. All that is required is to place a stack high enough so that the heated pollutant gases penetrate the inversion. This may be a local solution but not a regional solution, because the pollutants may be trapped by inversions covering larger areas.

Radiation Inversions

Radiation inversions (Fig. 4-13) are rather generally occurring inversions which when associated with a point source of air pollution, may produce a serious but localized problem. No highly specific topographic features are necessary, so that these can occur almost anywhere. The radiation inversion results from a cooling of the surface during clear nights. Radiation of heat into the atmosphere causes a cool mass of air to occur near the surface. Above 250 m the temperature of the air more normally approximates the adiabatic lapse rate, so that a cool air mass develops under a warm air mass. The inversion dissipates daily during midmorning because of heating of the surface by the sun. By midafternoon the surface temperature is above the normal adiabatic lapse rate.

FIGURE 4-13 Radiation inversion.

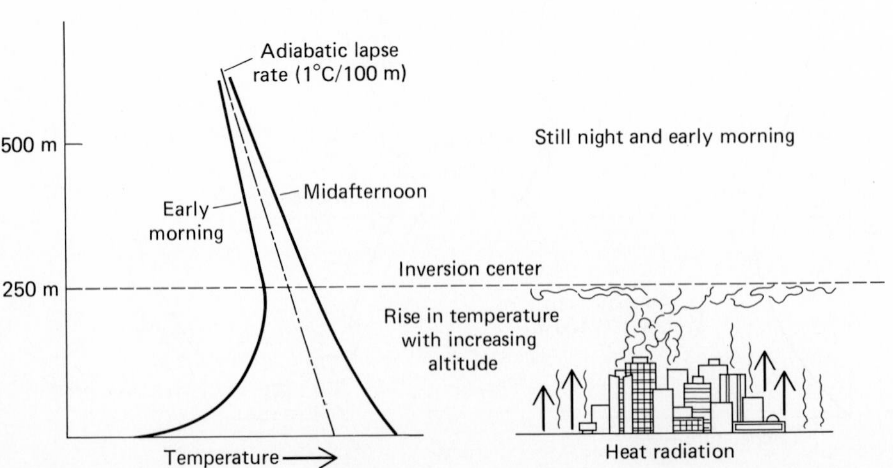

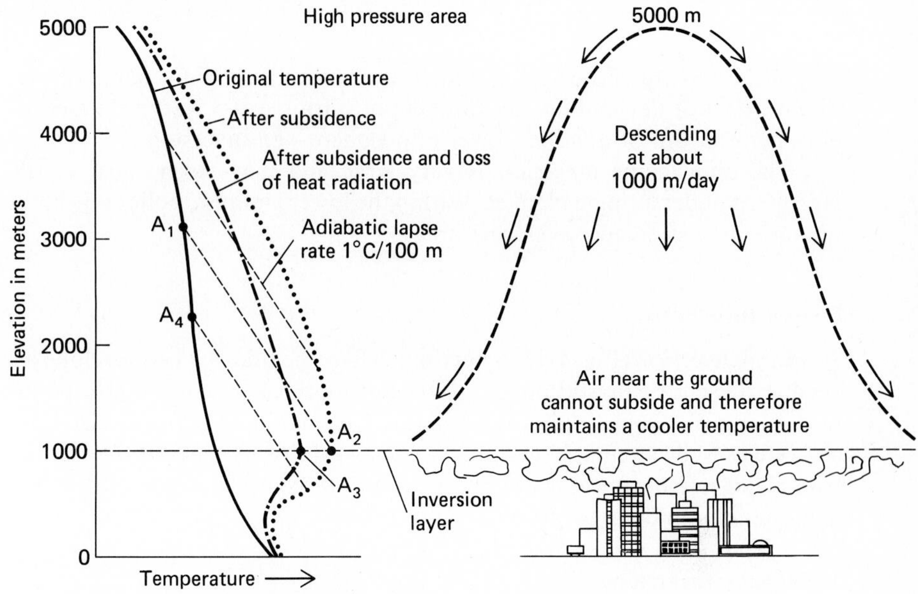

FIGURE 4-14 Anticyclone inversion.

Pollutants released at night and during the early morning may build up underneath the inversion and cause problems. Some may be brought down in localized areas during the up and down air movements associated with thermal activity during midday. These localized air-pollution injury pockets may be remote enough from the source to make it almost impossible to determine the actual source of the problem.

Anticyclone Inversions

The most serious type of pollution inversion on the eastern half of the North American continent occurs when an anticyclone or high-pressure system settles in for a few days (Fig. 4-14). Usually, highs and lows move across the United States from west to east rather quickly, but sometimes one gets held up by a major weather system off the Atlantic coast.

An anticyclone produces a subsidence inversion. The air temperature with altitude of a high-pressure system is slightly above what a normal adiabatic lapse rate would predict for various altitudes. The air in an anticyclone is descending and spreading out as winds from the center outward. The descending air is warmed along adiabatic-lapse-rate lines because of increasing pressure with decreasing altitude. Some of the increased heat is lost during mixing with cooler air, but the overall effect of subsidence is to increase the temperature of the descending air. Air near the top of the 5000 m anticyclone can descend for 4 to 5 days at 1000 m per day and can, theoretically, be warmed by up to 50°C. Air in the anticyclone at 1000 m can only be warmed 10°C. Obviously, the air closer to the ground can descend very little and therefore is warmed little above the original temperature. The product of descending and differential warming is a warmer upper atmosphere on top of a cooler lower atmosphere—an inversion.

The anticyclone inversion is very large and may cover most of the Northeast under a single blanket. Within the lower air mass, pollutants build up and cause problems over vast areas.

Frontal Inversions

A frontal inversion (Fig. 4-15) is very much like one side of an anticyclone inversion. Occluded or stationary fronts can more easily concentrate pollutants along the front to phytotoxic levels, but fast-moving fronts also concentrate as well as generate pollutants. Electrical storms generate ozone. The frontal inversion concentrates the electrical-discharged ozone and other, man-made pollutants to produce plant injury.

Turbulence Inversions

A turbulence inversion (Fig. 4-16) is a low-altitude inversion of air in the zone of turbulence caused by wind passing over irregular topographic features. The movement of air up and down by turbulence causes it to be warmed or cooled along the adiabatic lapse rate. If one averages all the adiabatic-lapse-rate lines for the turbulence zone, the interface of the upper point of the turbulence zone is cooler than the nonmixed air.

Turbulence inversion may concentrate and funnel polluted air from sources to remote areas along prevailing wind directions. It may also produce intermittent pollution problems due to irregular wind directions.

FIGURE 4-15 Frontal inversion.

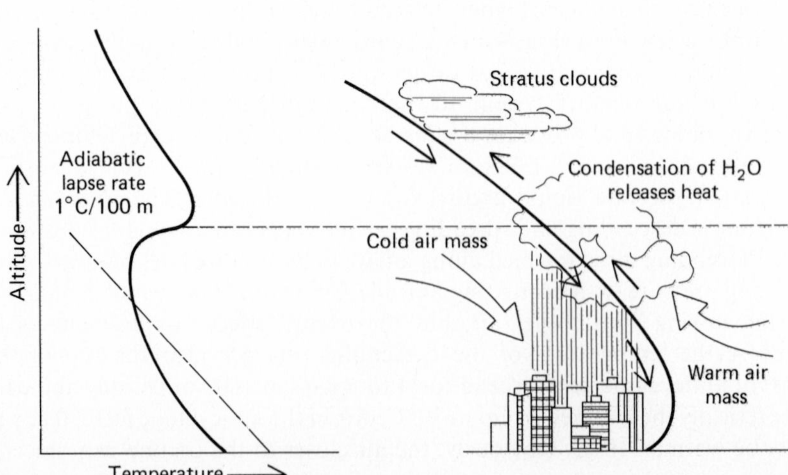

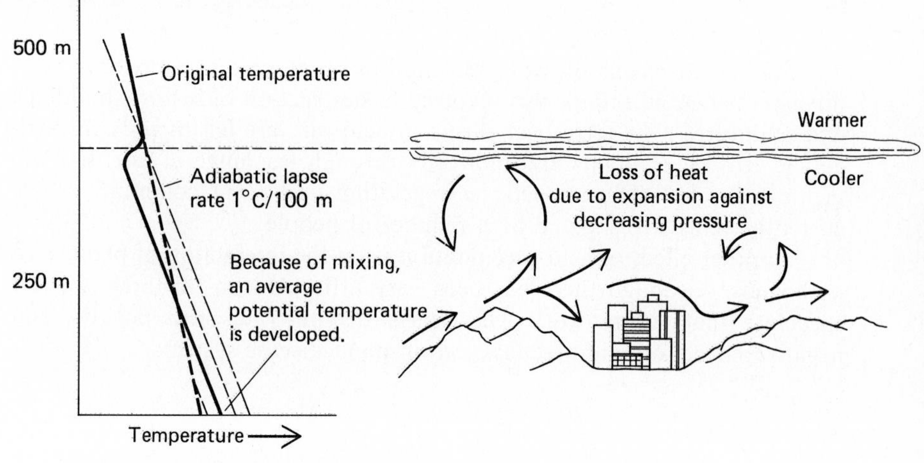

FIGURE 4-16 Turbulence inversion.

REAL-WORLD INTERACTION OF POLLUTANTS AND WEATHER SYSTEMS

Although it is easier to separate the individual phytotoxic gases into specific concentrations and types of symptoms produced, it is more realistic to think of pollution as being caused by a mixture of pollutants. It is also easier to separate the inversions into types, but realistically they may superimpose and negate or enhance the concentrating effects. It is the synergistic and negating interactions of individual pollutants and types of weather systems that make real-world diagnosis of pollution problems much more difficult than one may initially be led to believe.

Conflicting reports of enhanced symptoms and repression of symptoms in plants exposed to mixtures of ozone and SO_2 show the complexity of the problem. Do not be fooled into hasty diagnosis of pollution cause-and-effect relationships. Biologists need to learn much more about the interrelationships among the causes of problems they see in the real world.

ACID PRECIPITATION

An air pollution topic of considerable concern today revolves around the fallout of pollutants with rain. Dilute solutions of sulfuric and nitric acid from SO_2 and NO_2 pollution are the main sources of acidification. Rain, in an unpolluted atmosphere, would generally have a pH value of 5.6 due to dissolved carbonic acid from CO_2 in the atmosphere. Industrial pollution has caused the pH to fall to 4.5 and even lower in areas of eastern North America and western Europe. Rain storms as low as 2.4 have been recorded. This is the pH equivalent of vinegar.

Significant effects of acid precipitation on aquatic systems have been documented. Acid rain is also involved in destruction of historic buildings and sculptures. The terrestrial effects of acid rain are felt directly through cuticle erosion of plants and indirectly through leaching of essential elements from soil. The impacts on vegetation and plant communities are currently being investigated by a number of people.

Another effect of acid precipitation is on the interaction of plants and pathogens. Although there has been very little work in this area, we can speculate that future work will clearly demonstrate both positive and negative effects of acid precipitation on plant disease systems.

REFERENCES

COSTONIS, A. C. 1973. Injury to eastern white pine by sulfur dioxide and ozone alone and in mixtures. Eur. J. For. Pathol. *3*:50–55.

DAVIS, D. D., and F. A. WOOD. 1972. The relative susceptibility of eighteen coniferous species to ozone. Phytopathology *62*:14–19.

DOCHINGER, L. S., and C. E. SELISKAR. 1970. Air pollution and chlorotic dwarf disease of eastern white pine. For. Sci. *16*:46–55.

EVANS, L. S., N. F. GMUR, and F. DaCOSTA. 1978. Foliar response of six clones of hybrid poplar to simulated acid rain. Phytopathology *68*:847–856.

HEPTING, G. H. 1968. Diseases of forest and tree crops caused by air pollution. Phytopathology *58*:1098–1101.

JACOBSON, J. S. and A. C. HILL, eds. 1970. Recognition of air pollution injury to vegetation: a pictorial atlas. Air Pollution Control Association, Pittsburgh, Pa.

KOZLOWSKI, T. T. 1980. Impacts of air pollution on forest ecosystems. BioScience *30*:88–93.

LOOMIS, R. C., and W. H. PADGETT. 1973. Air pollution and trees in the east. USDA For. Serv. State Private For., Northeast Area and Southeast Area. 28 pp.

PETTERSSEN, S. 1958. Introduction to meteorology, 2nd ed. McGraw-Hill Book Company, New York. 327 pp.

SCORER, R. 1968. Air pollution. Pergamon Press, Oxford. 151 pp.

SHRINER, D. S. 1978. Effects of simulated acidic rain on host-parasite interactions in plant diseases. Phytopathology *68*:213–218.

SMITH, W. H. 1971. Lead contamination of roadside white pine. For. Sci. *17*:195–198.

WALTHER, E. G. 1972. A rating of the major air pollutants and their sources by effect. J. Air Poll. Contr. Assn. *22*:352–355.

WOOD, F. A. 1968. Sources of plant-pathogenic air pollutants. Phytopathology *58*:1075–1084.

PART TWO

BIOTIC AGENTS OF TREE DISEASES

The organisms of this planet have evolved relationships one with another. Plants are the predominant autotrophs capable of photosynthetically capturing the sun's energy for biochemical synthesis. Heterotrophic organisms are dependent upon the complex carbon compounds synthesized by the autotrophs. In turn, autotrophs are dependent upon heterotrophs to break down the structure and release the elements locked by previous populations of both heterotrophs and autotrophs. All the chemical elements of biological synthesis are cycled again and again through the various organisms and recycled back through the soil and air to be reused in biological synthesis.

It is appropriate to keep this balance of nature in mind when considering biological agents of disease. Disease is not an imbalance in nature but rather a very normal part of the cycling and recycling of elements.

Part 2 will deal with the various biotic agents capable of parasitizing and causing diseases of trees. Chapters on nematodes, viruses, bacteria, fungi, and higher plant parasites will present the reader with an initial basic understanding of the biological characteristics of pathogens. The pathogenic aspects of the parasites are separated into mode of parasitic action, basic disease cycle, characteristic symptoms of disease, methods for diagnosis of disease, specific examples of disease agents, and aspects of control.

An understanding of the natural relationship between pathogen and host is incomplete without an understanding of environmental interactions. These three elements—host, pathogen, and environment—are interrelated in the development of biotic plant disease.

When confronted with a plant disease problem, the average person expects the plant pathologist (the plant doctor) to prescribe and the local nursery (the drug store) to supply a pesticide that will control the pathogen. We have a few prescriptions of this type for forest and ornamental crops, but the majority of tree disease problems cannot be handled in this way. Most tree disease agents are more appropriately controlled, not exterminated, by manipulation of the environment or the host. Therefore, the chapters that follow treat all three elements of the tree disease triangle.

5

NEMATODES AS PLANT PARASITES AND AGENTS OF TREE DISEASES

- Types of nematodes
- Mode of action of plant-parasitic nematodes
- Nematode disease cycle
- Symptoms of nematode disease and injury
- Methods for diagnosis of nematode problems
- Examples of nematode problems of trees
- Control of nematode diseases

Nematodes are the only group of plant-parasitic animals studied by plant pathologists. Other plant-parasitic animals, such as insects and mites, are usually left to the entomologists.

At the present time, nematodes are not considered highly significant limiting factors in the growth of ornamental or forest trees. In fact, nematode problems occur only where there are nematologists (people who study nematodes), and there are only a few of these people in forest pathology.

Why devote a chapter to such a limited subject? If we can learn from the successes and mistakes of plant pathologists with regard to agricultural crops, we can see that there was the same skepticism regarding nematode importance when there were very few nematologists in that field. Intensive agricultural practices introduced new and different problems, one of which was nematodes.

We can expect and already see similar problems in some forest nurseries. Nematodes will also become a significant factor in plantations where the same species are used in second and third rotations.

TYPES OF NEMATODES

Nematodes are unsegmented round worms, in the phylum Nemathelmenthes, with well-developed digestive and reproductive systems (Fig. 5-1).

FIGURE 5-1 Plant-parasitic and free-living nematodes as seen with the 100× magnification compound microscope.

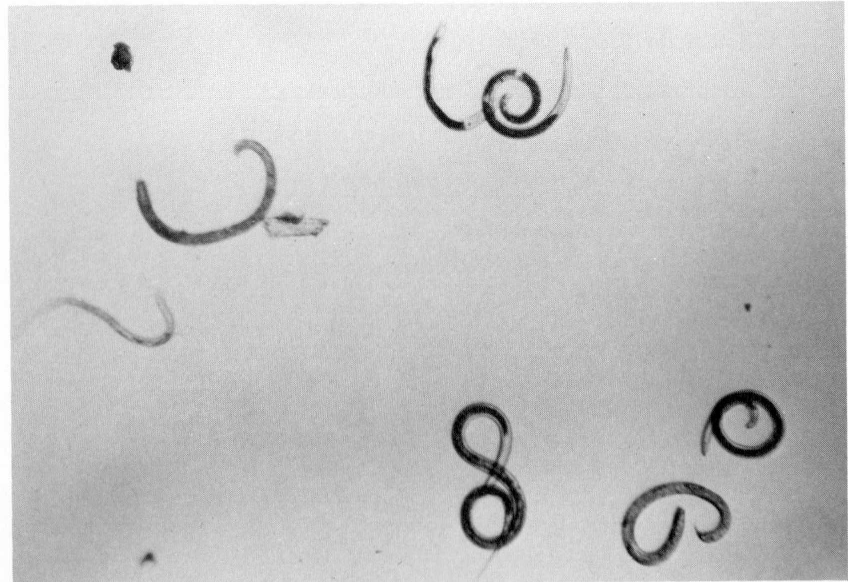

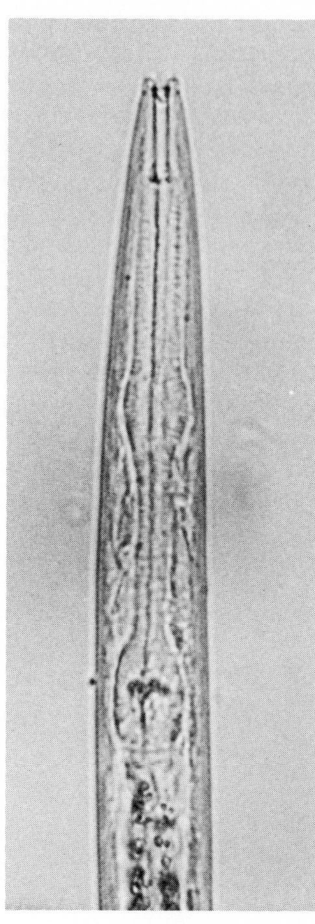

FIGURE 5-2 Free-living nematode with a cavity mouth part.

FIGURE 5-3 (Below) Plant-parasitic nematode with a stomatostylet.

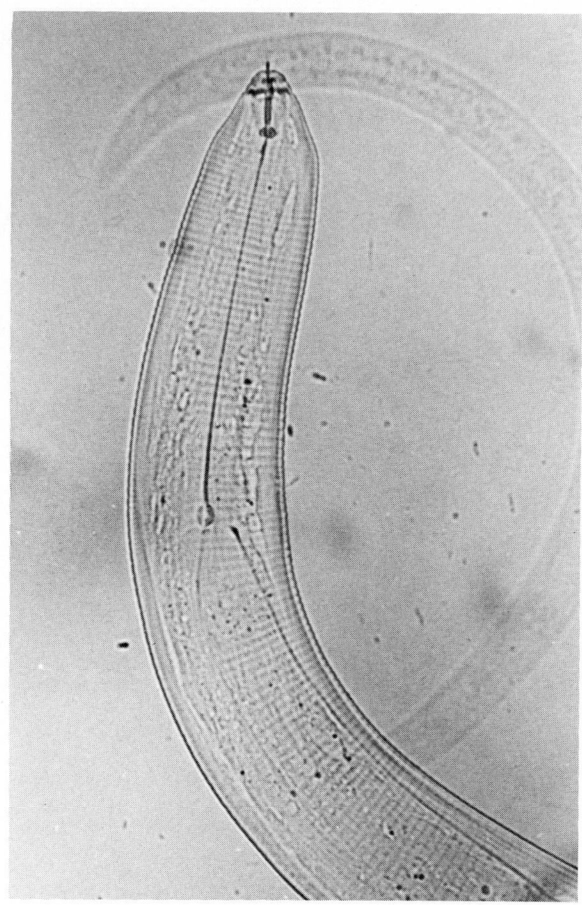

Nematodes can be placed into four groups. The majority fall into a group of small, free-living saprophytic nematodes characterized by a hollow cavity like a mouth for picking up organic matter and feeding it into the esophagus (Fig 5-2). A second group of variable-sized nematodes are animal parasites with mouths adapted for suction attachment to animal cells. A third group are predators on small invertebrates in the soil and have mouth parts with modified teeth-like projections. The group we are interested in is a fourth group, 0.5 to 2.5 mm in length. They are plant parasites characterized by a stylet or spear-like feeding apparatus.

There are two types of plant parasitic nematode stylets. The stomatostylet is the most common (Fig. 5-3). The odontostylet is a characteristic feature of some of the larger species of nematodes.

Plant parasitic nematodes are split into two groups based on feeding habits (Fig 5-4). Some have a browsing feeding habit. They pierce a cell, remove the nutrients, retract the stylet, and move on to another location to

FIGURE 5-4 Types of plant-parasitic nematodes.

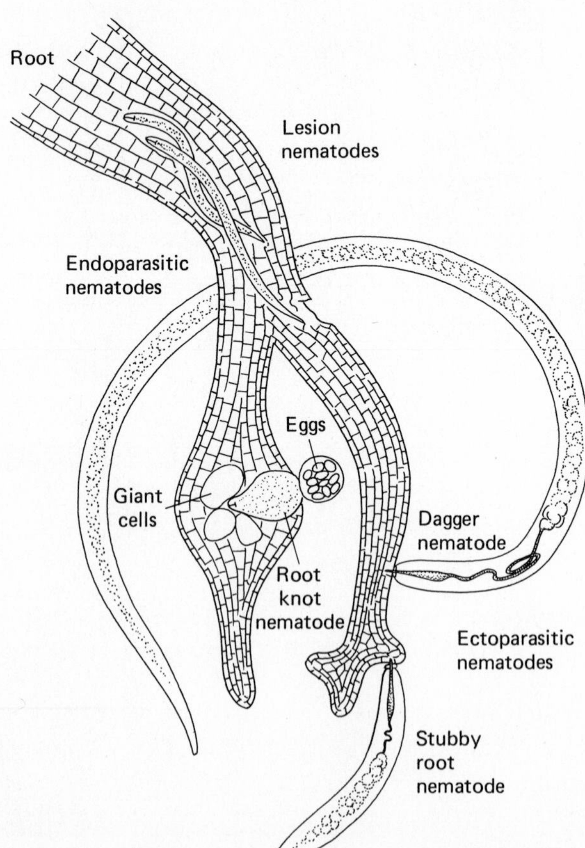

pierce another cell. Another group has a sedentary feeding habit, which involves staying in the general location of their first feeding. They may nurse the cells along and remove only a small amount of nutrient, or they may continue to penetrate deeper and deeper into the root as cells die. The former group are called ectoparasites and the latter group endoparasites.

MODE OF ACTION OF PLANT-PARASITIC NEMATODES

The stylet is a hollow pin-like structure used to pierce plant cell walls. Salivary juices are injected into the cell and soluble nutrients removed by means of the stylet.

Plant-parasitic nematodes usually feed on the root systems of plants. They may cause disease or injury directly by their feeding.

A second role of nematodes is to produce wounds for infection by soil fungi and bacteria. The destructive effects of the root-knot nematode (*Meloidogyne* spp.) and the *Fusarium* wilt pathogen on mimosa trees are more dramatic than either alone.

A third role of nematodes is as vectors of viruses. Ash ringspot virus is transmitted by the dagger nematode (*Xiphinema americanum*).

A fourth role of nematodes is the parasitism of mycorrhizal fungi. (Mycorrhizae are discussed in detail in Chapter 9.) The protective barrier and enhanced uptake of mycorrhizal roots is destroyed by the lance nematode (*Hoplolaimus galeutus*).

The fifth role of nematodes is as a contributing factor in declines. (Declines are discussed in Chapter 18.) The stubby root nematode (*Trichodorus christiei*) is associated with littleleaf disease of southern pines.

NEMATODE DISEASE CYCLE

The plant-parasitic-nematode life cycle involves an egg, one larval stage within the egg, three larval plant-feeding stages, and an adult plant-feeding stage. Some species produce male and female individuals. Others are hermaphrodites with both male and female organs in the same individual. A third type reproduces parthenogenically or in the absence of fertilization of eggs.

Those nematodes with odontostylets shed the stylet with each molt. One or two embryonic stylets to replace the shed stylet can often be seen in second- and third-stage larvae. Nematodes with stomatostylets maintain their stylets through the successive molts.

Under ideal conditions the life cycle from egg to adult and reproduction of up to 500 eggs is completed in about 1 month, so more than one generation per growing season is possible, even in northern climates.

Nematodes move in the film of free water surrounding soil particles. On their own, they can move only a few centimeters per year. One factor in parasitic nematode buildup in agricultural or intensively managed forest soils is increased potential for spread of the parasite on cultivation equipment and on transplants.

SYMPTOMS OF NEMATODE DISEASE AND INJURY

Root destruction by nematodes produces symptoms in plants similar to nutrient deficiencies. Symptoms of nematodes on plants include growth reduction, imbalance of root to shoot ratio, sparse yellow foliage, reduced size of foliage, premature leaf drop, and abnormal wilting during hot, dry periods. Root symptoms include root lesions, impaired root growth, root rot, excessive root branching, and galls.

METHODS FOR DIAGNOSIS OF NEMATODE PROBLEMS

Nematode diseases are detected initially by the symptoms produced on aboveground portions of plants. Examination of roots often shows characteristic symptoms of nematode activity, such as lesions or galls. To finally characterize a problem as being caused by nematodes requires extraction and identification of nematodes from the soil and/or roots.

Isolation of nematodes from soil is generally done by mixing a sample of soil in water. The slurry of soil and water is passed through a number of screens to remove the larger soil particles. Most plant parasitic nematodes are not small enough to pass through a 320-mesh (44-micron) screen, so nematodes and some soil particles in that size range are trapped on a 320-mesh screen. The concentrated nematode sample is washed from the 320-mesh screen and examined directly, or after further concentration by use of a centrifuge or Bierman funnel.

The Bierman funnel consists of a beaker in which the nematode slurry is placed (Fig. 5-5). A layer of cotton cloth or facial tissue is placed over the top of the beaker, which is then inverted into a water-filled funnel. Nematodes in the beaker wiggle their way through the tissue and settle to the neck of the funnel, where they are collected in a small volume of water a few hours later.

Any nematode with a stylet is potentially a plant parasite but the question arises as to which plants are parasitized. Some nematodes have very narrow host ranges and others very large. The species of the genus *Meloidogyne* (root-knot nematodes) are parasitic on at least 1500 species of plants, including 100 forest trees. Some stylet-bearing nematodes feed on fungi exclusively. Others may feed on both higher plant roots and fungi.

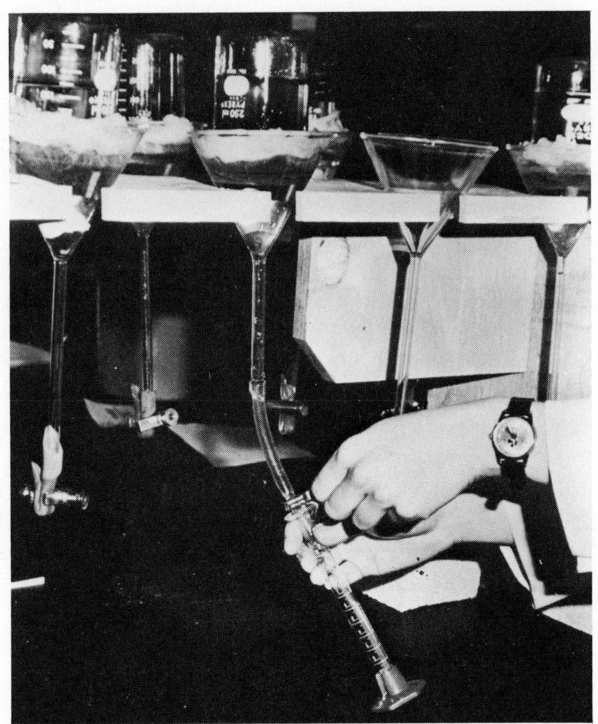

FIGURE 5-5 Bierman funnel for concentrating nematodes.

Demonstration of parasitism of a particular plant by a nematode species requires controlled experimentation. The suspected parasite is introduced into the soil of a potential host grown in pots in a greenhouse. Extraction of nematodes from the pot at a later date should recover the introduced nematode. Evidence of larval stages indicates that the nematode can complete its life cycle. Evidence that the population increased is used to postulate parasitism of the particular host.

You can readily see the major difficulty in demonstrating the parasitism of nematodes of large trees based on this type of test. Because of this, very little information on the role of nematodes on mature trees is available.

An example of the circumstantial evidence for nematode interaction in mature trees is a study of maple decline in Connecticut. Larger populations of plant-parasitic nematodes were found in soil beneath declining trees than in soil beneath healthy trees. However, it is not possible to predict a cause-and-effect relationship on the basis of this evidence alone.

Proof of parasitism is not proof of pathogenicity. It is even more difficult to demonstrate pathogenicity, or the capacity to cause disease. Modification of Koch's rules of pathogenicity (discussed in Chapter 1) are needed. The first requirement of constant association can be demonstrated by extraction of a large population of parasitic nematodes from the soil of diseased plants. Because plant parasites can be extracted from almost any soil, we need some quantification of the term "large." As a general rule, populations of fewer than 200 nematodes per half liter of soil from the root zone do not cause problems, but populations in excess of 1000 per half liter of soil may result in a problem.

The second step, isolation and growth in pure culture, is a bit more difficult. The nematode is a complex organism with other microorganisms maintained internally and externally. Surface sterilization of the outer skin does not affect the internal microorganisms. So it is almost impossible to isolate the nematode in pure culture. Growth in pure culture is equally difficult because nematodes need plant cells to feed upon. Nematodes may be grown on tissue culture of a plant or fungus, so that we can approach but not actually achieve pure cultures.

The third step, production of disease symptoms following inoculation with a "pure culture" of nematodes, and the fourth step, re-isolation, follow the procedures for proof of pathogenicity already described. Plant damage should be correlated with an increase in the nematode population.

EXAMPLES OF NEMATODE PROBLEMS OF TREES

Plant-parasitic nematodes are serious pathogens of agricultural crops and forest nursery crops. We also can expect, even though there is little evidence, that they play a role in reducing productivity of forest tree populations. The effects of nematodes on the diseases of natural populations of plants will be difficult to demonstrate. We find plant parasites present in almost any soil, but interactions of soil microorganisms and intermixing of roots of various plants retard the development of large populations of specific plant parasites.

Root-Knot Nematode

Over 1500 species of plants, including 100 species of trees, both conifers and hardwoods, are parasitized by the root-knot nematode (*Meloidogyne* spp.). Feeding by the female causes giant cells to form in the root. Cell walls between adjacent root cells dissolve, nuclear division increases, and cells enlarge. The female *Meloidogyne* nematode is an endoparasite and spends her entire life in one location within the root gall, feeding upon the few giant cells formed. Eggs form parthenogenically in the adult female and she swells to a pear shape. The eggs extruded from the female remain together in a saclike membrane on the surface of the root gall. The young that hatch often penetrate the root in the same general area as the parent and further accentuate gall formation.

Dagger Nematode

Dagger nematodes (*Xiphinema* spp.) are very large plant-parasitic nematodes, at least 10 times larger than most others. The stylet is of the odontostylet type, with a central opening large enough to allow virus particles to be taken up during feeding. The stomato stylet of other plant-parasitic nematodes has too small an opening to allow the uptake of virus particles. Dagger nematodes are ectoparasites which in sufficient numbers cause extensive destruction of roots. Many ringspot viruses, including ash ringspot virus, are known to be transmitted by this nematode, so that their role as vectors of virus may be as significant as their direct parasitism.

Stubby-Root Nematode

The stubby-root nematode *(Trichodorus christiei)* is associated with many agricultural crops as well as short leaf pine showing littleleaf symptoms. It was found in 19 of 35 southern pine tree nurseries. The effect of this ectoparasite is to produce short stubby roots and an overall reduction in both root and top weights of southern pine seedlings.

Lesion Nematode

Lesion nematodes (*Pratylenchus spp.*) are endoparasitic nematodes of agricultural and forest crops. Populations of nematodes burrow within the host root, causing necrotic areas of root. The nematode does not feed on just a few cells as with *Meloidogyne,* but continues to expand the size of the lesion by feeding on living cells on the periphery of the lesion.

Pinewood Nematode

The pinewood nematode (*Bursaphelenchus xylophilus*) has just recently been identified in the United States, but in Japan it has caused problems for more than 60 years on red and black pines. In Japan, approximately 1 million cubic meters of wood was lost to this nematode in 1975. *Bursaphelenchus,* unlike the previous four nematodes, is associated with the tops of trees. The nematode is vectored by Cerambycid beetles, which also provide wounds for infection. Thousands of nematode larvae are found on and within the insect vector. The nematodes spread quickly through both large and small pine hosts in the resin canals while feeding on the surrounding epithelial cells. Symptoms of infection include a reduction in oleoresin exudation flow from wounds, followed by the yellowing of needles. The needles wilt and turn brown. Mortality occurs within 40 to 60 days after summer inoculation. How important this nematode will be in the United States is not yet known. It has been recovered from Japanese black pine and Austrian pine. White pine and jack pine are reported to be highly resistant.

Many additional nematodes are presently known to parasitize trees, but the five described above represent the range of interaction between nematodes and trees.

CONTROL OF NEMATODE DISEASES

Control of nematodes is generally through application of chemical nematocides and sterilants to the soil. Crop rotations, interplanting nonhost cover crops, fallow periods between crops, organic amendments, and nematode resistance are some of the control methods that will be used in the future. None of the present or future techniques will totally eliminate nematode problems, so we may learn to live with some nematode-induced losses.

REFERENCES

CHRISTIE, J. R. 1959. Plant nematodes, their bionomics and control. Univ. Fla. Agr. Exp. Sta., Gainesville. 256 pp.

GILL, D. L. 1958. Effect of root-knot nematodes on Fusarium wilt of mimosa. Plant Dis. Rep. *42:*587-590.

HOLLIS, K. P. 1963. Action of plant-parasitic nematodes on their hosts. Nematologica *9:*475–494.

KIYOHARA, T., and K. SUZUKI. 1978. Nematode population growth and disease development in the pine wilting disease. Eur. J. For. Pathol. *8:*285–292.

MAI, W. F., and H. H. LYON. 1975. Pictorial key to genera of plant-parasitic nematodes. Cornell University Press, Ithaca, N.Y. 219 pp.

MAI, W. F., J. R. BLOOM, and T. A. CHEN. 1977. Biology and ecology of the plant-parasitic nematode *Pratylenchus penetrans.* Pa. State Univ. Coll. Agr. Bull. 815. 64 pp.

MAMIYA, Y. 1976. Pine wilting disease caused by the pine wood nematode. JARQ *10:*206–211.

NICKLE, W. R. 1972. Nematode parasites of insects. Proc. Annu. Tall Timbers Conf. Ecol. Anim. Control Habitat Manage., pp. 145–163.

PETERSON, G. W. 1962. Root lesion nematode infestation and control in a plains forest tree nursery. USDA For. Serv. Rocky Mount. For. Range Exp. Stat. Res. Note 75. 2 pp.

RIFFLE, J. W. 1971. Effect of nematodes on root-inhabiting fungi. *In* E. Hacskaylo, ed., Mycorrhizae. First North Am. Conf. Mycorrhizae, Proc. Apr. 1969. USDA Misc. Pub. 1189. pp. 97–113.

RIFFLE, J.W. 1973. Effect of two mycophagous nematodes on *Armillaria mellea* root rot of *Pinus ponderosa* seedlings. Plant Dis. Rep. *57:*355–357.

RUEHLE, J. L. 1964. Nematodes, the overlooked enemies of tree roots. Proc. Int. Shade Tree Conf. *40:*60–67.

RUEHLE, J. L. 1964. Plant-parasitic nematodes associated with pine species in southern forests. Plant Dis. Rep. *48:*60-61.

RUEHLE, J. L. 1966. Nematodes parasitic on forest trees. I. Reproduction of ectoparasites on pines. Nematologica *12:*443–447.

RUEHLE, J. L. 1969. Forest nematology—a new field of biological research. J. For. *67:*316–320.

RUEHLE, J. L. 1969. Influence of stubby-root nematode on growth of southern pine seedlings. For. Sci. *15:*130–134.

RUEHLE, J. L. 1971. Nematodes parasitic on forest trees. III. Reproduction on selected hardwoods. Nematologica *3:*170–173.

RUEHLE, J. L., and D. H. MARX. 1971. Parasitism of ectomycorrhizae of pine by lance nematode. For. Sci. *17:*31–34.

VIGLIERCHIO, D. R. 1971. Nematodes and other pathogens in auxin-related plant-growth disorders. Bot. Rev. *37:*1–21.

6

VIRUSES AS AGENTS OF TREE DISEASES

- Types of viruses
- Mode of action of plant-parasitic viruses
- Virus disease cycle
- Symptoms of virus diseases
- Methods for recognition and verification of virus diseases
- Examples of virus diseases of trees
- Control of virus diseases

Little research has been done on forest tree viruses. A summary of virus-like disorders of forest trees in 1964 listed 53 suspected viruses. Only 24 were reported from North America. At least five of the 24 are presently known or suspected to be caused by a mycoplasma-like organism (MLO), which will be described in Chapter 7. Seven of the 24 are viruses of *Prunus* spp. Concern with viruses of cultivated cherries demonstrated that wild *Prunus* spp. are often symptomless carriers of the viruses of cultivated cherries. Very few of the remaining 12 suspected virus diseases have been studied thoroughly enough to be called virus diseases today.

TYPES OF VIRUSES

Viruses are infectious agents composed primarily of nucleic acid and protein. They are considered lifeless by some because they do not possess the complement of normal metabolic enzymes used by organisms for synthesis and degradation of organic compounds. Others consider them primitive organisms because they possess functional nucleic acid, which is essential for continuity of genetic information for all living organisms. One can derive additional arguments for each point of view, but the final solution to the question is dependent upon which criteria you wish to define as essential for life.

Viruses are a rather diverse group of disease agents. The simplest viruses consist of a short strand of RNA. The most complex incorporate nucleic acid and proteins within a membrane capsule derived from the host-cell plasma membrane.

Some viruses, such as bacteriophage viruses, are rather complex in structure and function. The phage has a DNA core, a protein coat, and a limited number of enzymes combined into a complex structure.

The plant viruses we have information on are generally simple in structure and function. The infectious nucleic acid material, usually a single strand of RNA, is contained within a protein shell composed of large numbers of identical protein subunits.

Most plant viruses operate as individual infectious units. Infection of a plant cell by one virus prohibits infection by others. But some plant viruses require other viruses to infect. For example, tobacco satellite virus (SV) cannot develop in tobacco plants in the absence of infection by tobacco necrosis virus (TNV). Apparently, SV lacks the ability to synthesize a necessary protein and therefore relies on TNV for that function.

The most simple plant virus, a viroid, is an infectious agent composed only of a short piece of RNA. It was reported for the first time early in the 1970s as the cause of spindle tuber of potato. The RNA strand of viroids is so short that it does not have enough nucleic acid to code for even the smallest

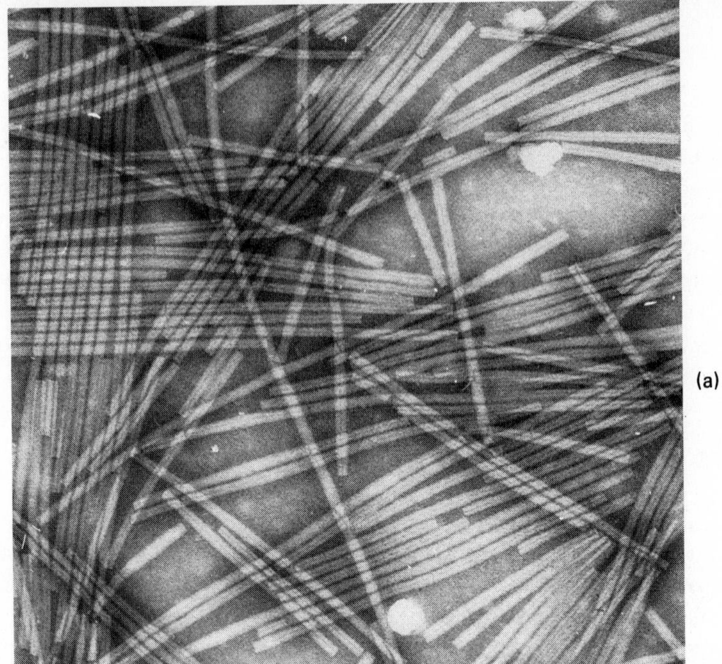

(a)

FIGURE 6-1 Examples of plant viruses as observed with the electron microscope. (a) Rigid rods of tobacco mosaic virus (b) Flexuous rods of aphid borne cowpea mosaic virus (c) Pohyhedrons of ash ringspot virus. (Photographs compliments of Dr. Garry Lahey.)

(b)

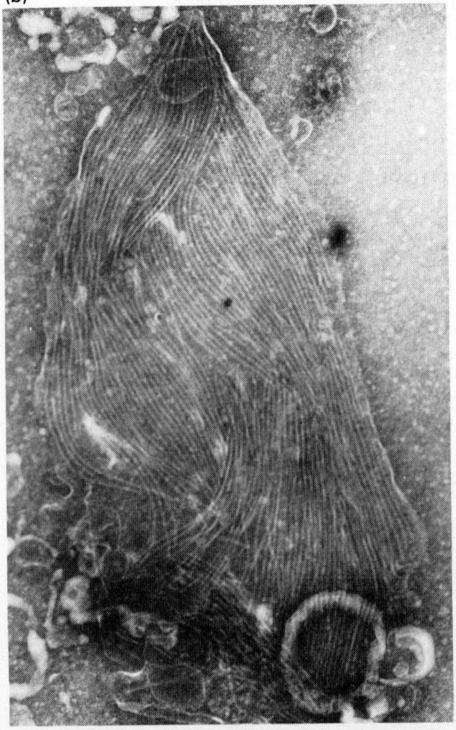

(c)

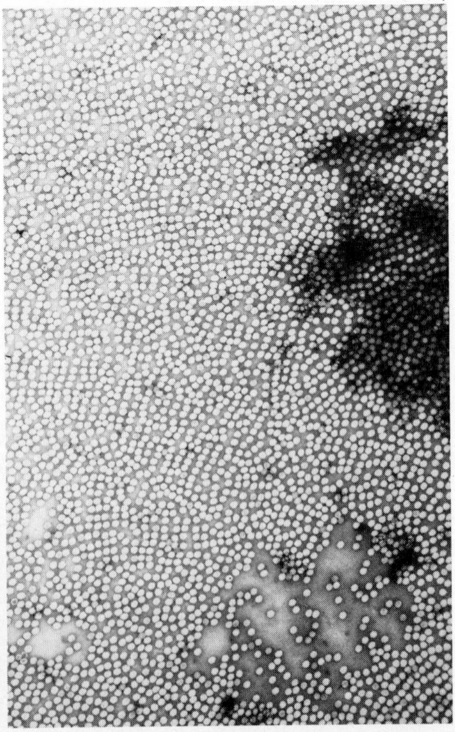

known protein. Obviously, it would be of interest to determine how such a short piece of RNA can affect plant function.

Plant viruses as observed with the electron microscope appear as rigid rods, flexuous rods, or polyhedron shapes (Fig. 6-1). Different viruses may be of different sizes, but all the infectious particles of a given virus will generally be of the same size. Some viruses are multicomponent, meaning that the infectious agent is made up of more than one particle.

The hosts for viruses appear to be limited only by our ability to recognize them. There is no reason to expect that any taxonomic group of organisms is free of viruses. The lack of viruses in certain taxonomic groups of plants, such as gymnosperms, ferns, and mosses, probably simply reflects the fact that they have not been extensively looked for in these groups.

Virus diseases of plants have been recognized since 1892, when Iwanowski demonstrated that the infectious agent of tobacco mosaic virus (TMV) would pass through a bacteria-proof filter (Fig. 6-2). Figure 6-3

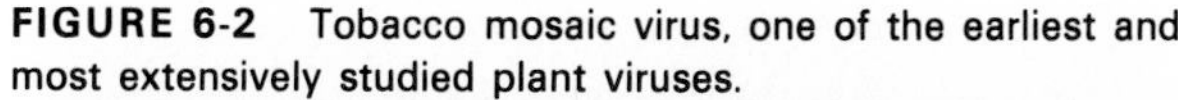

FIGURE 6-2 Tobacco mosaic virus, one of the earliest and most extensively studied plant viruses.

shows an example of tulip breaking, a virus disease of tulips recognized many years before viruses were known to exist. But it was not until the 1930s when the electron microscope was invented that viruses could be seen.

In 1935, Stanley demonstrated that viruses were composed of crystalline protein. It was three years later, when Bawden and Pirie showed that Stanley's virus protein was not a pure substance but one also containing a nucleic acid, that the true structure of viruses became known.

The naming of plant viruses is confusing. We use common names for viruses based upon the symptoms they produce. For example, ash ringspot virus (ARSV) produces chlorotic ring-like patterns on ash leaves (Fig. 6-4). The same symptoms are produced on ash leaves by tobacco ringspot virus (TRSV). It is a matter of choice which name to apply, so that we end up with numerous common names for the same virus. The name does not always fit the symptoms, either. For example, the ash ringspot virus may look more like chlorotic line patterns or chlorotic spots than chlorotic rings. Virologists cannot seem to agree upon a standardized system of nomenclature for viruses. Until they do, we will just accept the confusing common names.

FIGURE 6-3 Tulip breaking, a virus disease of tulips that was recognized long before viruses were known to exist.

FIGURE 6-4 Line patterns in ash associated with ash ringspot virus.

MODE OF ACTION OF PLANT-PARASITIC VIRUSES

Plant viruses are infectious agents which act like dominant gene-like factors within plant cells. They are obligate parasites and require living cells for their multiplication.

The process of virus replication within living cells disrupts normal metabolic functions. The single-strand RNA virus associates with ribosomes to produce a replicase enzyme which in turn mediates the production of complimentary strands of RNA from the virus RNA. The new RNA strands then produce complementary RNA strands identical to the original virus RNA. Nucleic acid bases for production of virus RNA and enzymes for RNA synthesis are derived from the plant cell. Production of virus RNA increases the demand on the cell for the metabolic synthesis of nucleic acids.

The virus RNA strand acts as a messenger RNA strand within the plant cell. In conjunction with plant ribosomes, the virus RNA strand also acts as a template for virus protein synthesis. Virus protein synthesis places additional demands on normal plant metabolism for the production of amino acids.

The assembly of protein coat subunits around virus nucleic acid appears to occur automatically. The assembled virus particles are sometimes enclosed within a vesicle. In some viruses they accumulate within plant organelles.

The overall effect of plant viruses is slowly to starve the cell and the organism of nutrients and energy. Plant viruses, in contrast to animal and bacterial viruses, seldom produce plant mortality. The usual effect is so subtle that it is difficult to distinguish morphological differences between virused and nonvirused plants.

Although viruses generally produce nonlethal effects on plants, it is usually possible to find some plants which are highly sensitive to virus infection. These plants produce necrotic spots, termed local lesions, where virus infection occurs. The necrosis of infected cells prevents further invasion and is therefore one type of resistance. The use of local lesion hosts in virus research will be discussed in the section on methods.

VIRUS DISEASE CYCLE

Virus particles enter cells through slight injuries or are injected by insect vectors.

Within living cells, as discussed above, the virus RNA molecule causes the cell to produce a complementary RNA strand, which then produces more virus RNA in the normal manner of nucleic acid synthesis. The virus RNA molecules also act as messenger RNA (mRNA). In conjunction with ribosomes, the virus RNA directs the synthesis of the virus protein coat (Fig. 6-5).

Virus RNA and coat protein somehow assemble into virus particles. The completed virus particles may accumulate in storage vacuoles within cell organelles or in the cytoplasm. These accumulations of virus particles form crystalline structures which are sometimes observable with the light microscope.

Viruses move from cell to cell through protoplasmic strands called plasmodesmata. They can also become distributed throughout the plant in the phloem or xylem vascular system.

Viruses require outside assistance to get from one plant to another. Horticultural activities such as grafting, vegetative propagation, and even cultivation sometimes move viruses from one plant to another. Other viruses rely on specific insect vectors, such as leafhoppers, aphids, white flies, thrips, and beetles.

Mites, nematodes, and fungi can also serve as vectors. Some viruses are carried in seeds from one generation to the next, and a few are carried in pollen. Even though pollen and seed transmission are infrequent means of virus dissemination, the occurrence is sufficient to initiate virus in a popula-

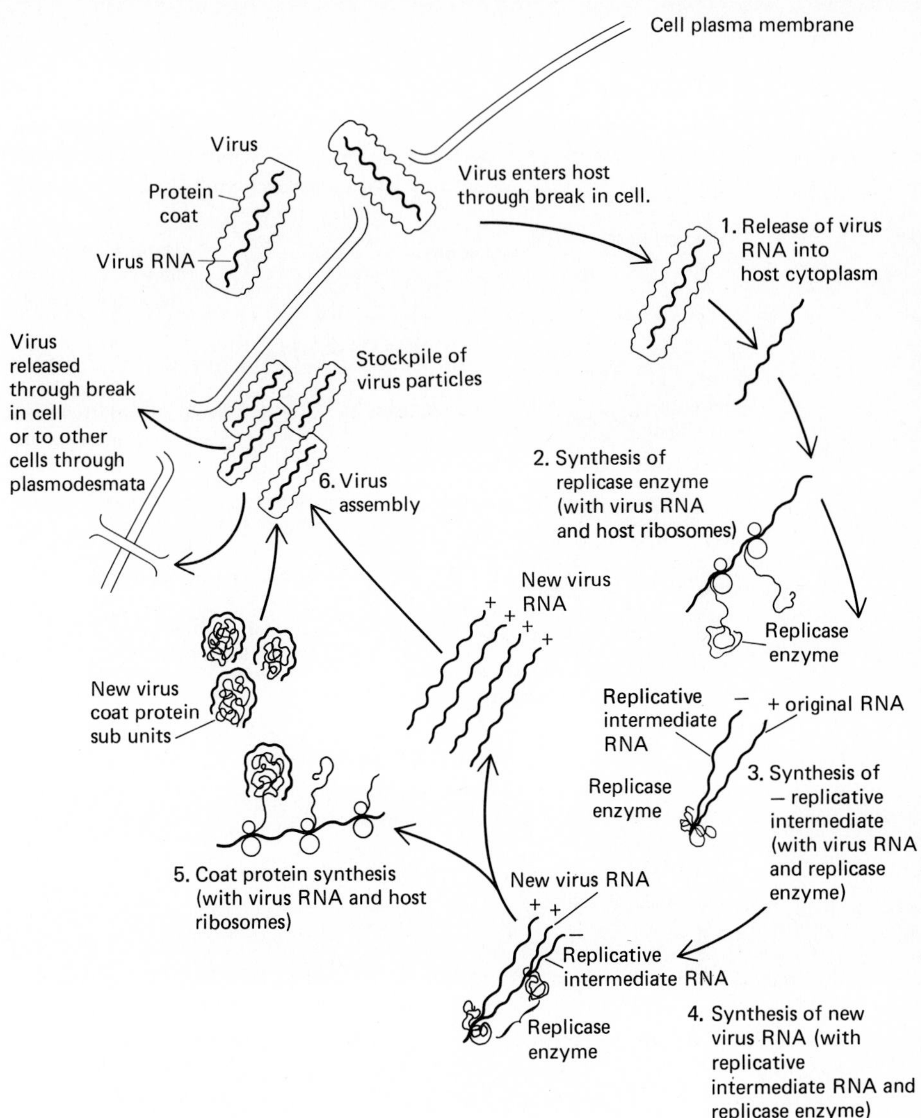

FIGURE 6-5 Tobacco mosaic virus "life cycle".

tion for subsequent short-range dispersal and intensification by insect and nematode vectors.

The need for vectors to move from plant to plant caused viruses to evolve varying types of associations, primarily with sucking insects such as aphids and leafhoppers. These associations with insects include mechanical transmission of viruses, which contaminate the insect mouth parts, to ingestion and multiplication within the insect prior to transmission to uninfected plants. A third type of virus–insect vector association involves ingestion by the insect and circulation without multiplication of virus within

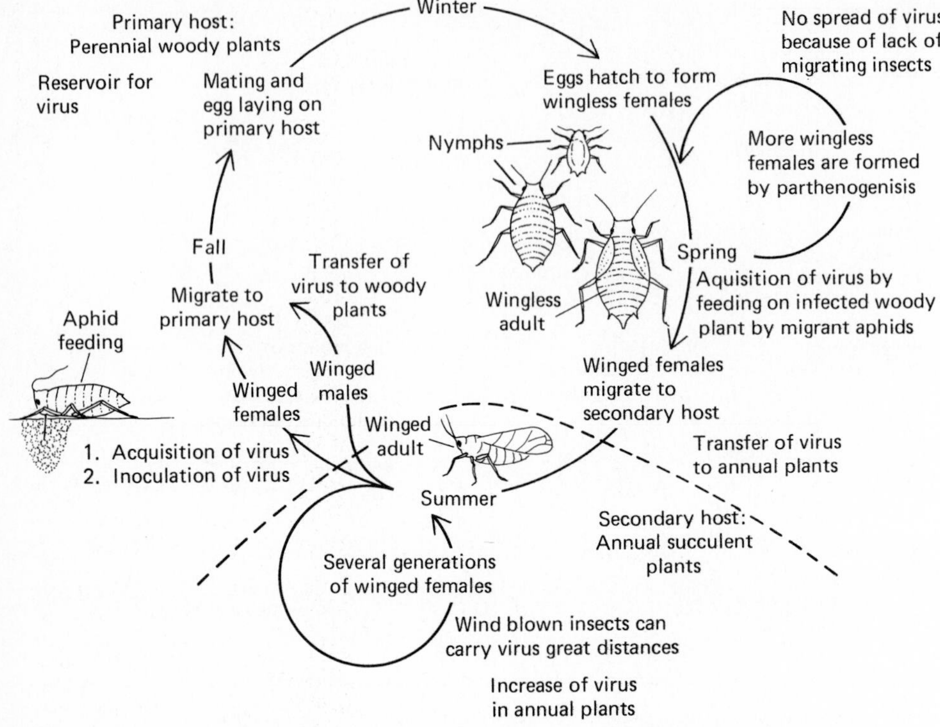

FIGURE 6-6 Aphid life cycle and its relationship to virus diseases of plants.

the insect, before infection of another plant takes place during subsequent feeding.

Aphids are ideal vectors of virus (Fig. 6-6). The aphids' perennial host provides a long-term virus reservoir. The aphids' succulent annual host is ideally suited for maximum virus increase. Seasonal shifts in aphid feeding ensure virus survival.

SYMPTOMS OF VIRUS DISEASES

The effects of viruses on their hosts result in a variety of symptoms and in some cases no visible symptoms. Typical symptoms involve flower color streaks; foliage ring spots, flecks, necrotic lesions, mosaic or mottling patterns (Fig. 6-7), chlorosis, veinal chlorosis, marginal chlorosis, line patterns, curling, and dwarfing; and stem morphology changes, excessive branching (witches' brooming) and rosettes. The subtle effects of symptomless or hidden viruses may produce only yield losses. If nearly all of the plants of a given species are affected, it is almost impossible to recognize a virus by its effect on the host.

The role viruses play in ecological community relationships is totally unexplored. Even though we have virtually no evidence as to the effects of viruses on natural communities, we can speculate from our knowledge of agricultural problems that they should have a profound effect.

METHODS FOR RECOGNITION AND VERIFICATION OF VIRUS DISEASES

We may think that a particular set of symptoms are caused by virus, based on the characteristic nature of virus symptoms, but other disease agents can cause some of the same symptoms. Therefore, transmission of the suspected causal agent by grafting plant tissue that shows virus symptoms to a healthy plant, followed by the development of symptoms in the formerly healthy plant, is a beginning step in the verification of virus etiology. Mycoplasma-like organisms (discussed in Chapter 7) can also be transmitted by grafting, so that additional steps are necessary. Increased understanding of other disease agents, plus improvements in techniques available to researchers, have expanded our methods for the identification of virus diseases. These methods will be discussed briefly next.

Transmission Studies

The suspected causal agent should be transmitted to and produce symptoms in healthy hosts, such as selected varieties of bean, cucumber, tobacco,

FIGURE 6-7 Bean yellow mosaic virus is common in garden beans.

FIGURE 6-8 Indicator plants used in virus transmission studies.

Datura, Gomphrena, and a number of other indicator plants (Fig 6-8). Based on the host range and symptoms produced on these herbaceous hosts, the suspected virus can be related to known viruses.

Transmission to herbaceous hosts usually involves extraction of plant juice from the diseased plant and mechanical inoculation of the test plant with the extract. Carborundum or other fine abrasives are used to produce small injuries for virus infection. Precautions to prevent denaturation of proteins and destruction of nucleic acid during extraction must be exercised. Buffers to control pH and additives to inhibit protease and nuclease enzymes are used.

Stability of Virus Extract

Various properties of the virus in crude plant sap, such as pH effects, thermal inactivation, and dilution end point, are determined. It is ideal to have some method of quantification of the amount of active virus in the extract, so that a herbaceous host which produces specific necrotic flecks at points of infection by the virus is desirable. This is called a local lesion host. Using the number of lesions produced per quantity of extract applied to the leaf as a quantitative indicator of virus, the investigator determines the optimum pH and pH range for the virus. Next, the minimum temperature necessary to inactivate the virus (no lesions on the local lesion host) is determined, together with an estimate of the number of infectious virus particles in the extract. This is done by determining how much dilution of the extract is necessary to prevent infection of the local lesion host. These properties are compared with the properties of known viruses to further the identification of the unknown virus.

Purification of Virus

The methods used for the purification of viruses are those of the protein chemist. Buffers are used to prevent denaturation of the virus protein and to enhance denaturation of specific plant proteins in the extract. Low-speed centrifugation is used to separate cell fragments from the virus in solution. High-speed centrifugation is then used to concentrate the virus into a pellet, which is then resuspended in buffer. Low-molecular-weight compounds, such as sugars, amino acids, and salts, are separated from high-molecular-weight substances such as proteins and viruses (protein and nucleic acid) by dialyzing the small molecules through a semipermeable membrane. The addition of ammonium sulfate to a partially purified protein solution causes precipitation of specific proteins, depending upon the concentration of ammonium sulfate used, so that some plant proteins in the extract are separated from the virus by this method.After each step, a bioassay of the extract is performed on the local lesion host to verify which fraction contains the virus. Virus in solution will absorb ultraviolet light at a wavelength of 260 nm. The amount of absorption of ultraviolet light in a spectrophotometer is therefore used as a qualitative and quantitative assay tool during the purification process.

Sedimentation Properties

Purified virus particles suspended in solution will settle in an ultracentrifuge at different rates depending on the mass of the particle. Usually, the centrifuge tube on which the virus preparation is placed contains a series of layered increasing concentrations of sucrose solutions from the top to the bottom. Higher concentrations of sucrose produce a more dense liquid through which the virus migrates more slowly. The gravimetric force on the virus particle caused by the high-speed centrifugation plus the interacting migration-retarding effect of increasing density of solution produce very good separation of particles according to mass and volumes. Identification of the virus layers is accomplished by scanning the centrifuge tube with a spectrophotometer. The virus layer is recognized by its absorbence at 260 nm and verified by infectivity on the local lesion host. The size of the virus particle can be calculated from the sedimentation properties.

Electron Microscopy

Crude extracts, partially purified virus preparations, or highly purified preparations from density gradient centrifugation can be examined with the electron microscope. Differential staining with electron-dense materials or shadow casting with heavy metals is necessary to observe the virus. The size and shape of the virus is determined using the electron microscope.

Serological Assay

Antigen-antibody reactions in a serological assay are used to determine the similarity of the newly observed virus to other, known viruses. The test is not absolute proof of identity but rather an indication of similarity. The procedure is as follows (see Figs. 6-9 and 6-10). A purified virus preparation of a known virus is injected into the bloodstream of an animal such as a rabbit. The foreign virus protein is an antigen which induces the rabbit to produce a specific protein antibody to inactivate the foreign substance. Antibodies bind to antigen in the first step of inactivation. More antibodies are produced than are necessary. Therefore, one can purify from the blood of the rabbit a protein fraction antiserum with a high concentration of the specific antibodies. The antigen–antibody interaction within animals is the basis of acquired resistance by exposure or immunization used by the medical profession. The virologist can place a portion of a purified antiserum preparation, made by him or purchased from a central supplier, in a small depression in a petri plate containing agar. A sample of the original virus is placed in another depression in the agar and a sample of the unknown virus is placed in a third depression. The proteins in the antiserum, as well as the virus, diffuse into the agar. Where they come in contact, the antibodies of the antiserum will chemically bind to the virus protein much as they do in the

FIGURE 6-9 Preparation of antibodies to a virus.

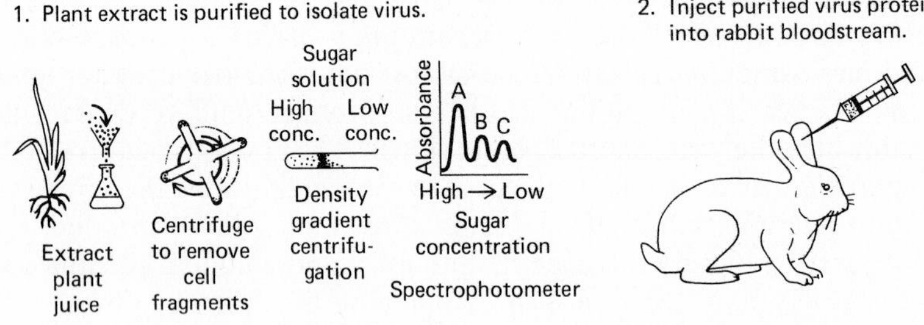

3. Antibodies are formed in the bloodstream to the foreign protein.

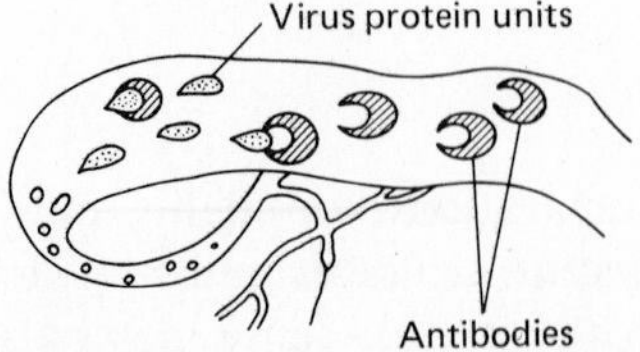

4. Bleed the rabbit and purify the fraction containing the antibodies.

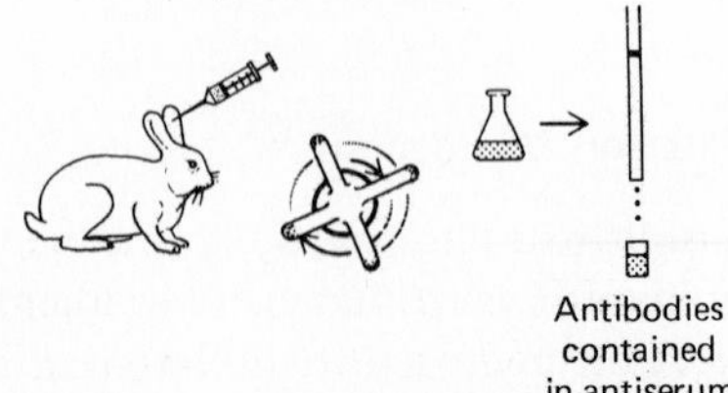

1. Extract sap from suspected virused plants.

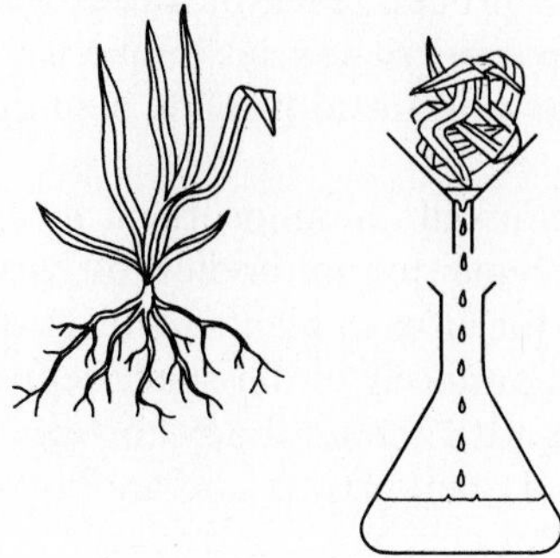

2. Place antiserum in center well and plant extracts in outer wells.

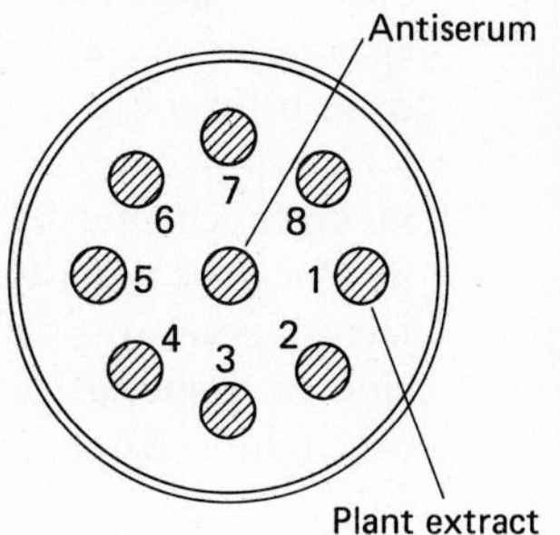

3. Examine plates for precipitation zones; extract from plants 2 and 5 contain the virus specific for the antiserum.

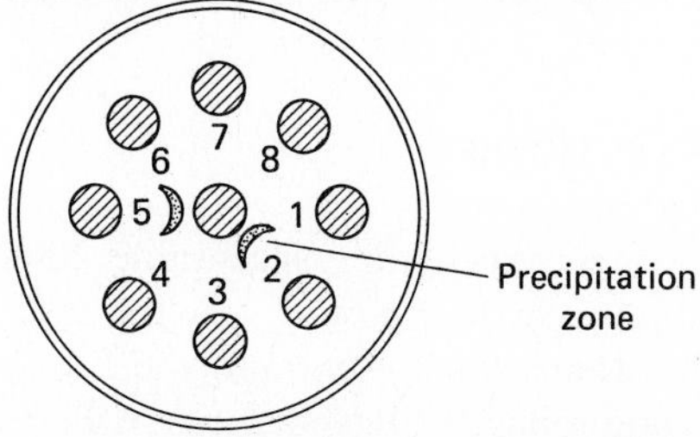

4. The precipitation is an antigen-antibody reaction similar to what takes place in the rabbit

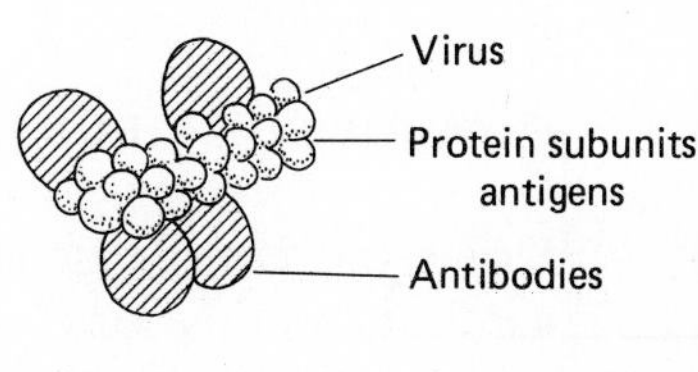

Figure 6-10 Serological test of plant extracts for a specific virus.

animal's bloodstream. A line of precipitated antibody–virus protein antigen will appear in the agar plate between depressions with serologically similar virus and animal antiserum.

Vectors

A final step in characterizing a virus is to identify the vectors. Once the virus is identified, the identification of the vector is somewhat easier. Known vectors of similar viruses are first checked to see if they are involved in the disease cycle in the plant under study. For example, serological and other properties of ash ringspot virus demonstrated a relationship to tobacco ringspot virus. Dagger nematodes (*Xiphinema* spp.) are known vectors of tobacco ringspot virus. Therefore, it was logical to check out nematodes as vectors of ash ringspot virus. The dagger nematodes were shown to be the vectors of ash ringspot virus.

Quick Methods for Recognition of Viruses

Until recently, virus identification was a slow process. Identifications were based primarily on characteristic plant symptoms. A new technique using enzyme-linked antibodies for specific viruses appears to provide a simple diagnostic tool for virus surveys.

A procedure known as ELISA (*e*nzyme-*l*inked *i*mmunosorbent *a*ssay) involves chemically binding a specific enzyme to antibodies of virus produced as described previously. The assay for virus in plants involves the introduction of plant sap and enzyme-linked antibody into a small depression in a special plate. The virus will react with the antibody to form a precipitate. After washing the nonprecipitated plant extract and antibodies from the plate, the specific substrate for the enzyme is introduced and allowed to react. A small quantity of virus–antibody–enzyme precipitate can then be visually recognized by the change in color of the substrate. The enzyme can produce a large quantity of chemical reaction, thereby enhancing recognition of very dilute concentrations of virus within plant extracts. In many instances, virus concentrations below levels necessary for the production of symptoms in plants can be recognized by the ELISA technique.

EXAMPLES OF VIRUS DISEASES OF TREES

Using the criteria described above as reasonably acceptable tests for verification of the virus etiology of a disease, we know of very few virus diseases of North American trees other than *Prunus* spp. and the fruit trees. These include tobacco mosaic virus and tobacco ringspot virus of white ash, elm mosaic virus of American elm, apple mosaic virus and cherry leafroll virus of birch, poplar mosaic virus, arabis mosaic virus, potato virus Y and tobacco necrosis virus of poplars, and oak mosaic virus of California oak.

The significance of these viruses and others to the health of tree populations is not yet known, but the ash and birch viruses are thought to play a role in the decline of the species (see Chapter 18).

The lack of well-documented examples of tree virus diseases is somewhat misleading. Very few people have ever really looked for viruses in forest trees. Van der Plank assumes that viruses are not present in wild populations of trees because the trees are resistant. In Chapter 18, I speculate that viruses are common but do not express symptoms readily except during periods of stress to the plant. Others suggest that trees such as poplars, which are vegetatively propagated, ultimately deteriorate with time because of the presence of viruses. Virus diseases of trees is a subject that we know very little about, primarily because there are very few people doing research on forest and shade trees. There is no reason to assume that viruses are any less important in trees than they are in other plants.

CONTROL OF VIRUS DISEASES

The control of plant viruses is generally that of prevention. Diseased plants usually do not recover and are therefore discarded. In agriculture, considerable effort is put into producing virus-free planting stock. A grower may spray with insecticides to reduce vectors once the crop is planted. Heat treatment is sometimes used by plant breeders to eliminate virus from foundation stock. Virus-free plants can sometimes be generated from apical meristems abscised from rapidly growing plants. Tissue culture media and hormones are used to induce rooting of the abscised meristems to develop virus-free plants.

REFERENCES

BALL, E. M. 1974. Serological tests for the identification of plant viruses. The American Phytopathological Society, Inc., St. Paul, Minn. 31 pp.

BAWDEN, F. C., and N. W. PIRIE. 1938. A plant virus preparation in a fully crystalline state. Nature *141*:513–514.

BERBEE, J. D., J. O. OMUEMU, R. R. MARIN, and J. D. CASTELLO. 1976. Detection and elimination of viruses in poplars. *In* Intensive plantation culture. USDA For. Serv. Gen. Tech. Rep. NC 21, pp. 85–91.

COOPER, J. I., and J. B. SWEET. 1976. The detection of viruses with nematode vectors in six woody hosts. Forestry *49*:73–78.

FORD, R. E., H. E. MOLINE, G. L. McDANIEL, D. E. MAYHEW, and A. H. EPSTEIN. 1972. Discovery and characterization of elm mosaic virus in Iowa. Phytopathology *62*:987–992.

GIBBS, A. J., B. D. HARRISON, D. H. WATSON, and P. WILDY. 1966. What's in a virus name? Nature *209*:450–454.

GOTLIEB, A. R., and J. G. BERBEE. 1973. Line pattern of birch caused by apple mosaic virus. Phytopathology *63*:1470–1477.

HARDCASTLE, T., and A. R. GOTLIEB. 1977. Detection of the yellow birch strain of apple mosaic virus (ApMv) using an enzyme-linked immunosorbent assay (Elisa). Proc. Am. Phytopathol. Soc. *4*:188.

HIBBEN, C. R., and R. F. BOZARTH. 1972. Identification of an ash strain of tobacco ring spot virus. Phytopathology *62*:1023–1029.

MATTHEWS, R. E. F. 1970. Plant virology. Academic Press, Inc., New York. 788 pp.

NAVRATIL, S., and M. G. BOYER. 1968. The identification of poplar mosaic virus in Canada. Can. J. Bot. *46*:722–723.

SELISKAR, C.E. 1964. Virus and virus-like disorders of forest trees. FAO/IUFRO Symp. Internationally Dangerous For. Dis. Insects, Oxford, July 20–30, 1964. 44 pp.

SMITH, K. M. 1972. A textbook of plant virus diseases. Academic Press, Inc., New York. 684 p.

STANIER, R. Y., M. DOUDOROFF, and E. A. ADELBERG. 1970. The microbial world, 3rd ed. Prentice-Hall, Inc., Englewood Cliffs, N.J. 873 pp.

STANLEY. W. M. 1935. Isolation of a crystalline protein possessing the properties of tobacco mosaic virus. Science *81*:644–645.

VAN DER PLANK, J. E. 1975. Principles of plant infection. Academic Press, Inc., New York. 216 pp.

VOLLER, A., A. BARTLETT, D. E. BIDWELL, M. F. CLARK, and A. N. ADAMS. 1976. The detection of viruses by enzyme-linked immunosorbent assay (Elisa). J. Gen. Virol. *33*:165–167.

WESTER, H. V., and E. W. JYLKKA. 1959. Elm scorch, graft transmissible virus of American elm. Plant Dis. Rep. *43*:519.

7

BACTERIA AS AGENTS OF TREE DISEASES

- Types of bacteria
- Mode of action of plant-pathogenic bacteria
- Disease cycle of plant-pathogenic bacteria
- Symptoms of bacterial diseases
- Methods for recognition and verification of bacterial diseases
- Examples of bacterial diseases
- Role of soil bacteria in nutrient cycling and indirectly in plant health
- Control of bacterial disease

Bacteria are found in a wider variety of habitats than are any other group of organisms. *Clostridium botulinum,* a soil-inhabiting bacterium, can survive the boiling temperatures used in food processing and develop even when sealed inside an airtight container. A toxin is produced under these conditions that is so potent that it would take only a few grams to kill every person on the planet.

Millions of bacteria occupy every cubic centimeter of topsoil. They are found inside and outside healthy as well as diseased plants and animals. Bacteria are found in association with infections caused by organisms such as fungi and nematodes and may influence the development of these disease-causing agents.

TYPES OF BACTERIA

Bacteria are unicellular procaryotic microorganisms (Figs. 7-1 to 7-3). They may be spiral, rod-like, or spherical in shape as seen with the light microscope. Other morphological shapes, such as spring-like, mycelial, and pleomorphic, are also possible. Special stain procedures demonstrate that some bacteria have flagella on the outer surface.

Electron microscopy reveals that bacteria lack mitochondria, golgi bodies, and the endoplasmic reticulum found in eucaryotic organisms. There is also no membrane around the nuclear region. Ribosomes and a nuclear region are the main features of bacteria as observed with the electron microscope.

FIGURE 7-1 Rod, spherical, and spiral bacteria as seen under high magnification with the light microscope.

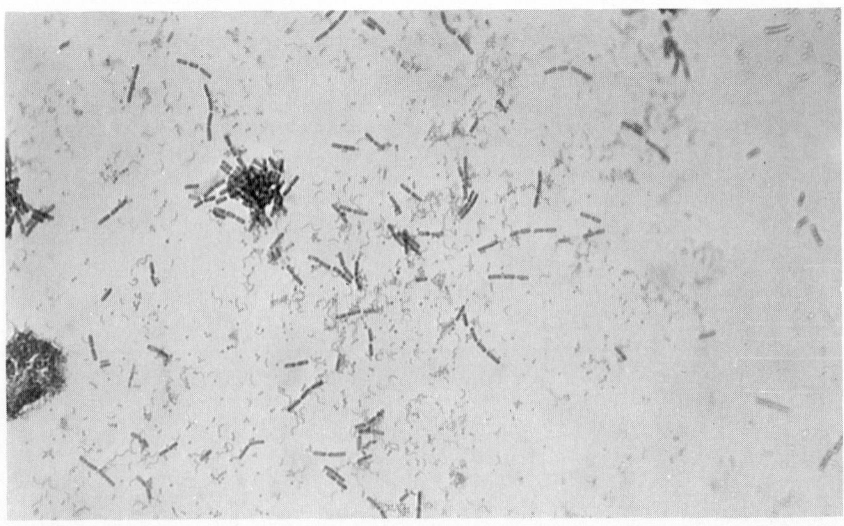

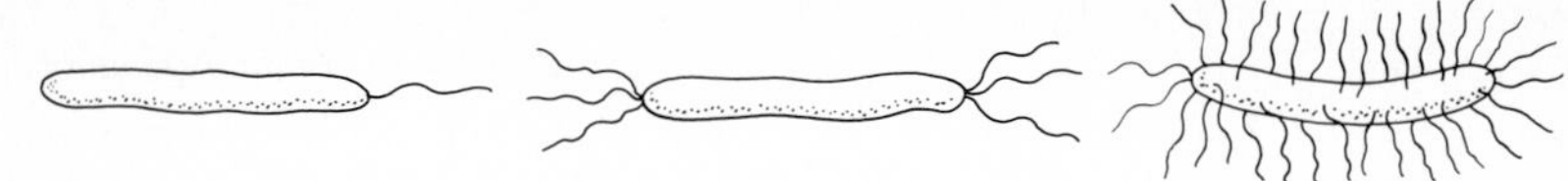

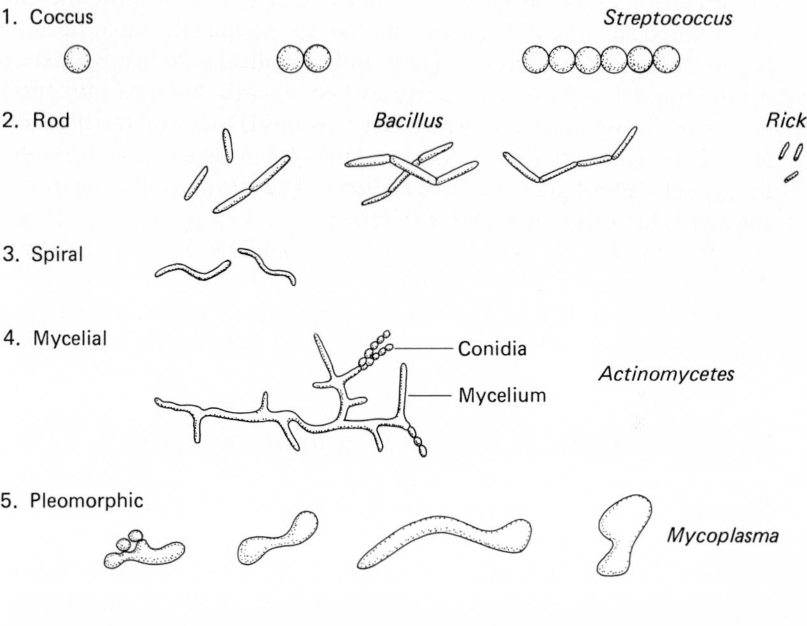

FIGURE 7-2 Bacterial morphology.

FIGURE 7-3 Comparison of procaryotic and eucaryotic cell morphology.

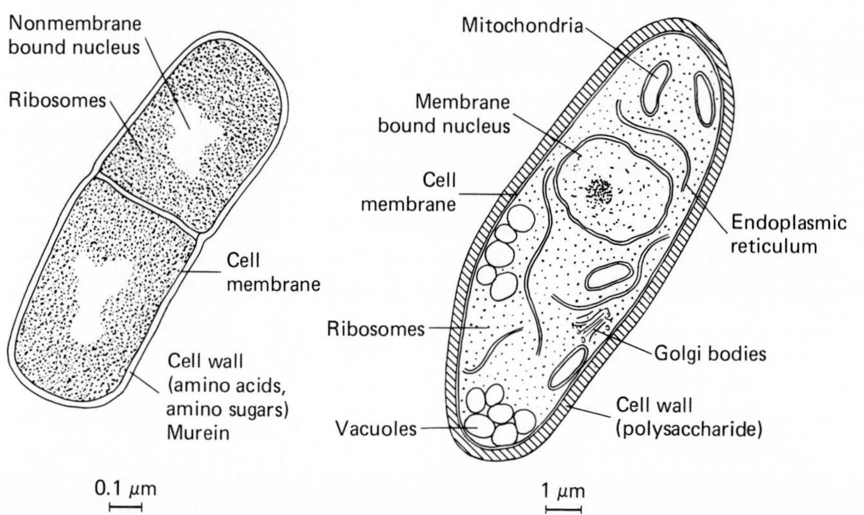

Species of higher organisms are characterized by groups of interbreeding morphologically similar individuals. Vegetatively propagating bacteria cannot be grouped by interbreeding criteria, and the range of morphological variation is too limited to adequately separate dissimilar groups. Biochemical, physiological, and ecological criteria are superimposed on the limited structural diversity to help classify bacteria into species.

A major separation into two large groups is made on the basis of a cell-wall Gram-stain procedure. The Gram-stain binds crystal violet with iodine to the cell walls of gram-positive bacteria. The crystal violet does not bind to the cell walls of gram-negative bacteria.

Morphological variation of bacteria is limited. Therefore, other criteria are utilized to characterize bacterial species. Oxygen, nitrogen, and carbon requirements; fermentation products; gas evolution when grown on specific media; liquefication of gelatin medium; and many other criteria are used.

MODE OF ACTION OF PLANT-PATHOGENIC BACTERIA

Plant-pathogenic bacteria produce disease by enzymatic maceration of plant cell walls and by secretion of toxic substances. The primary walls and middle lamella of succulent plant cells are dissolved by cellulase and pectinase enzymes, while the plasma membrane, which controls the osmotic concentration of the protoplasm, is disrupted by bacterial toxins.

Another mode of action of bacteria involves interference with cell division and differentiation. This mechanism is described in more detail for crown gall later in the chapter.

DISEASE CYCLE OF PLANT-PATHOGENIC BACTERIA

Bacteria have no means of actively penetrating and infecting plants. They rely on wounds made by other organisms or natural openings such as stomata or hydathodes in leaves or openings in flower parts for ingress into plants. Once inside, a general softening and dissolving of cell walls and membranes by bacterial extracellular enzymes allows bacteria access to additional tissues.

Bacterial dispersal from one plant to another generally requires splashing rain or an insect vector, because plant-pathogenic bacteria will not survive the desiccation of wind dissemination. Some bacteria (none of the plant parasites) survive in desiccating environments by producing an extracellular polysaccaride slime outer shell or by production of endospores.

Bacteria generally reproduce by transverse binary fission. Duplication of the single circular chromosome is followed by separation into two daughter cells.

Genetic recombination in bacteria has been studied extensively, but few of the studies have used plant-pathogenic bacteria. The genetic information in bacteria is carried on one circular DNA chromosome but may also be carried on one or more small strands of DNA called plasmids. Some plasmids function independently of the chromosomal DNA. Others are involved in replication of the chromosomal DNA and in mediating transfer of the genetic information from one bacterium to another.

Genetic recombination in bacteria occurs by one of five mechanisms.

1. *Plasmid transfer* from one bacterium to another can occur between and within species of bacteria. This extrachromosomal transfer of genetic information can be used in the laboratory to study gene functions and to produce new bacteria, which may be very useful in the production of industrial and pharmaceutical chemicals. It may also result in the formation of a new bacterium with resistance to antibiotics and other properties which may or may not generate future problems. There is, at this time, much controversy over the hazards and benefits of genetic research in the laboratory. In any case, it is important to recognize that bacteria may utilize these same mechanisms in the field when it is of a selective advantage.

2. *Conjugation* involves one cell, acting as a donor cell, which injects a copy of its chromosomal DNA strand into a receptor cell. The DNA is incorporated into the receptor genetic information so that some of the characteristics of the donor cell are expressed by the fission products of the acceptor cell. Plasmids mediate both the replication and transfer of the chromosomal DNA.

3. *Transformation* involves an incorporation of DNA from one bacterium into another in the absence of contact between the two bacteria. The donor bacterium lyses. Its DNA is released into the medium and incorporated into the receptor bacterium.

4. *Transduction* involves transfer of DNA from one bacterium to another by a virus (bacteriophage). Virus multiplication in one cell incorporates some of the bacterial DNA into the virus genome. Release of the virus following lysis of the cell and subsequent virus infection of another cell introduces the virus and donor-cell DNA into a receptor cell.

5. *Mutation* occurs in all organisms. These errors in the replication of genetic material are usually lethal or of a selective disadvantage to the organism. When the genetic change is a selective advantage, the mutation or recombinants produced through any of the foregoing mechanisms may provide the bacterium with a way of adapting to a changing environment.

SYMPTOMS OF BACTERIAL DISEASES

Bacteria produce diseases of various plant parts with characteristic symptoms for each. Bacterial pathogens of leaves and shoots produce water soaking initially, followed by necrosis and maceration of tissues. Bacterial infection of vascular tissue produces wilting. Mycoplasma infection of phloem causes yellowing and witches' brooming. Crown gall bacterial infection is characterized by hypertrophy of stem tissue. Bacterial infection of wood results in increased permeability due to perforation of pit membranes.

METHODS FOR RECOGNITION AND VERIFICATION OF BACTERIAL DISEASES

Bacterial infection of succulent tissues is often diagnosed based on characteristic symptoms. Microscopic examination of diseased tissue should reveal masses of bacteria streaming from infected tissue. Darkfield indirect illumination enhances the visibility of bacterial streaming.

Thorough diagnosis of bacterial diseases usually involves isolation on culture media. Infected tissue is shaken in sterile water or liquid media to separate the bacteria from the tissue. Dilutions of the initial bacterial suspension are mixed in agar media held at about 60°C. The media is solidified by cooling and the bacteria are allowed to develop for a few days. Colonies of shiny cream to light-colored spots are characteristic for bacteria. Individual colonies are subcultured to agar slants.

Identification of species of bacteria requires specific staining procedures to observe morphology and selected chemical tests to determine growth requirements. *Bergey's Manual of Determinative Bacteriology* (1974) is the standard reference used for identification of bacteria.

If the bacterial disease is not described in the literature it is necessary to go through Koch's proof of pathogenicity, as described in Chapter 1.

Specialized methods are necessary for the recognition of mycoplasma diseases. These are described for elm phloem necrosis later in the chapter.

EXAMPLES OF BACTERIAL DISEASES

There are fewer than 200 species of plant-pathogenic bacteria, with only a few causing problems in trees. A few selected plant pathogenic bacteria are listed in Table 7-1. *Rhizobium* is included in the table to show its taxonomic relationship to plant-pathogenic bacteria. Nitrogen fixation by *Rhizobium* and other bacteria is discussed later in the chapter. The question marks after *Rickettsia* and *Mycoplasma* are there to emphasize our incomplete understanding of these two groups of plant pathogens. They have been identified primarily on the basis of electron microscopy of diseased plant tissue.

A cursory examination of the bacterial diseases of forest and shade trees

TABLE 7-1 Taxonomic Classification of Bacteria Associated with Plants

Class	Order	Family	Genus	Disease
Schizomycetes	Pseudomonadales	Pseudomonadaceae	*Pseudomonas*	Bacterial canker and gummonis of stone fruits
			Xanthomonas	Leaf blight of walnut
	Eubacteriales	Rhizobiaceae	*Agrobacterium*	Crown gall and hairy root
			Rhizobium	Nitrogen fixation
		Enterobacteriaceae	*Erwinia*	Fire blight, wet wood, and soft rot of vegetables
		Corynebacteriaceae	*Corynebacterium*	Ring rot of potato
	Actinomycetales	Streptomycetaceae	*Streptomyces*	Potato scab
	Rickettsiales	Rickettsiaceae	*Rickettsia ?*	Pierce's disease of grape, phony peach
Mollicutes	Mycoplasmatales	Mycoplasmataceae	*Mycoplasma ?*	Phloem necrosis
			Spiroplasma	Citrus stubborn

listed in a USDA handbook, *Diseases of Forest and Shade Trees of the United States*, shows that there are fewer than 100 reported for the 215 species of trees covered in the book. Closer examination of the references indicates that very few present-day forest pathologists are working on bacterial diseases, since many of the references are at least 10 years old. A possible conclusion from these observations might be that bacterial diseases do not represent a serious threat to forest and shade trees. But this does not reflect the true significance of bacterial pathogens. Most present-day forest pathologists are very inadequately trained in the modern techniques of bacteriology. They stay away from bacterial problems and rely on the poorly founded concepts of a former generation, which recognized bacterial diseases and studied them as well as was possible.

Bacteria have evolved to fit into almost every niche available, so there is no reason to assume that they have not developed pathogenic capacity for trees.

The tree will be divided into functional parts, and examples of bacterial disease will be presented to characterize the role that bacteria play in each of the parts.

FIGURE 7-4 Fire blight on apple causes a shoot blight recognized by the blackened wilted leaves and the curled shoot tip.

FIGURE 7-5 Canker caused by *Erwinia amylovora* is the site for overwintering of the bacterium.

Shoot Blights

The succulent tissues of newly forming shoots are ideally suited for bacterial invasion. Pectolytic enzymes of the pathogen readily macerate the succulent nonlignified cell walls.

A good example of a shoot blight disease is fire blight, caused by *Erwinia amylovora* (Figs. 7-4 and 7-5). This was the first bacterial disease to be studied by plant pathologists and represents an example of a disease agent that we exported to Europe. Fire blight is one of the most serious diseases of apple and pear trees. It is largely responsible for the elimination of major pear production in the Northeast and the establishment of pear growing as a major crop in the Pacific Northwest. In addition to apple and pear, quince, hawthorn, mountain ash, sweet cherry, and serviceberry are affected by the bacterium.

Even though this was the first bacterial disease of plants studied by plant pathologists, we still know very little about the disease. Work done about the turn of the century established the disease cycle (Fig 7-6): survival of the bacterium over winter in stem cankers (see Fig 7-5); oozing of bacteria from the cankers in the spring; dissemination of the bacterial inoculum from cankers to young shoots and flower parts by splashing rain and insects; secondary rapid spread of the bacterium to additional flowers by bees and other insects; infection through hydathodes of the flower and stomates of the leaf; and invasion and maceration of parenchyma and cambium tissues of the leaf, petiole, shoot, and branch.

The major deficiencies in our knowledge of the disease involve the role of environmental factors in disease intensification. In certain years, the disease is very destructive, causing major amounts of damage to many of apple and pear varieties. Then for a period of years, it is almost impossible to find new infections. The need for rain to enhance the infection and spread is well established, but the interactions among amount of inoculum, fluctuating weather conditions, flower emergence, and fruit set are not understood.

Control of fire blight is accomplished by pruning out the cankers to remove the source of inoculum to prevent future spread and spraying the trees with the antibiotic streptomycin to prevent infection in the spring. Pruning should be done in late summer or winter to avoid spreading bacteria by means of pruning tools. Another recommended precaution is to dip tools in Chlorox after each cut. Chemical sprays are also used, but the effectiveness of the protectants is difficult to assess because of the variation in spread of the pathogen due to varietal and environmental variables.

Species other than *E. amylovora* are probably involved with shoot blights in other plants. *Pseudomonas syringe,* a bacterium that causes a serious foliage disease of beans, is reported as causing shoot blight of cherry.

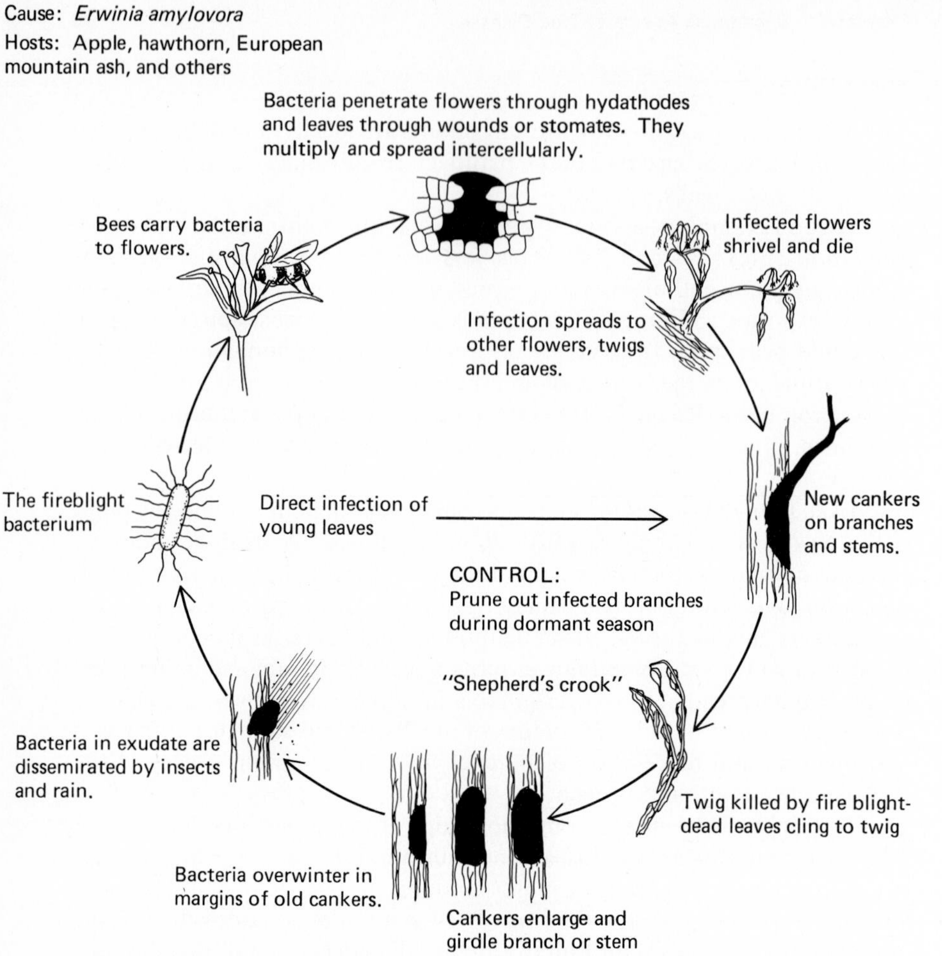

Cause: *Erwinia amylovora*
Hosts: Apple, hawthorn, European mountain ash, and others

Bacteria penetrate flowers through hydathodes and leaves through wounds or stomates. They multiply and spread intercellularly.

Infected flowers shrivel and die

Infection spreads to other flowers, twigs and leaves.

New cankers on branches and stems.

"Shepherd's crook"

Twig killed by fire blight-dead leaves cling to twig

Cankers enlarge and girdle branch or stem

Bacteria overwinter in margins of old cankers.

Bacteria in exudate are dissemirated by insects and rain.

The fireblight bacterium

Bees carry bacteria to flowers.

Direct infection of young leaves

CONTROL:
Prune out infected branches during dormant season

FIGURE 7-6 Fire blight disease cycle.

Foliage Blights

There are very few foliage diseases of trees caused by bacteria. Bacteria such as *E. amylovora,* which parasitize shoots, also cause foliage blotches, leaf spots, and death. *Pseudomonas syringe,* also mentioned above, is a foliage pathogen of citrus. Other Pseudomonads cause leaf spots of bigleaf maple, tung, California laurel, and red mulberry.

It is curious that so few bacterial foliage pathogens are reported for trees because serious bacterial foliage diseases of agricultural crops, ornamental shrubs, and vines are rather general in occurrence.

Stem Vascular Disorders

Vascular wilt caused by *Pseudomonas solanacearum* is a major disease of tomato, potato, tobacco, banana, and other agricultural crops. It causes a wilt of horsetail casuaria, an Asian tree introduced to Florida and southern California, but is not reported on any of the major tree species of North America. The pathogen invades the xylem vessels and causes plugging, toxin accumulation, and tylosis-type disruptions of the upward movement of water, somewhat similar to vascular diseases caused by such fungi as *Ceratocystis ulmi*, the Dutch elm disease fungus.

The number of examples of vascular xylem wilt bacteria of trees is very limited, but phloem bacteria are currently of great interest to forest pathologists. These are mycoplasmas—or are probably more properly referred to as mycoplasma-like organism (MLO), because they have not yet been characterized sufficiently. The first MLO was described in 1967 as the agent responsible for aster yellow, a disease assumed until that time to be caused by a virus. We now recognize many diseases of agricultural crops with mycoplasma etiology and four well-described mycoplasma diseases of North American trees: elm phloem necrosis, black locust witches' broom, pecan bunch, and ash witches' broom.

Mycoplasmas have been recognized to be disease-causing agents of animals since the early part of this century. Both parasitic and nonparasitic mycoplasmas are part of the normal microflora of the human mouth.

Mycoplasmas are prokaryotic microorganisms that have no cell wall. They usually require sterol in the growth medium and are sensitive to tetracycline antibiotics.

L-form bacteria also do not have a cell wall, but they can revert back to walled normal bacteria. A mycoplasma is always without a cell wall. Two

FIGURE 7-7 Mycoplasma-like organisms seen in the phloem with an electron microscope. (Photograph from C.R. Hibben and B. Wolanski. Phytopathology *61*:153.)

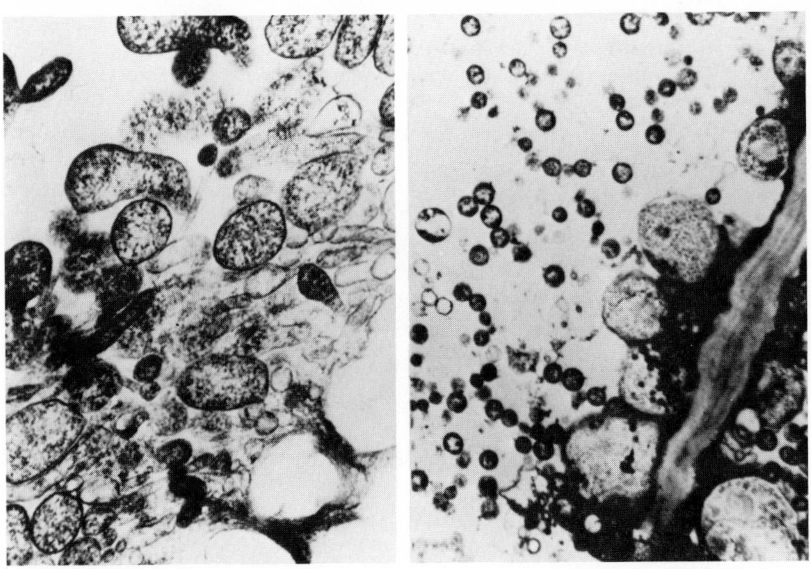

other bacteria lacking cell walls, bedsonia and the pitacycosis agent, are separated from mycoplasmas by their specific growth requirements.

Mycoplasmas observed with the electron microscope are generally variable in shape (pleomorphic) because the outer membrane is not rigid (Fig. 7-7). The three-layered outer unit membrane surrounds a protoplasm consisting of granular ribosomes and a nuclear region. Somehow the spiral mycoplasmas maintain a specific shape in the absence of a cell wall.

Growth on culture media, serological tests, and host response are the criteria used to identify species of mycoplasma. Most of the plant mycoplasmas have not been grown on culture media, so that they are tentatively identified as mycoplasmas based on morphology observed with the electron microscope and host responses.

Elm phloem necrosis is typical of mycoplasma diseases of trees. The disease agent infects an elm tree as a result of feeding by the leafhopper vector *Scaphoideus luteolus* (Fig. 7-8). Many of the mycoplasmas are transmitted by, and even multiplied within, the leafhopper vector. The leafhopper feeding in the phloem of the elm leaf injects the mycoplasma into the phloem. Mycoplasmas presumably multiply by fission or budding.

Chlorosis of foliage is a characteristic symptom of mycoplasma diseases (Fig. 7-9). In elm, the disease also produces a browning and wintergreen odor of the phloem in the lower stem. The browning and odor are diagnosed from chips of bark taken from the lower stem. Another symptom of mycoplasma diseases of trees, seen in some elms but more characteristic of other mycoplasma diseases of trees, is the production of clusters of short branches with chlorotic foliage from larger branches and main stems called witches' brooms (Fig. 7-10).

FIGURE 7-8 *Scaphoideus luteolus* leafhopper, the vector of phloem necrosis. (Photograph courtesy of the U.S. Forest Service.)

FIGURE 7-9 American elm dying of elm phloem necrosis in New York is characterized by a chlorotic thin crown.

FIGURE 7-10 Witches' broom associated with mycoplasma-infected white ash. (Photograph compliments of Dr. Craig Hibben.)

Elm phloem necrosis and presumably other mycoplasma diseases result in degeneration of feeder roots. Disruption of downward transport of photosynthetic products can be expected to cause degeneration of roots. The death of a tree is like a slow decline, because of the effects of the mycoplasma agent on roots. Elm phloem necrosis is sometimes controlled with insecticide sprays to prevent feeding by the vector.

Remission of symptoms of mycoplasma diseases can usually be accomplished by the introduction of tetracycline antibiotics into the plant. The antibiotic does not kill the pathogen but just slows it down or affects the host such that the witches' brooming and yellowing of the foliage is reduced. After one growing season the effects of the antibiotic wear off and the symptoms reappear.

Very little is known about the environmental and host factors affecting the development of mycoplasma diseases, so that additional approaches to control are not known.

Stem Cankers

Cankers or localized necrosis of vascular cambium of stems and branches can be caused by bacteria. The fire blight bacterium, already discussed as a shoot pathogen, also has a canker phase. Other bacteria, such as *Xanthomonas pruni* and *Pseudomonas syringe,* cause cankers of cherries but also cause shoot and foliage blights. Bacterial canker of poplars in Europe, caused by *Aplanobacter populi,* is probably the most important bacterial canker of trees. No one has seriously looked for the disease in North American poplars, but we can predict that importation of the pathogen, if it is not already here, may produce a future problem.

Stem Galls

The bacterium *Agrobacterium tumefaciens* induces a cancer-like uncontrolled proliferation of plant cells (Fig. 7-11). The bacterium survives in the soil and gains access to living tissue by means of wounds. Presumably, a tumor-inducing principle is released by the bacterium, which causes DNA replication to proceed at an uncontrolled rate. A tumor-inducing principle is associated with a plasmid in the bacterium. Plasmids are extranuclear DNA particles which are known to participate in a number of bacterial functions. Some plasmids function in conjugation and are known as F factors.

Agrobacterium causes disruption of control of DNA replication within the plant host. Host DNA replication is not synchronized with cell division, so that multiple sets of chromosomes accumulate within some nuclei. Hyperplasia (abnormal increase of cells) and hypertrophy (abnormal increase in cell size) result in gall formation.

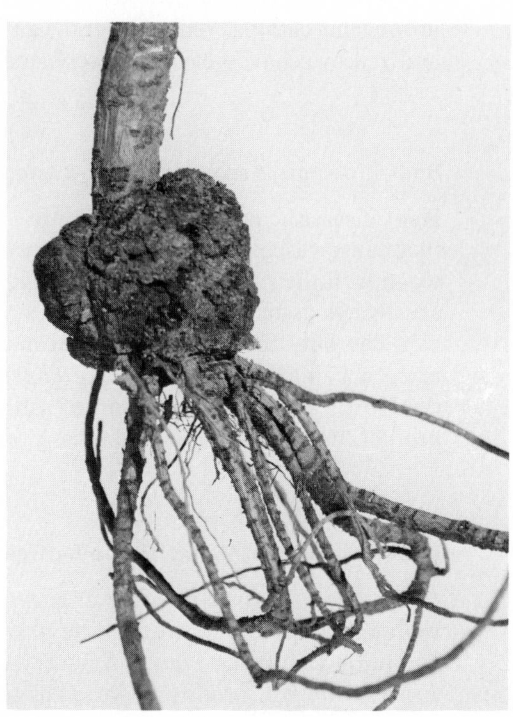

FIGURE 7-11 Crown gall on *Prunus.* (Photograph compliments of Dr. Wayne Sinclair.)

The bacterium sloughs off with a few outer cells and returns to the soil. The uncontrolled cell multiplication will continue on in the absence of the bacterium and may even occur at other locations in the plant remote from the original infection center. At one time it was thought that the secondary galls were caused by translocation of the bacterium within the plant, but it is now believed that an unknown tumor-inducing factor, probably a plasmid, is translocated, because bacteria cannot be recovered from the secondary galls. You can readily imagine the potential importance to cancer research if the exact function and nature of the tumor-inducing principle were known, and therefore, as might be expected, much work on crown gall is presently being conducted by medical research scientists.

A very interesting approach to the control of crown gall has recently been discovered and is already in widespread use. Another bacterium, *Agrobacterium radiobacter* strain 84, produces a bacteriocin molecule (agrocin 84) of unknown chemical identity, which presumably inhibits the *A. tumefaciens* tumer-inducing plasmid from transferring to the host plant. Even though the mode of action of the bacteriocin produced by *A. radiobacter* is still poorly understood, this means of control has been shown to be effective and economically acceptable to the nursery industry.

Most of the hardwoods are susceptible to crown gall but birch, hornbeam, catalpa, redbud, beech, ginko, sweetgum, tulip tree, magnolia, and zelkova, have never been reported as hosts.

Root Problems Associated with Bacteria

Root diseases caused by bacteria are very limited in number. The bacteria that cause vascular wilts and the crown gall bacterium infect trees through wounds in the roots and lower stem. Root rot fungi and nematode diseases are always associated with bacteria.

The only root disease of trees in which the bacterium is the primary agent is hairy root, caused by *Agrobacterium rhizogenes*. It is not known in the Northeast but is a problem of ash in the Great Plains states. Excessive proliferation of roots is characteristic of this disease.

Decomposition of Wood Tissue by Bacteria

Bacteria are involved in decay or enzymatic breakdown of woody tissues of both living trees and wood after the trees are cut. In the living tree, a condition known as wetwood is very common (Fig. 7-12). The bacterium *Erwinia nimipressuralis* is reported as the cause in elm and is assumed to be the agent in other hardwoods. *Corynebacterium humiferan* is another bacterium associated with wetwood in poplars.

Activity of the bacterium in the inner sapwood/outer heartwood region results in production and accumulation of water and methane. Pressure is built up which forces both gas and water from wounds. A stream of liquid and gas will sometimes spew forth from an increment borer wound when the core is pulled from the borer. Tree trimmers are sometimes doused by the foul-smelling liquid as they cut branches from the top of wetwood-infected elms or poplars. The characteristic odor of elm and poplar lumber as it is being cut is due mainly to the bacterial products.

Wetwood is reported to cause death in elms, but surveys of elm populations have, on at least two occasions, demonstrated that every tree in the population was infected with wetwood. Therefore, when a few trees die, it is almost impossible to demonstrate that death is caused by the bacterium.

Despite inadequate proof of pathogenicity for *E. nimipressuralis*, the belief among arborists that this is a potential disease agent causes them to attempt to prevent infection by painting wounds in trees with wound dressings. Another popular activity is to drill a hole in the elm in the zone of wetwood and insert a pipe to release the liquid and pressure and allow the cambium to heal over a festering wound. The value of either of these two activities is questionable.

FIGURE 7-12 Wetwood in slippery elm causes a black odoriferous liquid to flow from a pruned branch.

A serious defect resulting from wetwood occurs as wood from wetwood-infected trees is dried prior to its use as lumber. In the absence of carefully controlled drying conditions, the wood will irregularly collapse and warp, to form almost useless lumber. Warp and collapse associated with elm, poplar, and balsam fir lumber is very common, and therefore these species and others were not commonly used for lumber. Today, controlled drying schedules prevent warp and collapse, so that fir and poplar are now acceptable building materials.

Degradation of Wood Products

Bacteria invade logs submerged in holding ponds for long periods of time and logs that sink during the rafting of logs to mills. The bacteria cause degeneration of pit membranes between cells. No measurable reduction in the strength of wood cut from such invaded logs is evident, but the destruction of pit membranes allows liquids to move in and out irregularly and rapidly. This will become a problem if the wood is exposed to moisture in its final use, because decay fungi rapidly decay wood that is easily wetted.

Increased permeability of wood by bacterial deterioration of pits is a possible asset in wood that is to be treated with a wood preservative. Maximum uptake of preservative and deep penetration by the preservative are characteristics of ideal preservative treatments. But the bacteria do not uniformly invade the wood, and therefore the preservative treatment is not uniformly distributed throughout the wood.

Soft Rots of Fleshy Plant Products

A last role of bacteria that is very important in agricultural crops but is not a problem in trees is soft rot bacterial decomposition of fruits and vegetables. *Erwinia carotovora* can decompose a bushel of carrots or potatoes to a pile of viscous, foul-smelling mush in a short period of time.

ROLE OF SOIL BACTERIA IN NUTRIENT CYCLING AND INDIRECTLY IN PLANT HEALTH

Biosynthetic processes of plants and animals tie up carbon, nitrogen, and sulfur. Bacteria in the soil are essential degraders of organic material, thereby recycling the bound-up elements. It has been estimated that the supply of carbon dioxide on this planet would be exhausted within 20 years if it were not for the bacteria and other microorganisms that release carbon dioxide by degrading organic matter. Most students have been exposed to the carbon, nitrogen, and sulfur cycles in basic biology courses. I will attempt to relate the subject to plant health.

Carbon Cycle

Although we can speculate on what would happen if carbon were not recycled, there is no documented case where a shortage of carbon dioxide is a major plant problem. The addition of more carbon dioxide to the atmosphere of an enclosed chamber such as a greenhouse enhances the growth of plants—so we can say that plants could use more carbon dioxide but get along well enough at the present levels.

Nitrogen Cycle

Nitrogen recycling by bacteria, on the other hand, is variable from one soil to another, and plant health is delicately balanced by the variable supply of nitrogen. The nitrogen cycle (Fig. 7-13) basically involves oxidation of ammonia in two steps to nitrate. Plants, in turn, reduce the nitrate immediately upon intake to the amino form, attach the amino form of nitrogen to transport amino acids, and then further synthesize additional amino acids and other nitrogen-containing compounds by chemically transferring the amino group.

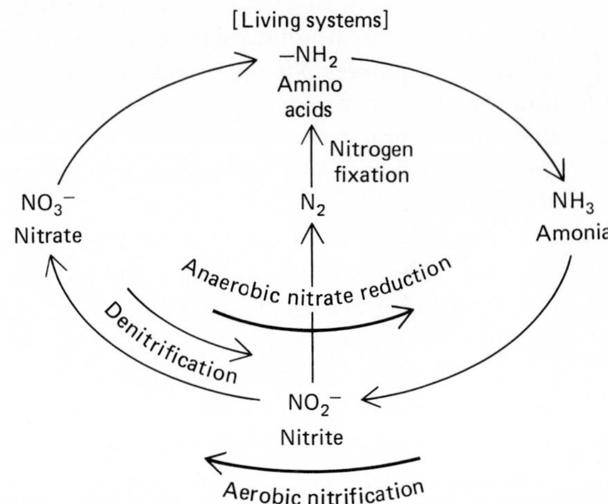

FIGURE 7-13 Nitrogen cycle.

The cycle thus far presented results in no increase or decrease of nitrogen in the biological system. In actual practice, the biologically cycling nitrogen is also interchanging with atmospheric nitrogen, the major element of our atmosphere.

Many aerobic bacteria can reverse the cycle by reducing nitrate to nitrite, but nitrite is toxic, so that the amount reduced is limited. A few normally aerobic bacteria such as *Pseudomonas* and *Bacillus*, can utilize nitrate under anaerobic conditions and reduce it beyond nitrite to molecular nitrogen. This type of reaction is one of the effects of anaerobic soil conditions which were discussed in Chapter 2.

The littleleaf disease of shortleaf pine is discussed at length in Chapters 16 and 18. Littleleaf sites are characterized by shallow topsoil on top of hardpan. It is important to recognize the possible role of anaerobic bacteria. High soil-moisture conditions during portions of the year reduce oxygen and probably cause the facultative anaerobic bacteria to shift to utilization of nitrate, thereby removing nitrogen from the soil in the form of nitrogen gas. Addition of nitrogen fertilizers reverses the yellowing decline symptoms of littleleaf disease, but fertilization has only temporary effects which are counteracted during the next period of high moisture.

Some bacteria are capable of nitrogen fixation. The importance of bacteria such as *Azotobacter* and *Clostridium* in nitrogen fixation is not fully understood. It is estimated that about 6 lb. of nitrogen per acre per year (6.7 kg/ha) are added to the soil by these bacteria. Some preliminary reports also indicate that these anaerobic nitrogen-fixing bacteria may play a role in supplying wood decay fungi with sufficient nitrogen to survive in the nitrogen-deficient environment of the xylem of trees.

A well-recognized contribution of nitrogen to the soil occurs as a result of bacteria in symbiotic relationships with legumes such as peas, beans, alfalfa, and clover. Nitrogen fixation by symbiotic association of bacteria and legume can fix as much as 400 lb of nitrogen per acre per year (448 kg/ha). Symbiotic nitrogen fixation results from invasion of tetraploid (four sets of chromosomes) cortical root cells of the legumes by species of *Rhizobium*.

The *Rhizobium* bacterium invades by penetration of a root hair. Bacteria within cortical cells stimulate the cells to divide, thereby forming a nodule. In the nodule is found hemoglobin, which neither the plant nor the bacteria can synthesize when grown separately. The hemoglobin is thought to play a role in reversible oxygen binding to maintain the appropriate reduction potential for fixation of nitrogen. After a period of time the bacteria disappear, presumably because they are absorbed by the host plant.

Nitrogen fixation by symbiotic associations of higher plants and bacteria also occurs by means of unknown bacteria associated with plants in the genera *Alnus, Myrica, Coriaria, Casuarina, Hippophae, Shepherdia, Discuria,* and *Ceanothus*. The amounts and importance of nitrogen fixation by these associations are not well known, but there is some speculation based on laboratory tests that alder mixed with Douglas fir affects soil nitrogen form and quantity. Damage by the root rot fungus *Phellinus (Poria) weirii*, which is discussed in Chapter 16, is reduced in mixed alder/Douglas fir stands.

Sulfur Cycle

Sulfur is another required element for plant growth that is cycled by bacteria in the soil. The normal bacterial cycling of sulfur from decomposing organic matter involves oxidation of hydrogen sulfide to sulfur and then further oxidation of sulfur to sulfate (Fig. 7-14). The plant takes up sulfate and reduces it for incorporation into the amino acid cysteine. Hydrogen sulfide oxidation results in the production of hydrogen ions. Therefore, sulfur or organic amendments commonly added to alkaline soils are used to increase acidity. Anaerobic bacteria utilize sulfate as an oxidizing agent. The product hydrogen sulfide can be smelled in mud at the bottom of ponds and bogs. The black color of mud is the accumulation of ferrous sulfide.

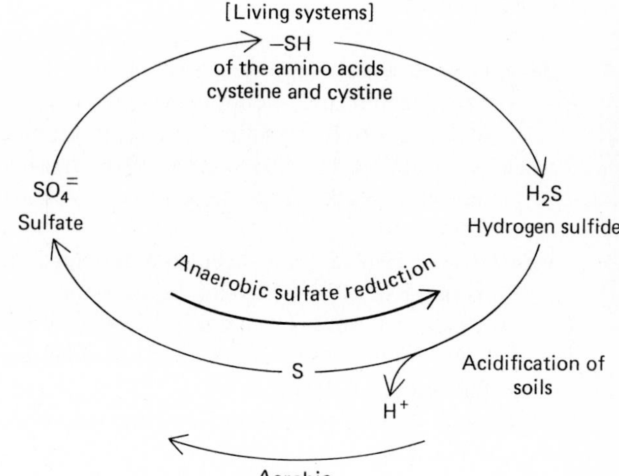

FIGURE 7-14 Sulfur cycle.

CONTROL OF BACTERIAL DISEASES

Plant-pathogenic bacteria are spread by insects, splashing water, and human beings. Insect control will prevent some bacterial diseases. Elimination of diseased individuals or diseased parts of individuals in a population avoids the short-range dispersal by splashing rain. Transport and storage of infected plants and plant parts and infested soil by human beings perpetuates and distributes the bacteria of agricultural crops. Quarantines on the movement of plant materials and soil do much to prevent this problem.

Direct control of bacterial diseases of plants by application of antibiotics is much less successful than is control of bacterial diseases of human beings and other animals, because plants do not readily absorb and efficiently distribute the antibiotics.

Manipulation of bacteria involved in nutrient cycling usually requires alteration of soil aeration. Obviously, one cannot change soil aeration over large forest areas. But it is possible to affect aeration around individual or groups of ornamental trees. Drainage tiles and gravel can be used to advantage in these situations.

Anaerobic bacterial production of plant toxic ammonia and hydrogen sulfide is a particularly serious problem when attempting to grow trees and other vegetation on reclaimed refuse dumps. Experimental tests using drainage tiles and planting on mounds of soil are being conducted because direct plantings into topsoil covering the refuse have failed. In the long run, it may be appropriate to screen various plant species for tolerance to ammonia and hydrogen sulfide. Development of tolerant plant materials for reclamation of refuse sites could become a major priority as cities and parks continue to expand onto these sites.

REFERENCES

AHO, P. E., R. J. SEIDLER, H. J. EVANS, and P. N. RAJU. 1974. Distribution, enumeration and identification of nitrogen-fixing bacteria associated with decay in living white fir trees. Phytopathology *64:*1413–1420.

BEER, S. V. and D. C. OPGENORTH. 1976. *Erwinia amylovora* on fire blight canker surfaces and blossoms in relation to disease occurrence. Phytopathology *66:*317–322.

BRAUM, A. C. 1962. Tumor inceptions and development in crown gall disease. Annu. Rev. Plant Physiol. *13*:533–558.

BUCHANAN, R. E., and N. E. GIBBONS. 1974. Bergey's manual of determinative bacteriology, 8th ed. Baltimore Williams & Wilkins Company, Baltimore. 1246 pp.

CARTER, J. C. 1945. Wetwood of elms. Ill. Nat. Hist. Surv. Bull. 23, art. *4*:407–448.

DUNEGAN, J. C. 1954. Blight of pears, apples, and quinces. USDA Leafl. 187. 2 pp.

GILMAN, E. F., I. A. LEONE, and F. B. FLOWER. 1977. Vegetating the completed sanitary landfill. Proc. Am. Phytopathol. Soc. *4*:188.

GREMMEN, J. and R. KOSTER 1972. Research on poplar canker (*Aplanobacter populi*) in The Netherlands. Eur. J. For. Pathol. *2*:116–124.

HARTLEY, C., R. W. DAVIDSON, and B. S. CRANDALL. 1961. Wetwood, bacteria, and increased pH in trees. USDA For. Serv. For. Prod. Lab. Rep. 2215. 34 pp.

HEPTING, G. H. 1971. Diseases of forest and shade trees of the United States. Agricultural handbook 386, USDA Forest Service. 658 pp.

HIBBEN, C. R., and B. WOLANSKI. 1971. Dodder transmission of a mycoplasma from ash witches' broom. Phytopathology *61*:151–156.

HOPKINS, D. L. 1973. Rickettsia-like bacterium associated with Pierce's disease of grapes. Science *179*:298–300.

HULL, R. 1971. Mycoplasma-like organisms in plants. Rev. Plant Pathol. *50*:121–130.

KNUTH, D. T., and E. McCOY. 1962. Bacterial deterioration of pine logs in pond storage. For. Prod. J. *12*:437–442.

LACY, G. H., and J. V. LEARY. 1979. Genetic systems in phytopathogenic bacteria. Annu. Rev. Phytopathol. *17*:181–202.

MARAMOROSCH, K., R. R. GRANADOS, and H. HIROMI. 1970. Mycoplasma diseases of plants and insects. Adv. Virus Res. *16*:135–193.

MOORE, L. W., and G. WARREN. 1979. *Agrobacterium radiobacter* strain 84 and biological control of crown gall. Annu. Rev. Phytopathol. *17*:163–180.

NIENHAUS, F., and R. A. SIKORA. 1979. Mycoplasmas, spiroplasmas, and rickettsia-like organisms as plant pathogens. Annu. Rev. Phytopathol. *17*:37–58.

PARKER, K. G. 1936. Fire blight: overwintering, dissemination, and control of the pathogens. Cornell Univ. Agr. Exp. Sta. Mem. 193. 42 pp.

SCROTH, M. N., S. V. THOMSON, D. C. HILDEBRAND, and W. J. MOLLER. 1974. Epidemiology and control of fire blight. Annu. Rev. Phytopathol. *12*:389–412.

SELISKAR, C. E., G. E. KENKNIGHT, and C. E. BOURNE. 1974. Mycoplasma-like organism associated with pecan bunch disease. Phytopathology *64*:1269–1272.

SELISKAR, C. E., C. L. WILSON, and C. E. BOURNE. 1973. Mycoplasma-like bodies found in phloem of black locust affected with witches' broom. Phytopathology *63*:30–34.

SINCLAIR, W. A., E. J. BRAUN, and A. O. LARSEN. 1976. Update on phloem necrosis of elms. J. Arboric. *2*:106–113.

SINCLAIR, W. A., and W. T. JOHNSON. 1972. Crown gall. Cornell Tree Pest Leafl. A-5. 12 pp.

SMITH, R. S. 1975. Economic aspects of bacteria in wood. *In* Biological transformation of wood by microorganisms, ed. W. Liese. Springer-Verlag, Berlin, pp. 89–102.

VAN LAREBEKE, N., G. ENGLER, M. HOLSTERS, S. VAN Den ELSAKER, I. ZAENEN, R. A. SCHILPEROORT, and J. SCHELL. 1974. Large plasmid in *Agrobacterium tumetuciens* essential for crown gall-inducing ability. Nature *252*:169–170.

WARD, J. C., R. A. HANN, R. C. BALTES, and E. H. BULGRIN. 1972. Honeycomb and ring failure in bacterially infected red oak lumber after kiln drying. USDA For. Serv. Res. Pap. FPL 165. 36 pp.

WHITBREAD, R. 1967. Bacterial canker of poplars. Ann. Appl. Biol. *59*:123–131.

WILSON, C. L., C. E. SELISKAR, and C. R. KRAUSE. 1972. Mycoplasma-like bodies associated with elm phloem necrosis. Phytopathology *62*:140–143.

8

INTRODUCTION TO FUNGI

- Fungi in the forest community
- Characteristics of fungi
- Taxonomic classification and life cycles of fungi
- Disease cycle of pathogenic fungi

Some estimates indicate that there are close to 100,000 species of fungi. Each has special environmental requirements and occupies a certain niche. Almost all niches and organic substrates have been experimented with and found suitable by some fungus. It has been said that if these meek organisms have not yet exactly inherited the earth, they do eventually inherit most of the living things on the planet as well as the goods and products made from them.

Only in the twentieth century have we been able to produce organic substrates that seem to defy the fungal ingenuity for variability. Some of plastics and styrofoams seem to be immune to the digestive enzymes of fungi and bacteria. We can all hope that the fungi and bacteria have not given up in this race with us and that they eventually come forth with a champion to degrade this portion of human residue.

FUNGI IN THE FOREST COMMUNITY

Fungi have many roles in the forest community (Fig. 8-1). For the most part, fungus activity is highly beneficial to both human beings and the forest community.

Saprophytic Fungi

The most important role of fungi is in the saprophytic decomposition of cellulose and lignin residues. In Chapter 7 it was noted that in the absence of carbon recycling, the supply of carbon dioxide would be used up in 20 years. How much of the carbon is recycled by fungi and how much is recycled by bacteria is not known. The fungi are well adapted for very rapid enzymatic breakdown of massive cellulose–lignin accumulations in woody stems. Fungal hyphae actively penetrate cell walls and permeate the whole substrate. Bacteria, on the other hand, passively move into a substrate and enzymatically macerate the tissues in localized areas.

Parasitic Fungi

A minor role of fungi in the forest community is the parasitic role. The basic difference between the parasite and the saprophyte is that the parasite encounters and deals primarily with the defenses of the living host, whereas the saprophyte has to compete with other saprophytes for dead organic matter.

A fraction of the parasites of the forest community are detrimental to their hosts; these are the pathogens. This small group of the total fungi is the group the forest pathologists have traditionally emphasized in their studies on wood decay, rust diseases, canker diseases, foliage diseases, wilt diseases, and root rot diseases. Much of the remaining chapters will deal with these problems.

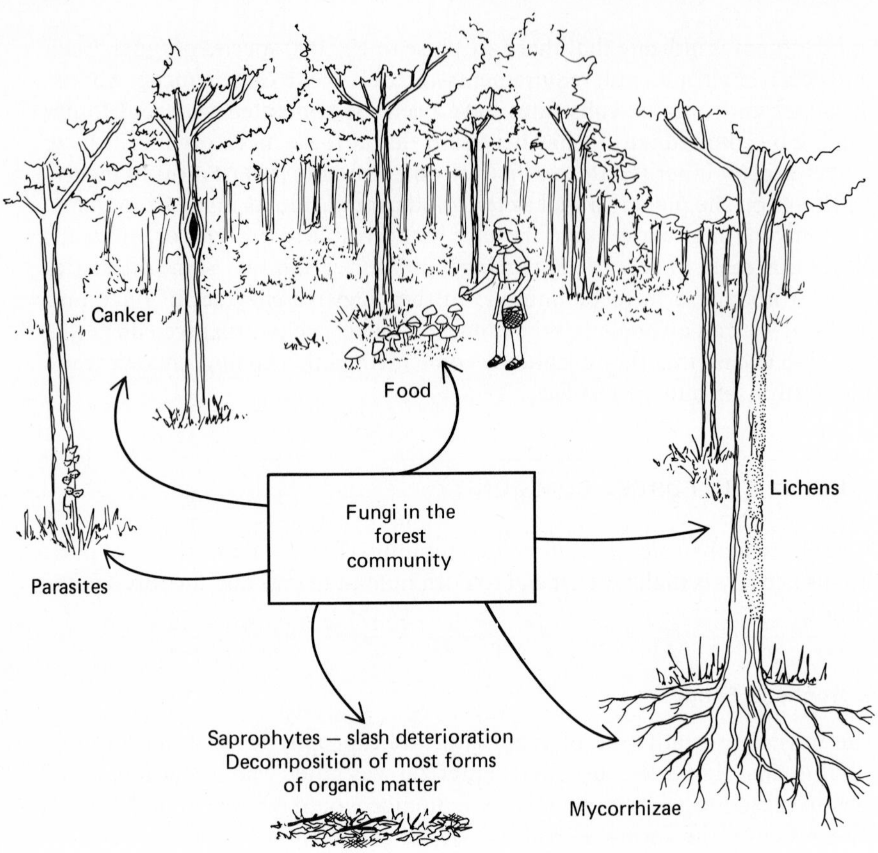

FIGURE 8-1 The roles of fungi in the forest community.

Fungi as Food for Other Organisms

In the forest community it is necessary to recognize that fungi serve as sources of food for both human beings and many other forms of life. Groups of invertebrates have specialized in the utilization of fungi for food. Instead of knocking off the next wood decay fungus fruit body you encounter, stop and observe the array of insect and other life that centers around the fruit body environmental niche.

Vertebrates, including man, recognize the culinary delicacy of fungi. The Vikings found another use for fungi. They ate a species of *Amanita* mushroom before going into battle. The mushroom produced hallucinogenic effects, which heightened their savagery in dealing with the enemy. Mushrooms in the genus *Amanita* have other noteworthy properties. Poisoning from eating mushrooms commonly occurs as a result of eating certain species of *Amanita.* A general rule to follow in selecting edible mushrooms is to choose those enclosed within a can or jar. If you are going to eat wild mushrooms, you had better have absolute confidence in the person who collected the mushrooms. Everyone should read the account of a family that was poisoned by mushrooms as described by Pilát and Ušák in the book *Mushrooms.* Mushroom poisoning is a very painful way to go.

Symbiotic Fungi in Lichens and Mycorrhizae

Fungi in the forest community are symbiotically associated with algae in lichens. Most biologists recognize the large foliose and fruiticose lichens that provide the array of colors and shapes to bark and limbs of trees, but few look closely at the bark of trees and recognize the subtle evidences of small crustose lichens. The fungus–algal association, lichens, are so common that we take them for granted. Human pollution of the environment results in changes in numbers and species of lichens. A possible future role of lichens could be as biological indicators of cumulative effects of low levels of air pollution.

To complete the roles of fungi, it is important to mention mycorrhizae. Mycorrhizae will be more fully discussed in Chapter 9, but for now it is important to recognize that fungi associated with plant feeder roots cause morphological changes in the roots. The morphologically modified roots called mycorrhizae are better nutrient- and water-absorbing organs. Most of the higher plants have developed this type of relationship with fungi.

CHARACTERISTICS OF FUNGI

Lack of chlorophyll, mycelial growth form, and reproduction by means of spores are the specific characteristics of fungi.

Heterotrophic Growth Requirements

Inability to synthesize their own food as chlorophyllous plants do forces fungi to live as heterotrophs on products synthesized by other organisms (Fig. 8-2). As already mentioned, most fungi consume the remains of dead organisms, while a few parasitize living organisms. Others straddle the fence, occasionally lending a hand to the demise of the host and then readily consuming the remains. Fungi that utilize dead substrates are called saprophytes. Those that parasitize only living hosts are called obligate parasites, and the fence straddlers are called faculative parasites or faculative saprophytes, depending upon their affinity for dead or living hosts, respectively.

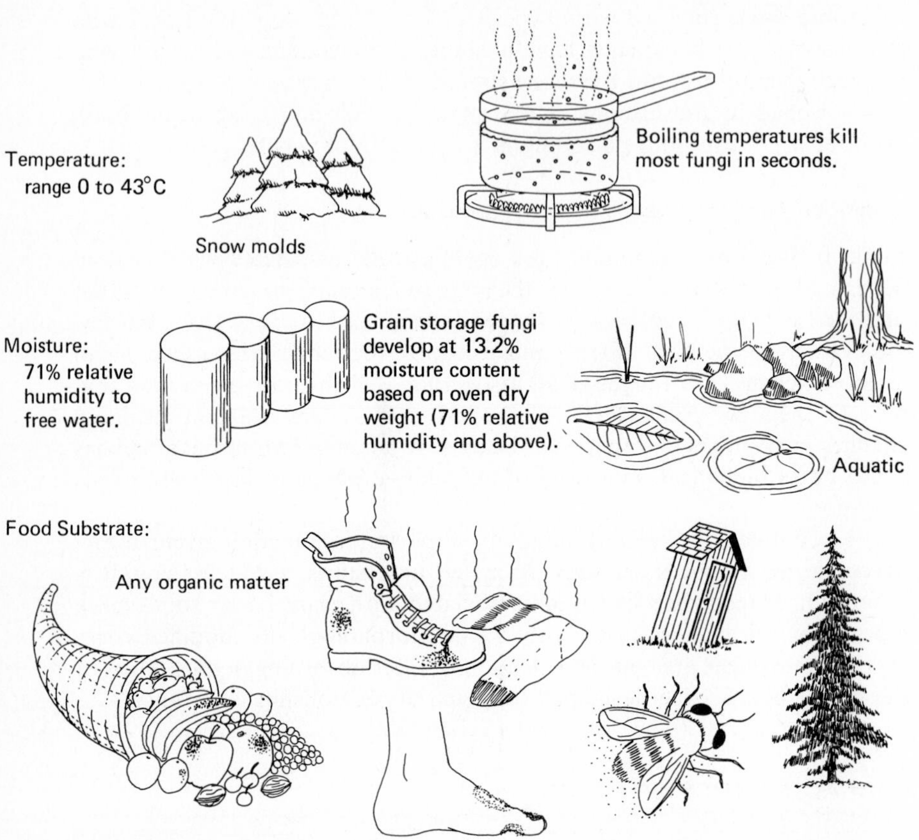

Note: Each fungus has its own specific requirements for temperature, moisture, and substrate. Oxygen: All require some oxygen (O_2), though some tolerate high concentrations of CO_2.

FIGURE 8-2 Requirements for growth of fungi.

Each fungus has its specific nutrient requirements and optimum environmental conditions for development. Cellulose, starch, and many other organic compounds are utilized by fungi for energy and for raw material for synthesis. Organic nitrogen in the form of amino acids or ammonia is the usual source of nitrogen. Other mineral requirements are similar to higher plants. Many fungi are deficient in the vitamins thiamine and/or biotin.

All fungi require oxygen, but some fungi can tolerate high levels of carbon dioxide. The yeasts used in the brewing industry generate alcohol in the absence of oxygen, but growth and development of the population capable of fermentation to 12% alcohol first took place in an environment with oxygen. The lack of oxygen during fermentation eventually results in the death of the yeasts. It is interesting to note that almost every civilization has exploited fungi for fermentation of local fruits, grains, potatoes, and many other substrates.

Most fungi require some free water to be able to grow in or on a substrate. A wooden table in a classroom will not be decayed by fungi because of lack of water. The same table placed outside will readily decay, because moisture, from rain and from the soil, absorbed by the wood is sufficient for fungal growth.

A few fungi, the most notable examples being the grain storage fungi, can grow in an environment where the substrate is in equilibrium with a relative humidity of 71%. At this relative humidity there is no free water. The grain moisture content is 13.2% based on an oven-dry weight.

You can readily recognize how difficult it is to prevent grain storage fungi from developing if the minimum requirements are a humidity of at least 71%. Moisture not dried out of the grain before it is placed in storage is a source of some moisture. A drop in temperature within a closed storage elevator or ship also will increase the relative humidity.

In large piles of grain, metabolism of the initial invading fungi adds moisture and heat. Additional fungi can become established as well as thermophilic bacteria. Eventually, if the grain is not dried and stored properly, spontaneous combustion takes over when the temperature exceeds that acceptable to the thermophiles. Today, we seldom hear of grain elevators burning because of spontaneous combustion, because the high value of grains has forced us to a better understanding of fungus activity. Every year, though, a few barns burn because high-moisture hay is invaded by storage fungi, which start the process toward spontaneous combustion.

Today, we recognize another problem associated with fungus invasion in stored grain. Toxic chemical products produced by grain storage fungi induce accumulative deleterious effects when ingested by livestock or human beings. These products, called aflatoxins, have, during the past 10 to 15 years, become serious concerns in animal feeds and in products processed for human consumption.

The other extreme in moisture requirements is represented by the aquatic fungi. Some fungi are totally aquatic in their existence. Leaves and other organic material that finds its way into well-aerated streams is decomposed by an interesting group of fungi called aquatic hyphomycetes. The curious characteristic of these fungi is their star-shaped spores, spores with long appendages presumably for easy dispersal in rapidly moving water. Lack of oxygen in stagnant water prevents development of these fungi, thereby allowing organic debris to accumulate.

Another requirement for fungus growth is an environment with a temperature range between freezing and about 43°C. Most fungi have an optimum near 24°C, but some operate at the extremes of the range.

Snow molds (Fig 8-3) can develop below the snow in a temperature very near the freezing point and avoid much competition by other microorganisms. These fungi will grow at higher temperatures, but at higher temperatures, they have to compete with a wide array of other microorganisms. Therefore, an ecologically competitive advantage is exploited by the snow mold fungi growing at or near the freezing point.

Another group of fungi find the ecological niche of higher temperatures a less competitive environment. A maximum temperature of about 55 to 60°C is the upper limit for a fungus, *Chaetomium thermophile,* which can be found in the waters of the cooler geysers. Temperatures at the boiling point will kill all fungi in a very few minutes. Thermophylic bacteria, on the other hand, can survive at temperatures in excess of 60°C, and some produce resting spores that can survive boiling temperatures.

Mycelial Growth Form

Fungi have a branching filamentous thread-like growth form. Individual threads averaging 1 to 5 μm in diameter are called hyphae (plural) or hypha (singular); see Fig. 8-4. A collection of hyphae produces a thallus or a colony of mycelium (singular) or mycelia (plural).

FIGURE 8-3 Mycelial felt of a snow mold looking like a thick spider web covering the forest floor. The snow melting back at the left exposes the fungus, which had developed under the protective snow covering.

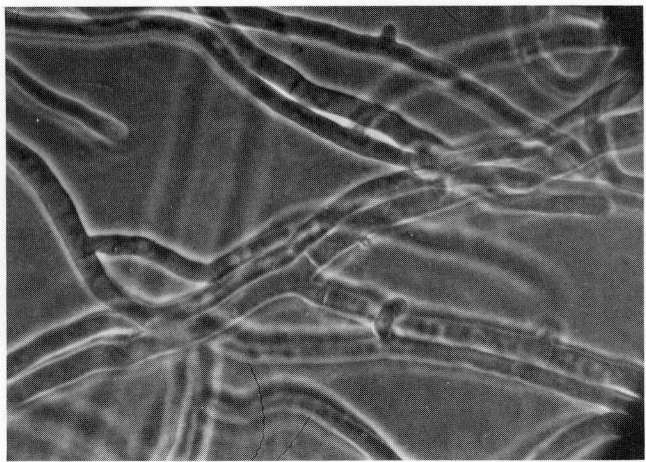

FIGURE 8-4 Fungus hyphae as seen under high magnification with a light microscope.

The thread-like hyphae grow at the tips and penetrate between and within cells of the substrate. Enzymes released from the fungus hyphae cause specific digestive reactions to take place which solubilize the substrate. The soluble form of the substrate is diffused through the cell wall of the fungus and is then further degraded within the fungus using the standard metabolic enzyme systems of living organisms.

Microscopic hyphae of some fungi aggregate and produce the more conspicuous fungal structures we are more familiar with, such as mushrooms.

There is not much variation in hyphae from one fungus to another. Except for a few fungi in the class Phycomycetes, all fungi have hyphae made up primarily of chitin, glucans, and mannans. Chitin is a long-chain polymer made up of *N*-acetylglucosamine building blocks. Chitin is also a major component of the insect exoskeleton. Glucans are 1, 3- and 1, 6-linked glucose molecules. Mannans are polymers made up of mannose and glucose.

Hyphae may be pigmented or without pigment (hyaline); they may have one continuous cytoplasm with many nuclei and organelles and no cross walls (coenocytic), or they may be split up into numerous cells by cross walls (septate). Cytoplasmic movement between cells is not eliminated in septate hyphae because the septa have a central pore which allows some movement of organelles and materials from one cell to another. The pore may become plugged if a cell is injured, thereby preventing the death of the entire hypha.

A septum with a central pore with swollen margins is called a dolipore septum. The dolipore septum is thought to be characteristic of the Basidiomycete class of fungi and is, therefore, one hyphal character with

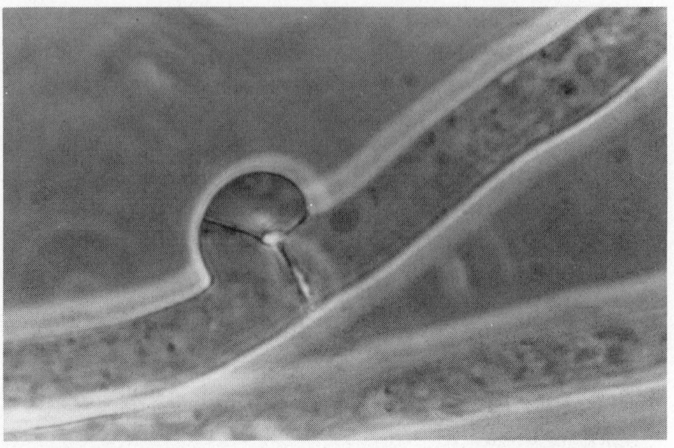

FIGURE 8-5 Clamp connection as seen under high magnification with a light microscope. (Photograph compliments of Dr. Edson Setliff.)

potential application in identifying fungi. The dolipore septum is best seen with the electron microscope.

Another hyphal characteristic unique to some of the Basidiomycetes is the clamp connection (Fig. 8-5). A hyphal branch from the terminal cell turns back and anastomoses with the cell just behind the septum. One nucleus is transferred through the hyphal branch to the cell just behind the terminal cell. In this way, each cell maintains two nuclei. Many Basidiomycetes do not have the clamp connection. The reason for its existence is still not known.

Reproduction by Means of Spores

Spores are the reproductive structures of fungi (Fig. 8-6). They differ from seeds of higher plants in that a spore cannot be subdivided into various tissues, such as embryo and endosperm. Spores are microscopic, one- to many-celled, and may be of diverse shapes and colors.

The macroscopic fungal fruiting structures that we often use to recognize fungi are evolutionary developments for more efficient spore production. Examples of the types of spore-producing structures of the Basidiomycete, Ascomycete, and Phycomycete classes, as well as the artificial Fungi Imperfecti class, will be presented.

Fungi generally produce extremely large numbers of spores because of the random or only partially directed dispersal of the spores to other sites. I have sampled and counted spores from the fruiting body of the wood decay fungus *Phellinus tremulae (Fomes igniarius* var. *tremuloides).* A figure in excess of 100,000 spores per square millimeter of undersurface per day was

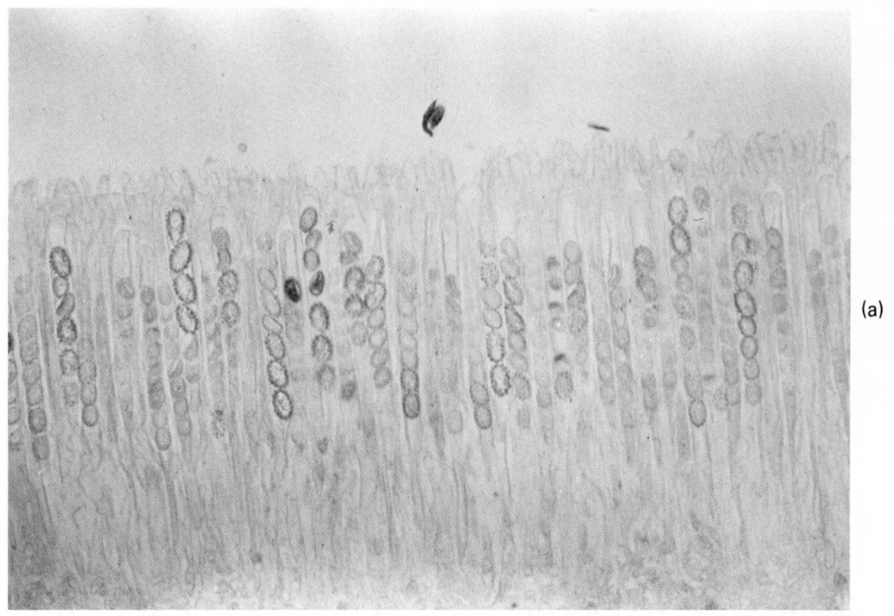

(a)

FIGURE 8-6 Examples of fungus spores as seen with the light microscope. (a) Rough-walled light-colored single-celled spores. (b) Dark-colored single-celled spores. (c) (page 110) Various fungus spore shapes.

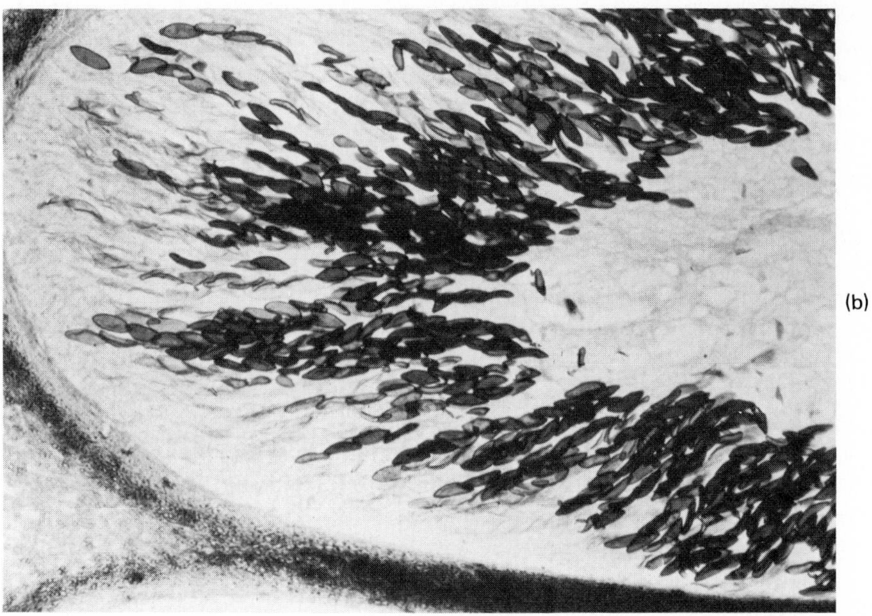

(b)

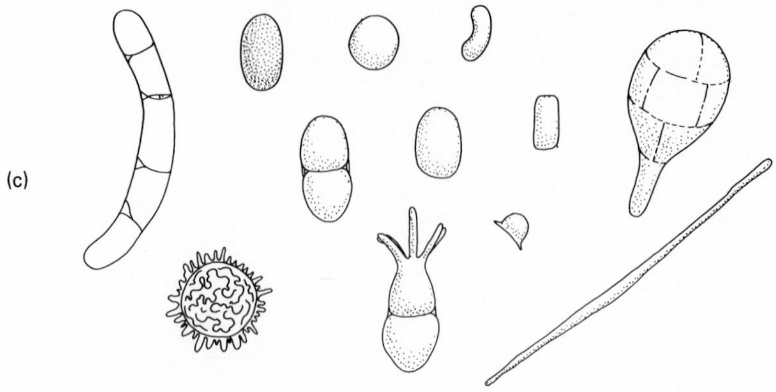

calculated. You can estimate the seasonal spore production of such a fungus using a 4 by 10-cm lower surface for the period between April 15 and October 15. Very few of these spores actually infect another tree. Most are lost on the forest floor.

TAXONOMIC CLASSIFICATION AND LIFE CYCLES OF FUNGI

The plant pathologist needs to be familiar with the details of fungus structures, life cycles, and taxonomy. The diagnosis of disease is often based upon identification of a specific fungus observed fruiting on the diseased tissue. Control procedures are more effective when applied to certain stages in the life cycle of a fungus, so an understanding of the life cycle is required.

Fungal populations are genetically variable and changeable, just like other populations of organisms. Therefore, we must understand the sexual activities of fungi to comprehend why disease-resistant varieties of plants do not always remain resistant.

Mycology is a subject that is not easily presented in a single chapter. You will become more proficient in the details presented here as we apply them to specific diseases in subsequent chapters. As these chapters are presented, it may be worthwhile to refer back to this chapter to see where the specific details fit into the scheme of things.

The taxonomic classification of any group of organisms provides a compilation of standardized names for organisms. This facilitates communication among people working with similar organisms. The rules for naming and grouping organisms into a taxonomic classification were devised by man. Ideally, they should represent natural biological groupings, but seldom is the ideal realized. Therefore, we continue to try and improve our system of organization.

To someone not intimately involved in taxonomic considerations, changes aimed at improving the system may seem more like changes to confuse than to improve. The biologist, continually confronted with changes, can resist the changes or adopt the new names, but in any case must be aware of both the old and the new systems.

The taxonomic treatment of fungi in this text will be a mixture of classification systems because there are convenient aspects of each for the introduction of fungi. Anyone greatly interested in fungi will quickly discover why the older system is inadequate.

The reader interested in current concepts in taxonomy of fungi is directed toward the taxonomic review of *The Fungi,* Vols. IVA and IVB, edited by Ainsworth, Sparrow, and Sussman (1973).

Basidiomycetes

The Basidiomycete class of fungi produce basidiospores on basidia (Fig. 8-7). The basidium is a specialized terminal hyphal cell originating from a fertile layer called a hymenium or from a specialized fertile cell called a teliospore.

Within the basidium, two nuclei (n + n) fuse to form a diploid nucleus ($2n$) which immediately undergoes meiosis to form four haploid nuclei ($1n$); see Fig. 8-8. Each of the nuclei migrates to a small projection, sterigma, from the basidium, and a spore forms by what appears to be a blowing out of the tip of the sterigma. When mature, the basidiospore is forcibly ejected from the sterigma to be picked up by wind currents and disseminated to another host or substrate.

FIGURE 8-7 Dark brown basidiospores formed on basidia lining both sides of a gill of a mushroom as seen under high magnification with a light microscope.

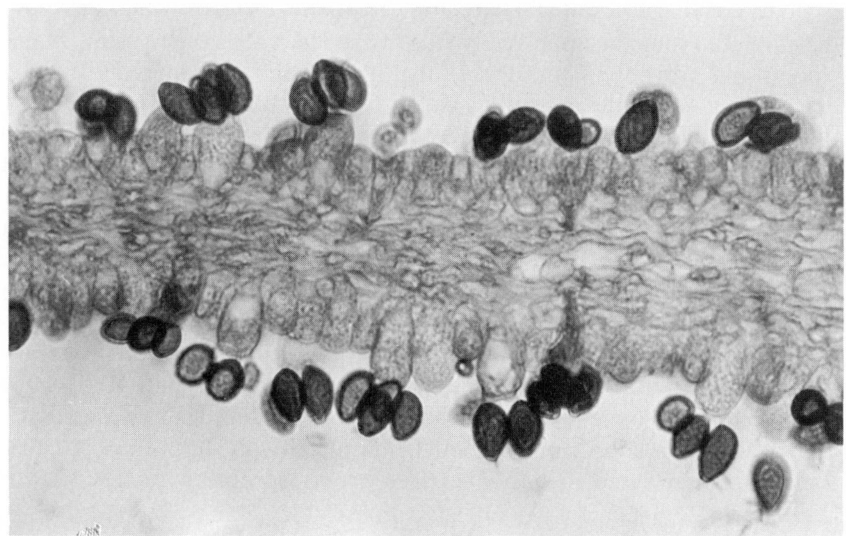

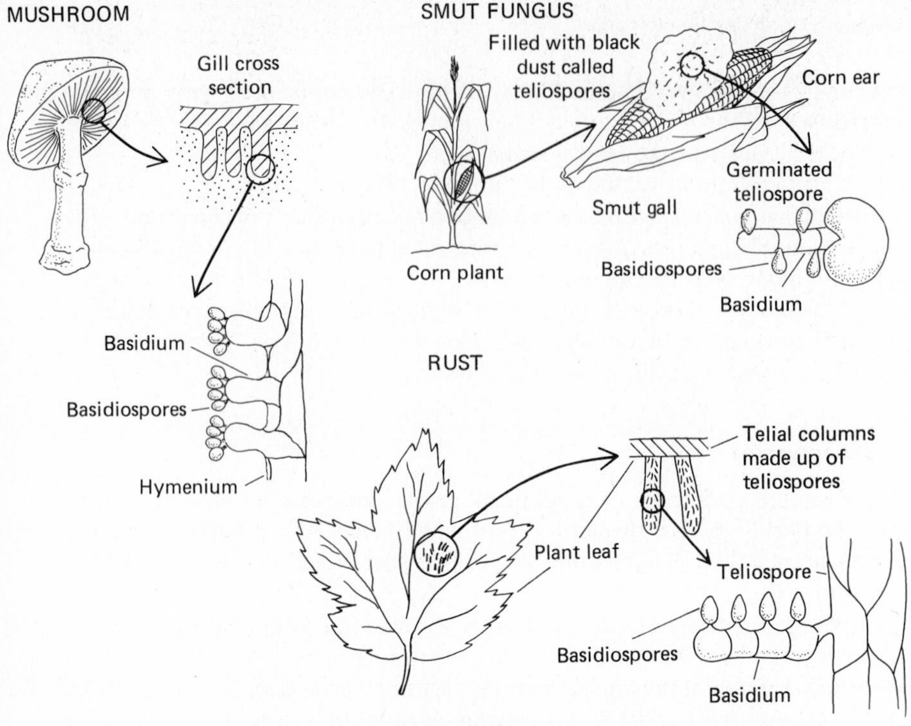

FIGURE 8-8 Basidiomycete reproductive structures.

A few Basidiomycetes produce their basidia enclosed within a closed chamber. In this case, the basidia disintegrate and the basidiospores are disseminated upon breakage of the chamber. The puffballs are good examples of the latter type and mushrooms of the former type.

The class Basidiomycetes is subdivided into two subclasses. Those with septate basidia are called Heterobasidiomycetes. Those with nonseptate basidia are Homobasidiomycetes. Rusts are Heterobasidiomycetes. Mushrooms and polypore wood decay fungi are Homobasidiomycetes.

Earlier, when describing the clamp connection, it was pointed out that each cell contains two nuclei. The standard basidiomycete life cycle begins when one of the basidiospores containing a single haploid nucleus ($1n$) (the product of meiosis) germinates to produce hyphae. Nuclear division, mitosis, results in multiplication of the single genome within the developing thallus. If the hyphae come in contact with another thallus developing from another basidiospore and the two strains are compatible, a fusion of the hyphae may occur to introduce a second genome to the thallus. The introduced nucleus in each thallus multiplies rapidly and migrates throughout the thallus out to the growing tips. The hyphae, which up until fertilization produced only simple septa, may then begin to produce clamps each time a new cell wall is laid down. The nuclei mitotically divide at the same time (conjugate nuclear division). Therefore, from the time when fertilization takes place until some time in the future when the fungus produces a fruiting body and sporulates, the two genomes ($n + n$) are maintained within each cell. The completion of the life cycle to production of spores following meiosis has already been discussed.

Some Basidiomycetes produce asexual spores following mitotic division. These vegetatively produced spores called oidia or conidia are very much like the conidia of the Ascomycetes, which will be discussed next.

Ascomycetes

The Ascomycete class of fungi are characterized by the production of ascospores within asci. Asci are sac-like structures usually containing eight ascospores.

The life cycle of Ascomycetes usually starts with germination of haploid ($1n$) ascospores (Fig. 8-9). Fertilization, the bringing together of two $1n$

FIGURE 8-9 Ascomycete reproductive structures.

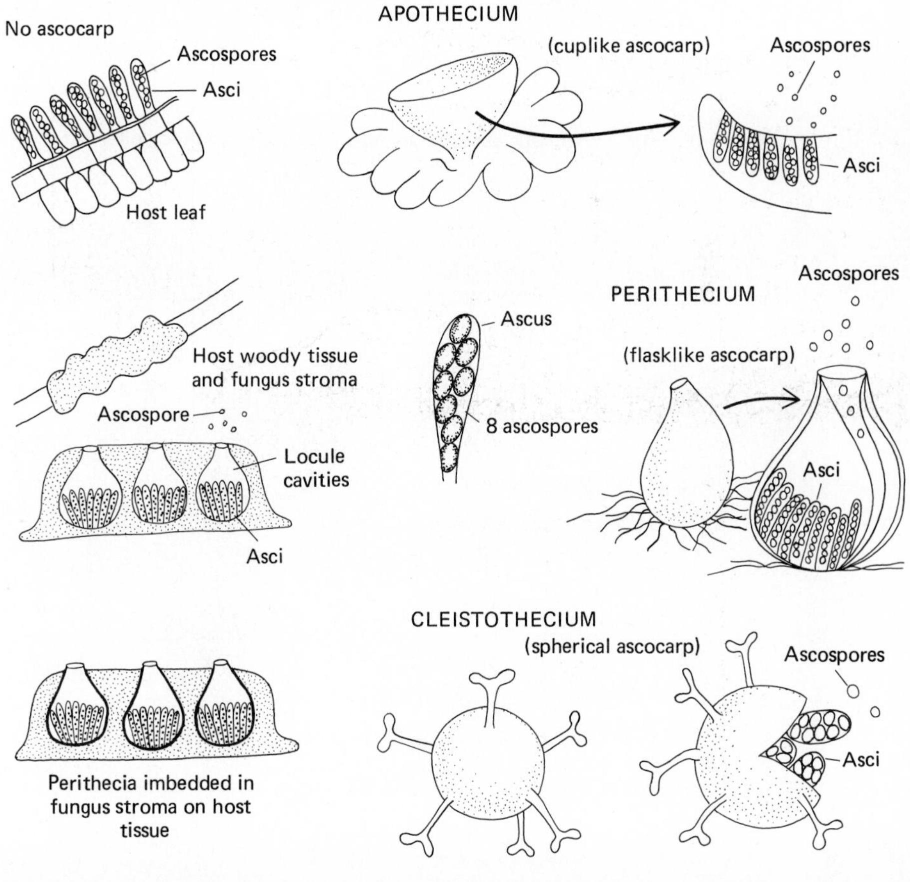

genomes, is somewhat different from that in Basidiomycetes. Ascomycetes may produce two morphologically distinct fertilization structures, ascogonia and trichogynes. The nuclei of the trichogyne are transferred to the ascogonium and pair with nuclei of the ascogonium. Pairs of nuclei migrate into developing ascogenous hyphae. A crozier or hook forms on the tips of the ascogenous hyphae. The crozier is very similar to the clamp connection of the Basidiomycetes. Other forms of fertilization may occur.

Two nuclei of the terminal cell fuse. Meiosis results in four haploid nuclei. Each nucleus undergoes mitotic division to produce eight nuclei. The terminal cell elongates during the meiotic and mitotic divisions. The cell contents of the terminal cell become incorporated into the spores that form around each nucleus.

The asci of Ascomycetes are aggregated into a number of different morphological structures (Figs. 8-10 to 8-13). Asci that are born free on the surface of the substrate are characteristic of the Hemiascomycetidae. Those born within spherical structures (cleistothecia), those born within flask-like structures (perithecia), and those born within cup-like structures (apothecia) belong to the subclass Euascomycetidae. Those that form asci in cavities within fungus stroma are in a subclass Loculoascomycetidae.

FIGURE 8-10 Asci with ascospores borne free on the surface of a leaf.

FIGURE 8-11 Cleistothecia of *Microsphaeria alni* with asci and ascospores.

FIGURE 8-12 (a) Perithecia of *Nectria galligena* as seen under the dissecting microscope. (b) Cross section through a perithecia as seen with a compound microscope. Ascospores occur in groups of eight. The ascus wall is not readily visible in this section.

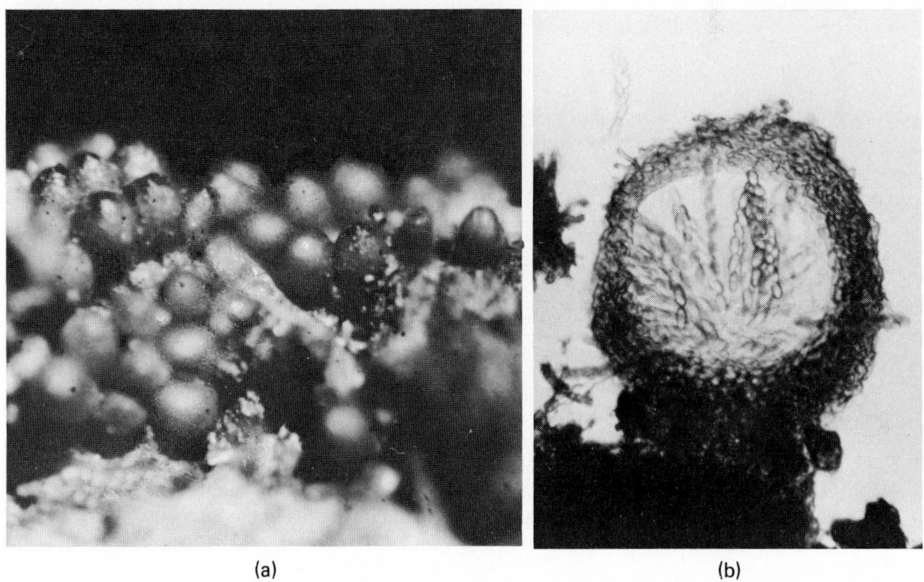

(a) (b)

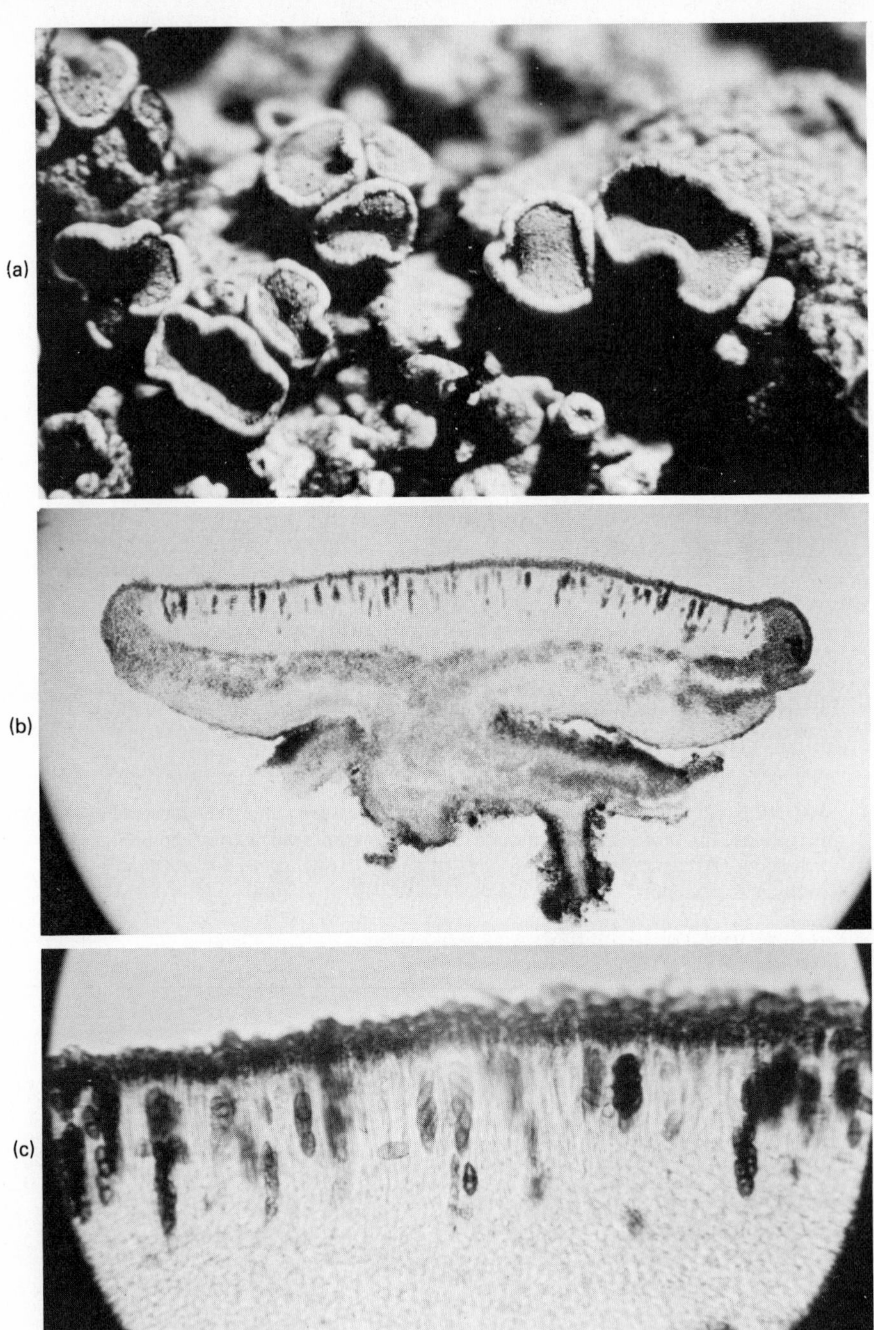

FIGURE 8-13 Apothecia of a lichen as seen with the dissecting microscope. (b) Cross section through an apothecium as seen under low magnification with a compound microscope. (c) Higher magnification of the apothecium cross section showing clusters of ascospores. The asci are not readily visible in this section.

A difficult concept for most beginning biologists to understand about the fungi is that many fungi have more than one spore stage. The ascospores of the Ascomycetes, called sexual or perfect spores, are produced following meiotic nuclear division. Ascomycetes and other fungi also can reproduce asexually. The asexual or imperfect spores of Ascomycetes, called conidia, are produced following mitotic division.

To make things even more confusing to the beginner, we have developed a second classification system for fungi that produce conidia. A given fungus may therefore have more than one name. One name is for the sexual stage and the other is for the asexual stage. (You will note two names for many canker and foliage fungi discussed in later chapters.)

Some fungi produce only asexual spores, or at least we have never found a sexual spore stage. Those fungi that do not produce sexual spores and the asexual spore stages of Ascomycetes are grouped into the Deuteromycetes or Fungi Imperfecti. Some mycologists and pathologists prefer to present the Fungi Imperfecti as a separate class, but I prefer to place the Fungi Imperfecti as an artificial yet convenient grouping within the Ascomycetes.

The major characteristic of the Fungi Imperfecti is the production of conidia on conidiophores (Figs. 8-14 and 8-15). The conidiophore is a modified hypha which gives rise to spores either by segmentation or by a blowing out of spores from the tip or sides. Conidiophores bearing conidia can be aggregated inside flask-like structures called pycnidia. They may be

FIGURE 8-14 Fungi Imperfecti reproductive structures.

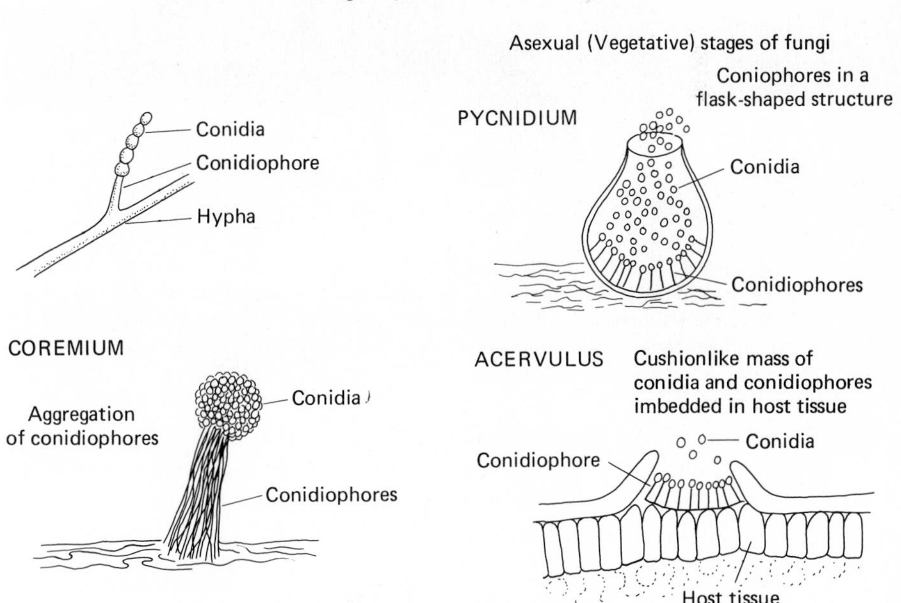

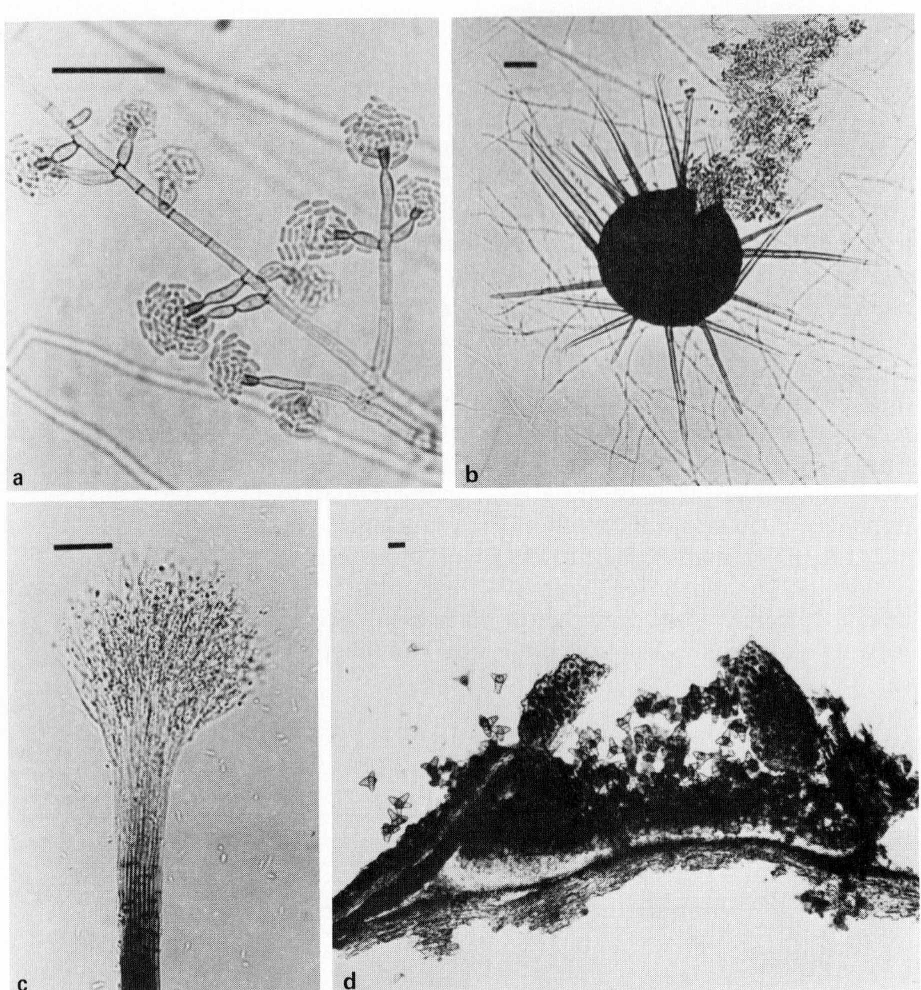

FIGURE 8-15 Photomicrographs of Fungi Imperfecti reproductive structures (the bar in each photograph is 25 μm) (a) Conidiophores and conidia of *Phialophora americana.* (b) Pycnidium of *Chaetomella* sp. (c) Coremium of the Imperfect stage of *Ceratocystis ulmi.* (d) Acervulus of *Asterosporium asterospermum.* (Photographs compliments of Dr. Chun-Juan Wang)

aggregated into various shaped cavities of host and fungus tissue or they may be born unenclosed. These three groups represent the artificial orders: Sphaeropsidales, Melanconiales, and the Moniliales, respectively. A fourth group, Mycelia Sterilia, of the Fungi Imperfecti, contains a few fungi that produce no conidia or any other type of spore.

Phycomycetes

The third class of fungi are the Phycomycetes, which some mycologists prefer to break up into three classes. Phycomycetes are characterized by the production of either oospores or a zygospore for a sexual stage and sporangia and sporangiospores for the asexual stage (Fig. 8-16).

The subclass Oomycetidae produces oospores following the fusion and nuclear transfer from an antheridium to an oogonium. The antheridium is smaller and different in shape from the oogonium, so that we recognize the Oomycetes as having sexual fusion of dissimilar gametangia.

The subclass Zygomycetidae is characterized by the production of a zygospore following the fusion of similar gametangia.

Meiosis takes place in zygospores or oospores to release haploid spores.

The asexual stages of Phycomycetes are characterized by the formation of sporangiospores by cleavage of the cell contents of a terminal cell called a

FIGURE 8-16 Phycomycete reproductive structures.

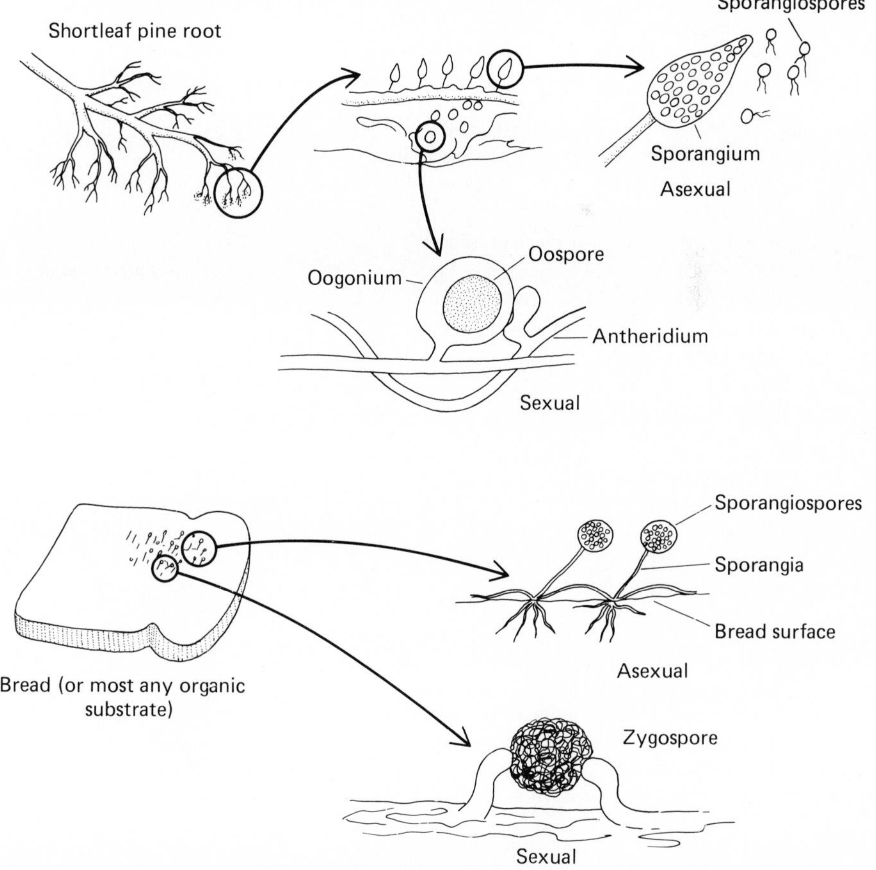

TABLE 8-1 Taxonomic Outline of Fungi

Mycota (division)	Characteristics	Examples
Myxomycotina (subdivision)	Amoeboid growth form Reproduction by sporangia	Slime molds
Eumycotina	Presence of mycelium	
Phycomycetes (class)	Nonseptate hyphae, sporangia, and sporangiospores	
Oomycetidea (subclass)	Sexual stage of antheridia and oogonium	*Phythopthora*, late blight of potato
Zygomycetidae	Sexual-stage zygospore	*Rhizopus* mold
Ascomycetes	Septate hyphae, asci, and ascospores	
Hemiascomycetidae	Asci born free	
Endomycetales (order)		Yeasts
Taphrinales		*Taphrina* leaf blisters
Euascomycetidae	Asci born in ascocarps	
Plectomycetes (series)	Cleistothecium ascocarp	
Erysiphales		Powdery mildews
Eurotiales		*Penicillium*
Pyreonomycetes	Perithecium ascocarp	
Sphaeriales		*Hypoxylon and Ceratocystis*
Hypocreales		*Nectria*
Discomycetes	Apothecium ascocarp	
Helotiales		Cup fungi and most lichens
Pezizales		
Tuberales		
Loculoascomycetidae	Asci born in locules within fungus stroma	
Myriangiales		
Microthyriales		Many foliage diseases
Hysteriales		
Pleosporales		
Dothidiales		
Deuteromycetidae (Fungi imperfecti)	Asexual stage of Ascomycetes, conidia on conidiophores	
Sphaeropsidales	Conidiophores in a pycnidium	*Cytospora*
Melanconiales	Conidiophores in an acervulus	*Gloeosporium*
Moniliales	Conidiophores born free	*Verticillium*
Mycelia Sterilia	No conidia	*Rhizoctonia*

TABLE 8-1 Taxonomic Outline of Fungi (*cont.*)

Mycota (division)	Characteristics	Examples
Basidiomycetes	Septate hyphae (sometimes clamps) Basidiosphores born on basidia	
Heterobasidiomycetidae	Septate basidium	
Tremellales		Jelly fungi
Uredinales		Rust fungi
Ustilaginales		Smut fungi
Homobasidiomycetidae	Nonseptate basidium	
Exobasidiomycetes	Basidia born free	
Hymenomycetes	Basidia arising from a fertile layer called a hymenium	
Polyporales		*Phellinus, Inonotus (Fomes, Polyporus)*
Agaricales		Mushrooms
Gasteromycetes	Basidia born enclosed within fungus tissue	
Lycoperdales		Puffballs
Phallales		Stink horns
Nidulariales		Bird's-nest fungi

sporangium. Some sporangiospores in the Oomycetidae have flagella for motility in water. Sporangiospores in the Zygomycetidae are without flagella and are dispersed in a dry environment.

The third subclass of the Phycomycetes includes a group of aquatic fungi, called Chytrids, which are of interest to aquatic biologists but play no role in woody plant systems.

A summary of the orders of fungi with typical examples are presented in Table 8-1.

DISEASE CYCLE OF PATHOGENIC FUNGI

The disease cycle of a pathogenic fungus is an interaction of the life cycles of both the plant and the fungus. Placement of spores or other vegetative reproduction propagules in an infection court on the host is called *inoculation. Incubation* of the spore involves uptake of water, possibly nutrients, and initiation of metabolic processes leading toward *germination* or the transition from the resting spore to dynamically growing mycelial phase of the fungus. *Infection* occurs when the fungus germ tube makes contact with the host cells and begins deriving nutrients from the enzymatic digestion of host material rather than utilizing stored reserves within the spore. *Invasion* is the ramification of fungus hyphae within the host tissues. *Disease symptoms* result from fungus invasion and host response. Further invasion may occur to intensify the disease symptoms. The fungus must go through a metabolic transition from vegetative expansion to growth termination and sporulation somewhere after invasion. In the case of foliage fungi, this transition may occur in just a matter of days, whereas in heart rot decay fungi it may take decades. Fungi that cause diseases of crops in temperate regions must accommodate their life cycles to survive the freezing conditions of winter. Fungi of other climates must survive high temperatures or drought. Therefore, some type of *dormant stage* is necessary. Some fungi combine sporulation and dormancy; others just stop mycelial growth and then begin mycelial growth again when temperature conditions are more favorable. Other fungi produce sclerotia. Sclerotia are tight aggregates of vegetative fungal cells surrounded by an impervious outer layer. Sclerotia may remain dormant in the soil until stimulated to grow by the presence of a susceptible host plant. Each fungus must have some method of *dissemination* of propagules from the diseased plant to another healthy plant. Some fungi have adopted insects as specific vectors for dissemination of the spores; others use wind or splashing water. The random or specifically directed placement of the fungus spore in the infection court completes and begins the disease cycle again.

Each type of disease discussed in future chapters will include a discussion of a typical disease cycle. You will soon recognize how little we really know about the disease cycles of fungi that cause diseases of trees. Hopefully, some of you may be stimulated to investigate some of the unknown aspects of diseases of trees. The rest of you will at least be reasonably capable of evaluating what is supposedly known.

REFERENCES

AINSWORTH, G. C., F. K. SPARROW, and A. S. SUSSMAN, eds. 1973. The fungi: an advanced treatise. Vol. 4A, 621 pp.; Vol. 4B, 504 pp. Academic Press, Inc., New York.

ALEXOPOULOS, C. J. 1962. Introductory mycology. John Wiley & Sons, Inc., New York. 613 pp.

BURNETT, J. H. 1968. Fundamentals of mycology. St. Martin's Press, Inc., New York. 546 pp.

CHRISTENSEN, C. M. 1961. The molds and man, an introduction to the fungi, 2nd ed. University of Minnesota Press, Minneapolis. 238 pp.

DEVERALL, B. J. 1969. Fungal parasitism. Studies in biology 17. St. Martin's Press, Inc., New York. 57 pp.

HUDSON, H. J. 1972. Fungal saprophytism. Studies in biology 32. Edward Arnold (Publishers),Ltd., London. 67 pp.

PILÁT, A., and O. UŠAK. 1951. Mushrooms. Spring Books, London, pp. 59–61.

SKYE, E. 1979. Lichens as biological indicators of air pollution. Annu. Rev. Phytopathol. *17*:325–341.

TALBOT, P. H. B. 1971. Principles of fungal taxonomy. St. Martin's Press, Inc., New York. 274 pp.

9

FUNGI AS SYMBIONTS OF TREE ROOTS: MYCORRHIZAE

- Types of mycorrhizae
- Mode of action of mycorrhizae
- Mycorrhizal association cycle
- Importance of mycorrhizae
- Recognition of mycorrhizae
- Examples of mycorrhizae
- Establishment of mycorrhizae on seedlings

Mycorrhizae (fungus roots) are the normal means by which almost all higher plants take up water and minerals from the soil. Only a few higher plant genera are known to lack mycorrhizal associations. One might question the insertion of this topic in a book on disease concepts, because mycorrhizae represent supposedly mutualistic symbiosis. The mycorrhizal fungus parasitizes the cortical cells of the root, but the presence or absence of mycorrhizae will dramatically affect seedling form and size and tree development.

Mycorrhizae are usually emphasized for their beneficial effect on root function. They are also important as barriers to infection by other destructive root pathogens in the nursery and in the field.

TYPES OF MYCORRHIZAE

Mycorrhizae are split into three groups, based upon the type of interaction of fungus and plant root. In one type, endomycorrhizae (Fig. 9-1), the fungus invades the cortical cells of the roots. Endomycorrhizae are characterized by the obligate relationship of orchids with the fungus *Rhizoctonia* spp. and a nonobligate but very normal relationship between grasses, legumes, some hardwood trees, and many other plants with *Endogone* spp.

Another type, ectomycorrhizae, is a fungus–host relationship where the fungus surrounds the roots, forming a mantle, and invades the roots but does not penetrate the cortical cells. A Hartig net (named for Robert Hartig, who first described the association) of hyphae forms between the cortical cells (Fig. 9-2). Most conifer and some hardwood mycorrhizae are of this type (Fig. 9-3).

The third type is called ectendomycorrhizae, which combines a Hartig net and a thin mantle with invasion of cortical cells (Fig 9-4). Ectendomycorrhizae are found in well-fertilized nurseries. Both conifers and broadleaf plants have this type of mycorrhizae.

The characteristics of the three types of mycorrhizae are summarized in Table 9-1.

TABLE 9-1 Characteristics of Mycorrhizal Types

	Mantle	Hartig net	Intracellular fungus invasion	Intercellular fungus invasion
Ectomycorrhizae	Yes	Yes	No	Yes
Endomycorrhizae	No	No	Yes	No
Ectendomycorrhizae	Yes	Yes	Yes	Yes

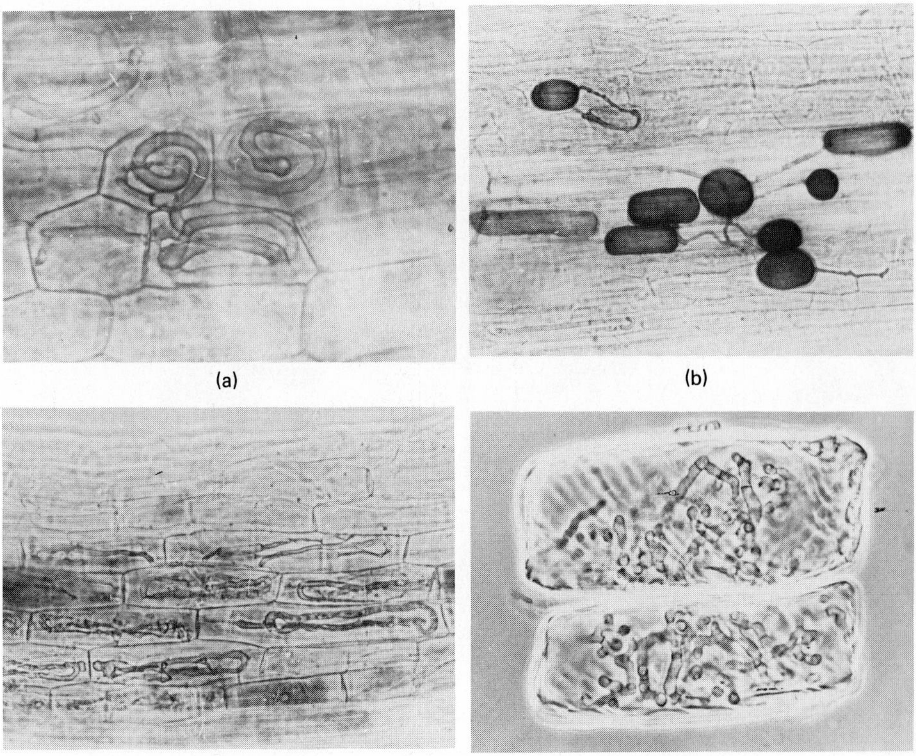

(a) (b)

(c) (d)

FIGURE 9-1 Endomycorrhizal root showing (a) Coiled hyphae. (b) Vesicles. (c) and (d) Intracellular hyphae. (Photographs compliments of Dr. Hugh Wilcox.)

FIGURE 9-2 Ectomycorrhizal root cross section, showing a Hartig net and mantle. (Photograph compliments of Dr. Hugh Wilcox.)

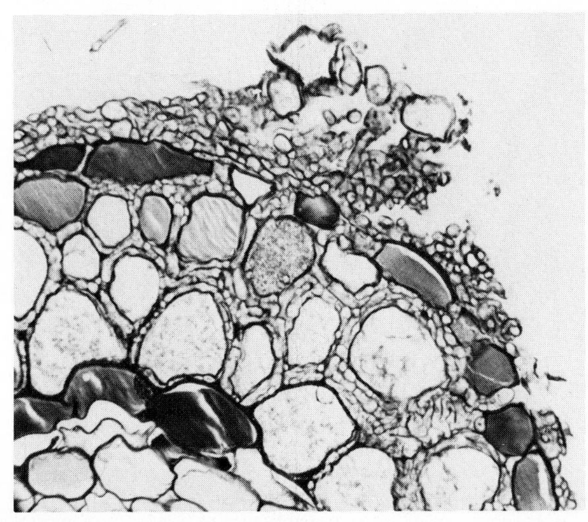

FIGURE 9-3 Examples of ectomycorrhizal roots of red pine. (Photographs compliments of Dr. Hugh Wilcox.)

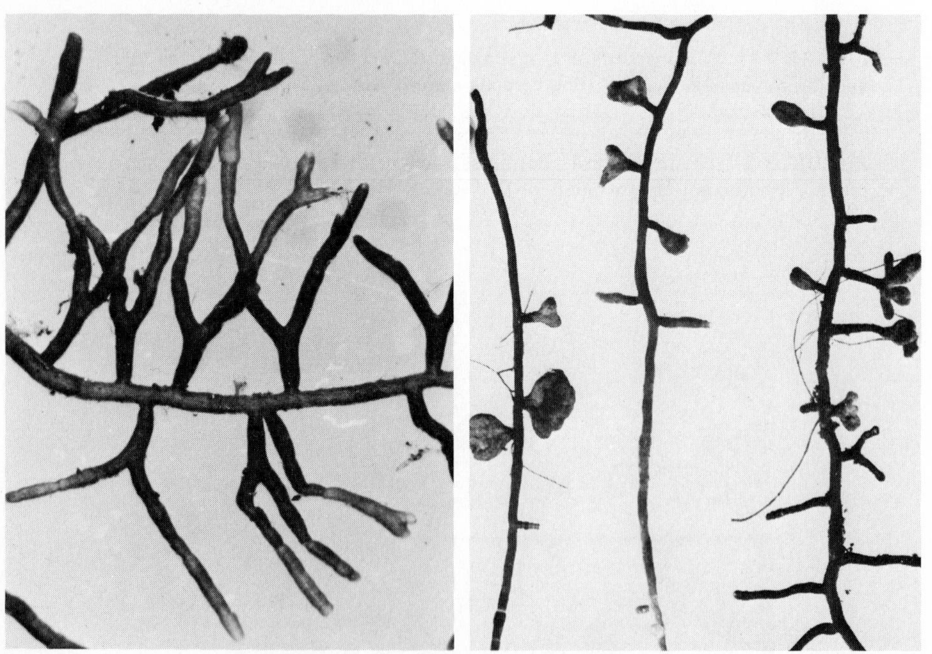

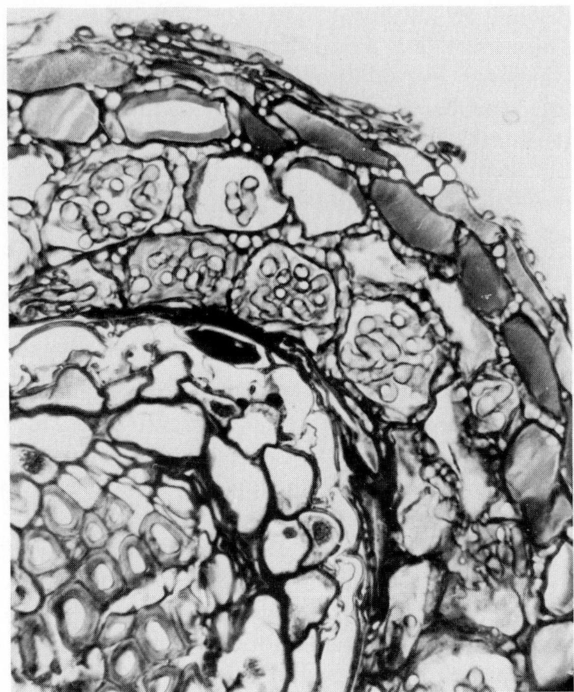

FIGURE 9-4 Ectendomycorrhizal root cross section, showing both intra- and intercellular invasion of cortical cells and a thin mantle. (Photograph compliments of Dr. Hugh Wilcox.)

MODE OF ACTION OF MYCORRHIZAE

The mycorrhizal association of fungi and plant roots is a two-way mode of action. Fungus invasion causes a major proliferation of short roots. In turn, the increase in roots increases the moisture- and mineral-element-absorbing surface and provides the plant with a better capacity to survive and grow.

In addition, the fungus-root association is a more efficient mineral-element-absorbing organ. More nitrogen and phosphorus are absorbed from the soil and accumulated in plants with mycorrhizae, as compared to plants without mycorrhizae. The fungus capacity for extraction of elements from the soil organic matter is assumed to be part of the increased efficiency.

The fungus parasitizes the plant for most of its carbohydrate and vitamin needs. According to some mycorrhizal specialists, the fungus may actually be obligately parasitic on plant roots.

Why are some fungi mycorrhizal, some pathogenic, and some saprophytic? These differences are probably based on the mode of action of the organism. The saprophytic fungi cannot penetrate the physical and chemical defense barriers of living roots. The parasitic fungi penetrate and cause a series of disruptive reactions leading toward the death of the host. The mycorrhizal fungi, like other obligate parasites, induce a limited host chemical defense reaction which is not totally disruptive to the host and is tolerated by the fungi.

If we could identify the important chemical reactions in mycorrhizal fungus root interactions, we might be able to utilize these natural chemical defense mechanisms for the prevention of plant diseases.

MYCORRHIZAL ASSOCIATION CYCLE

Most of the known ectomycorrhizal fungi are homobasidiomycetes, so initial infection of roots is probably by basidiospores. Reasonable mycorrhizal colonization of fumigated nursery beds usually occurs during the first growing season. The source of inoculum appears to be basidiospores from mushroom fruit bodies associated with tree roots around the nursery.

Mycorrhizal roots develop and are functional for a single growing season. Presumably, the fungus survives in the root association and reinfects new roots produced along the expanding root system of the following year.

Fruit bodies of Agarics (gilled mushrooms) and Boletes (pored mushrooms) are formed above ground. Hyphal strands connect the fruit body with the mycorrhizal root nutrient source.

The endomycorrhiza is initiated, in some instances, by germination of an *Endogone* chlamydospore. The chlamydospore is a relatively large, thick-walled, resting fungus spore produced inside the cortical cells or on the surface of endomycorrhizal roots. These spores can be collected from soil through special washing and sieving techniques. They appear to be well distributed in the upper layers of soil.

It is not known how *Endogone* chlamydospores are disseminated to infect additional plants. One can infest a soil by adding spores to soil or introduction of soil from an area where plants were recently grown. Planting cover crops such as grasses or legumes will also initiate a population of *Endogone* chlamydospores.

We do not know much about the life cycle of *Endogone* spp. because the fungus is difficult to grow in culture. Little is known beyond observations of the fungus root association.

Ectendomycorrhizae are not well understood either.

The fungi of the ectendomycorrhizal association can be grown in culture. They have not been identified taxonomically, because they do not sporulate in culture. Apparently initial invasion of root tips is like that of ectomycorrhizae, but eventually endomycorrhizal invasion of cortical cells occurs. There seems to be a progression involving initially ecto, then ectendo, and then primarily endo-like invasion of lateral roots.

The factors affecting sporulation, dissemination and infestation of soil by ectendomycorrhizae, are not known. Higher levels of soil fertility appear to favor ectendomycorrhizal associations, because this type of association is reported most commonly from relatively fertile conifer nurseries.

IMPORTANCE OF MYCORRHIZAE

Physiological and silvicultural aspects of mycorrhizae are important to foresters and others working with plants. Experiments in which seedlings are grown with and without the mycorrhizal fungi demonstrate the effect of mycorrhizae on plant growth and vigor. Plants with mycorrhizae are much larger and much more vigorous that nonmycorrhizal plants.

Mycorrhizal benefits to plants include (1) increased mineral and water availability resulting from nutrient exchange with the fungus, and increased root surface area; (2) resistance to root pathogens through mantle formation and reduced carbohydrate levels in the plant root; and (3) increased populations of nonpathogenic microorganisms in the rhizosphere of the plant, resulting from mycorrhizal activities.

It is apparent that mycorrhizal associations on plant roots are beneficial to the plant's well-being. Thus foresters, wishing to maximize timber and/or fiber yield in the shortest possible time, and other land managers, interested in the reclamation of highly disturbed sites, would be wise to investigate the effects of mycorrhizae on seedling and mature plant development as a natural means of increasing productivity and establishing new forests. Arborists should also consider mycorrhizae in the establishment and maintenance of ornamental plantings.

RECOGNITION OF MYCORRHIZAE

Ectomycorrhizae are rather variable in color and form. Some mycorrhizae are just single short stubs, others are the more characteristic fork-branched short roots, and others are coralloid clusters of branching roots.

Ectendomycorrhizae have similar gross morphological characteristics, but endomycorrhizae have no specific gross morphological characteristics.

More detailed observations on mycorrhizae require a microscope. Mycorrhizae are cleared for microscopic examination by boiling in dilute potassium hydroxide and then staining with aniline blue (cotton blue) or other stain to observe the fungus. Embedding in paraffin or other medium and sectioning with a microtome are sometimes necessary for very detailed observations of fungus–host interaction.

Proof of mycorrhizal association for specific fungi is accomplished in the laboratory using aseptically grown seedlings. A pure culture of the suspected mycorrhizal fungus is introduced into the rooting medium of the aseptically grown seedling, and the development or lack of development of a mycorrhizal association is evaluated after a period of time.

The usual method of obtaining pure cultures of fungi for such tests of mycorrhizal capacity is from basidiospores or from pieces of fungus tissue of fruit bodies observed in association with tree roots. If one attempts to culture from actual mycorrhizal roots, a different collection of nonsporulating mycorrhizal fungi and many common saprophytic soil fungi are recovered.

Identifiable basidiomycete fungi are seldom cultured from mycorrhizal roots, but those found fruiting in the forest often will form mycorrhizae in laboratory cultures. A vast array of unidentified fungi cultured from roots form equally good mycorrhizae. What are the actual roles and importance of these groups of fungi? It can only be left as a question at this time.

EXAMPLES OF MYCORRHIZAE

Specific examples of mycorrhizae are very numerous. One can get names of some know mycorrhizal associations from Hepting's book, *Diseases of Forest and Shade Trees of the United States* (1971). The rest are scattered in the literature.

One specific mycorrhizal association is being exploited extensively now to enhance survival of conifers planted in highly deficient sites. This is just the beginning of a promising future for mycorrhizae in tailormaking seedlings for specific sites.

The fungus that is being studied in detail throughout the United States and in other countries is *Pisolithus tinctorius*. This puffball-like mushroom is commonly observed in association with and isolated from trees growing on very deficient sites, such as mine spoils.

The fungus can be grown in quantity on sterilized grain and introduced into well-fumigated nursery beds at planting time. Seedlings in such infested beds develop *P. tinctorius* mycorrhizae rather than *Thelephora terrestris* mycorrhizae because the massive inoculum of *P. tinctorius* outcompetes the normal *T. terrestris* recolonization of the bed.

Survival and growth comparison of seedlings with *P. tinctorius* and *T. terrestris* show striking superiority of the *P. tinctorius* seedlings in deficient sites but not on highly fertile sites.

ESTABLISHMENT OF MYCORRHIZAE ON SEEDLINGS

Nature usually does an acceptable job of bringing together compatible fungi and roots in mycorrhizal associations. Lack of mycorrhizae accounts for a limited number of examples of nonsurvival of conifers. One such example was the attempt to establish conifer nurseries in the central plains. Only after introduction of some soil from forested areas were the prairie nurseries able to produce acceptable seedlings.

Natural infestation of fumigated nurseries from spores produced on fruit bodies around the nursery has been taken for granted. But there are potential advantages in controlling the types of fungi producing mycorrhizae. Introduction of grain infested with cultures of mycorrhizal fungi appears to work if the fumigation is properly done and the moisture and temperature conditions are favorable for the fungus. Some work is going into developing less bulky inoculum by using spores rather than infested grain.

Because the topic of mycorrhizae is just starting to get the practical attention it deserves, we will probably see additional methods for controlling various types of mycorrhizal fungi. Maybe we will learn to recognize good mycorrhizal sites for establishment of nurseries or maybe we will learn to modify present nursery sites to exploit appropriate natural inoculum.

Establishment of mycorrhizae on ornamental plant materials has not been a concern of the producers of such materials. Similarly, the role of mycorrhizae on ornamentals has not been seriously studied by researchers. One might speculate that concern and research on the topic of mycorrhizae could improve the survival and development of some of the ornamental materials presently being utilized.

Mycorrhizae and their role in agricultural production is presently a fruitful topic of research. With increasing costs of fertilizers, the use of mycorrhizal inoculation to improve plant vigor looks interesting.

REFERENCES

GERDEMANN, J. W., and J. M. TRAPPE. 1974. The Endogonaceae in the Pacific Northwest. Mycol. Mem. 5. 76 pp.

HACSKAYLO, E. 1971. Mycorrhizae. USDA For. Serv. Misc. Publ. 1189. 255 pp.

HACSKAYLO, E. 1972. Mycorrhizae: the ultimate in reciprocal parasitism. BioScience *22*:577–582.

HATCH, A. B. 1936. The role of mycorrhizae in afforestation. J. For. *34*:22–29.

HEPTING, G. H. 1971. Diseases of forest and shade trees of the United States. Agriculture handbook 386, USDA Forest Service. 658 pp.

MARKS, G. C. and T. T. KOZLOWSKI, eds. 1973. Ectomycorrhizae: their ecology and physiology. Academic Press, Inc., New York.

MARX, D. H. 1972. Ectomycorrhizae as biological deterrents to pathogenic root infections. Annu. Rev. Phytopathol. *10*:429–454.

MARX, D. H. and W. C. BRYAN. 1975. Growth and ectomycorrhizal development of loblolly pine seedlings in fumigated soil infested with the fungal symbiont *Pisolithus tinctorius.* For. Sci. *21*:245-254.

MOSSE, B. 1973. Advances in the study of vesicular–arbuscular mycorrhizae. Annu. Rev. Phytopathol. *11*:171–196.

SINCLAIR, W. A. 1974. Development of ectomycorrhizae in a Douglas-fir nursery. II. Influence of soil fumigation fertilization and cropping history. For. Sci. *20*:57–63.

SLANKIS, V. 1974. Soil factors influencing formation of mycorrhizae. Annu. Rev. Phytopathol. *12*:437–457.

WILCOX, H. E., and R. GANMORE-NEUMANN. 1974. Ectendomycorrhizae in *Pinus resinosa* seedlings. I. Characteristics of mycorrhizae produced by a black imperfect fungus. Can. J. Bot. *52*:2145–2155.

10

FUNGI AS AGENTS OF TREE DISEASES: FOLIAGE DISEASES

- Types of foliage diseases
- Mode of action of fungus-caused foliage diseases
- Disease cycle of a fungus foliage pathogen
- Symptoms of foliage diseases caused by fungi
- Methods for recognition of specific foliage diseases
- Examples of foliage diseases caused by fungi
- Control of foliage diseases caused by fungi

Foliage diseases are of concern because of their effects on photosynthetic activity and, therefore, on the growth of the plant. But the effects of reduced photosynthetic activity on plant vigor may influence the susceptibility of the plant to invasion by other fungi and insects, thereby compounding the potential losses. Another effect of foliage diseases that is difficult to quantify is the psychological trauma caused by such an obvious disorder on the homeowner with a limited number of well-cared-for plants.

TYPES OF FOLIAGE DISEASES

Foliage diseases are caused by a wide array of biotic and abiotic agents, such as fungi, insects, mites, air pollutants, viruses, mycoplasmas, bacteria, nematodes, and other factors (Fig. 10-1).

Foliage diseases caused by fungi are usually separated into groups based on the taxonomic position of the causal fungus. Basidiomycete rust fungi represent one group that will be discussed in Chapter 11. This chapter will emphasize foliage diseases caused by Ascomycete fungi or imperfect stages of Ascomycete fungi.

A foliage disease and tuber rot of potato, late blight, caused by the Phycomycete fungus *Phytophthora infestans,* produced one of the most devastating epidemics in recorded history. But foliage diseases of trees caused by Phycomycete fungi are almost unknown.

FIGURE 10-1 Nipple gall mite on sugar maple.

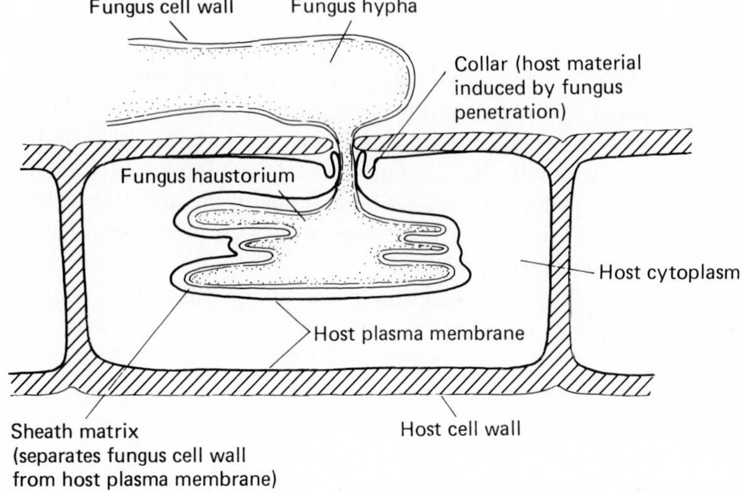

FIGURE 10-2 The relationship between the fungus haustorium and the host cell.

MODE OF ACTION OF FUNGUS-CAUSED FOLIAGE DISEASES

Some fungi that parasitize leaves produce intercellular or surface mycelium which extracts nutrients from living cells through haustoria (Fig. 10-2) that penetrate the cell wall but not the plasma membrane. Nutrients are absorbed through the plasma membrane of the plant cells into the haustorium. Parasitism of this type usually does not kill the cells immediately but rather causes the infected cells to act as a pumping system for nutrients from the plant. If one traces movement of elements by introduction of radioactive phosphorus or carbon to a plant, one finds that elements are preferentially translocated to the infection site.

Other fungi parasitize leaves by direct penetration of parenchyma and mesophyll cells and extraction of available nutrients. The rapid death of cells occurs.

The leaves being parasitized probably react with specific chemical responses to both types of infections. Studies on host response to foliage infection and invasion of agricultural crops show specific and genetically controlled reactions. There is no reason to believe that similar specific and genetically controlled reactions do not occur in trees also, but forest pathologists have not yet explored these possibilities to any great extent.

DISEASE CYCLE OF A FUNGUS FOLIAGE PATHOGEN

The disease cycle of a typical foliage pathogen of trees involves the production of ascocarps either during the fall or early spring on infected needles or leaves. Ascospores are shot from these ascocarps about the time that new foliage is coming out in the spring. Those few infections that take place produce large numbers of conidia which intensify the infection. The fungus overwinters as saprophytic mycelium or ascocarps in diseased leaves or in twig cankers to complete the cycle.

Ascomycete fungi are well suited as foliage pathogens because they generally have a sexual ascospore stage and an asexual conidial stage. The ascospores are formed in asci inside some type of ascocarp. Mycelium in infected tissue or the ascocarp may provide the fungus with a stage for survival during adverse environmental conditions such as winter. The ascospore or sexual stage also allows for genetic recombination to take place. Genetic recombination is essential for diversity in the parasite, and diversity is important for parasitic virulence on a host with wide genetic variability and for the ability to survive under varying environmental conditions.

FIGURE 10-3 Infection by foliage fungi is by direct penetration or through stomates.

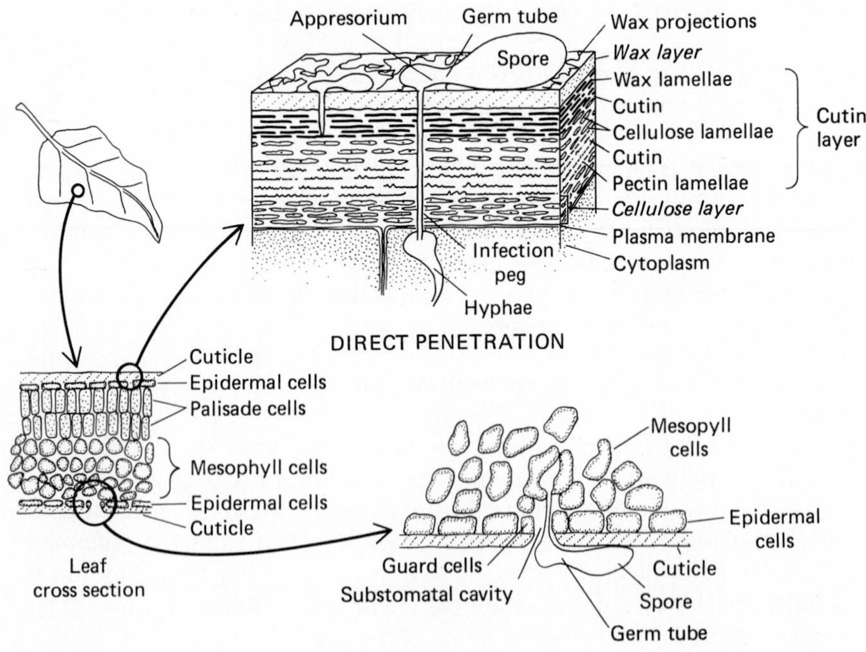

The asexual spore stage provides the fungus with the capacity to rapidly increase its numbers once a virulent ascospore infects a susceptible host. In the immediate area of infection, thousands of vegetatively produced conidia are formed. These need disperse only a short distance to find an infection point on susceptible tissue—often the same host. The selection and increase of the more virulent lines provide foliage diseases with explosive potential for causing disease epidemics.

Ascomycete fungi, except for mildews, can survive both as parasites and as saprophytes. This is necessary for survival on fallen or dead leaves of the host during the dormant season. The obligate parasitic mildews survive as cleistothecia rather than hyphae in dead host material.

Some Ascomycetes simplify their life style by infrequently utilizing the sexual stage. Therefore, many foliage diseases are better known by their Fungi Imperfecti names.

For foliage pathogen disease buildup, wet conditions are necessary. Moisture is required for the triggering of ascospore or conidia release as well as for germination and infection. Foliage diseases are most serious during years when cool, wet conditions occur.

The infection process of foliage fungi is by direct penetration of the leaf surface or through stomates (Fig. 10-3). The diversity of structure and chemical composition of the cuticle and leaf cell wall prevents direct penetration by all but a few fungi. The foliage fungus must penetrate these layers either by force or enzymatic action or by both to get to the cytoplasm, where it derives its nourishment.

SYMPTOMS OF FOLIAGE DISEASES CAUSED BY FUNGI

Infection by foliage fungi can result in total necrosis and shriveling of leaves or discrete localized necrotic areas.

Massive infection or invasion at yearly intervals may cause the tree to shed its leaves and produce a second crop. Reduced terminal growth and tip dieback sometimes occur. Adventitious buds may produce new shoots and foliage, forming a clustering of branches into witches' brooms.

Associated with necrotic symptoms, one usually finds signs of fungus fruiting structures.

METHODS FOR RECOGNITION OF SPECIFIC FOLIAGE DISEASES

Foliage problems caused by fungi are not always easily distinguished from those caused by insects, mites, air pollutants, virus, mycoplasmas, bacteria, nematodes, and other causes. Insect problems of foliage are usually characterized by some type of distinctive feeding symptom, the presence of larva or other insect stage, or the fecal pellets of former insect activity. Mites sometimes cause distinctive color and morphological changes, particularly in hardwoods. In conifers the symptoms of mites are subtle and difficult to recognize with certainty. Shortening and curling of some of the needles in a fascile and chlorosis is sometimes associated with mites. Air pollutants produce flecking that resembles the early symptoms of some fungal diseases. The hypersensitive reaction of a virus infection results in flecking. Other viruses cause irregular ring spots. Mycoplasma-induced chlorosis and witches' brooming of twigs may occasionally resemble fungal diseases. Bacterial leaf spots are distinguished from fungal diseases based on the presence of bacteria as seen with dark-field illumination of the microscope or by isolation on agar media. There are very few nematode diseases of foliage, and these are readily diagnosed by the presence of the nematode. Other natural phenomena, such as normal fall coloration and drop of conifer needles, or man-made problems resulting from phytotoxicity of chemicals applied to plants, are sometimes confused with foliage diseases.

Foliage diseases may be confused with late spring frosts. Frosts occur on leaves throughout the crown or in the upper crown of trees, while foliage diseases are usually more serious in the lower crown. The microclimatic conditions, particularly high relative humidity of the lower crown, favor infection by foliage pathogens.

Fungus induced foliage diseases usually produce distinctive symptoms and signs for easy identification. It is often necessary to make a freehand section and observe the fungus microscopically to make a positive identification. Using a fungus–host index, it is usually possible to identify the disease.

If one cannot find reference to a previous association of the observed fungus and host in the literature, it is advisable to prove pathogenicity using Koch's postulates. Some foliage fungi are obligate parasites, so that growth on culture media may not be easily accomplished. In this case it is necessary to inoculate with spores obtained directly from a host plant.

EXAMPLES OF FOLIAGE DISEASES CAUSED BY FUNGI

Hardwood Foliage Diseases

Foliage diseases have traditionally been assumed to be of little importance to hardwoods except where they are used as ornamentals. As an ornamental, any defect that reduces the appearance quality of a tree is of importance, not so much to the tree but to the owner or observer of the tree.

The reason foliage diseases of hardwoods are seldom important is that the tree appears to produce much more leaf surface area than it can efficiently fully utilize for photosynthetic production.

Foliage diseases can be expected to increase in importance with intensive management of hardwoods. In Europe, *Melampsoroa* rusts and *Marsonina* leaf spot are some of the most serious problems for their extensively used poplar varieties.

If approximately 60% of the leaf area is destroyed by either fungus or insects, growth reduction will occur and sometimes another crop of leaves will be formed.

If defoliation takes place early in the season, there may be a slight overall reduction in growth that year, but the tree will refoliate and generally suffer no consequences from the defoliation. But if defoliation occurs during the middle part of the summer, the tree will attempt to refoliate and may not become fully hardened-off for cold weather when fall comes. Under these conditions, there may be injury to the next year's buds by frost and an overall reduction in food reserves stored in the plant for initiation of growth the next year. Late summer or fall defoliation in the absence of refoliation does not seriously affect food reserves, so that refoliation rather than defoliation is the serious aspect of the problem. Reduction in food reserves seems to predispose a tree to attack by *Armillariella mellea*, a root-infecting fungus discussed in Chapters 16 and 18.

The fungi that cause foliage diseases will often parasitize young shoots and twigs. The twig and shoot blight foliage fungi are very similar to canker fungi in their interaction with the host cambium. Twig cankers and dead shoots sometimes provide a method of overwintering for the fungus.

Septoria leaf spot and canker *Septoria musiva* causes a foliage and canker disease of eastern cottonwood. The pathogen is of minor concern except when cottonwoods and hybrid poplars are grown in intensive culture.

The future for intensive culture for poplars in the United States looks rather promising. The trees grow very rapidly on good sites, producing up to 2.5 cm of radial increment per year. They can be readily propagated as clones by cuttings.

Septoria foliage and canker problems have been recognized on clones of both cottonwood and hybrid poplar. Growth loss is related to disease intensity. Variation in susceptibility suggests possible genetic control of resistance.

The easiest thing to do is to select a resistant plant for production of cuttings, but this type of practice will lead to disaster. It is important to recognize how the resistance is inherited in the host as well as the pathogen before moving into any improvement program. We have often seen resistant varieties of agricultural crops eventually destroyed by devastating diseases. Selection for resistance generally narrows the genetic variability. As discussed in Chapter 20, it is important to maintain genetic variability to avoid rapid disease development.

Another way to minimize loss due to *Septoria* and other similar diseases is to evaluate the effects of growing site and stand density on disease development. Environmental conditions influencing both the plant and the pathogen may be utilized to advantage in reducing disease losses. With *Septoria* canker in poplar, wider plant spacing was associated with increased cankering. In other diseases it may be just the opposite.

Anthracnose Anthracnose is one of the most common foliage diseases of hardwoods. Species affected by anthracnose include sycamore, white oak, walnut, maple, ash, hickory, elm, birch, catalpa, basswood, plane tree, tulip tree, and horsechestnut. Generally, the causal fungus is a species of *Gnomonia* (sexual stage), *Gloeosporium* (asexual stage). Typical symptoms are irregular dead blotches that merge together to cause death of most of the leaf and often result in defoliation (Figs. 10-4 and 10-5). If defoliation does not occur, the leaves remain distorted. On sycamore and white oaks, shoot infection causes the death of twigs and the emergence of lateral buds to form

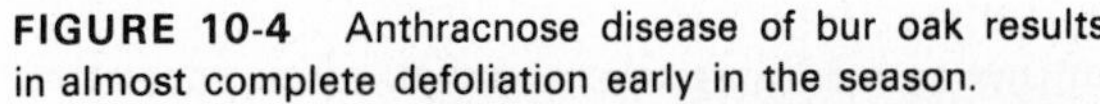
FIGURE 10-4 Anthracnose disease of bur oak results in almost complete defoliation early in the season.

FIGURE 10-5 Closeup of anthracnose-affected bur oak foliage.

witches' brooms on the heavily infected trees. The best control measure for anthracnose is to rake and destroy the infected leaves in the fall. This cuts into the disease cycle by eliminating the overwintering stage. In the field there is presently no control, since raking leaves or applying fungicides is impractical. Probably, very little damage occurs in natural stands of hardwoods, but with future intensification in the planting and management of hardwoods, we can expect foliage diseases to develop to the point where more effort will be necessary to reduce growth losses. The best approach with intensively managed hardwoods may be to select for genetically controlled resistance.

Powdery mildew Powdery mildew is another common foliage disease of many trees, shrubs, agricultural crops, and grasses. In trees and in shrubs the disease often develops late in the summer or fall and does not cause significant damage to the plants. Symptoms include a characteristic white cast to the leaves due to the mycelium of the fungus growing entirely on the outer surface (Figs. 10-6 and 10-7). Black cleistothecia often appear as distinct dots on the white surface.

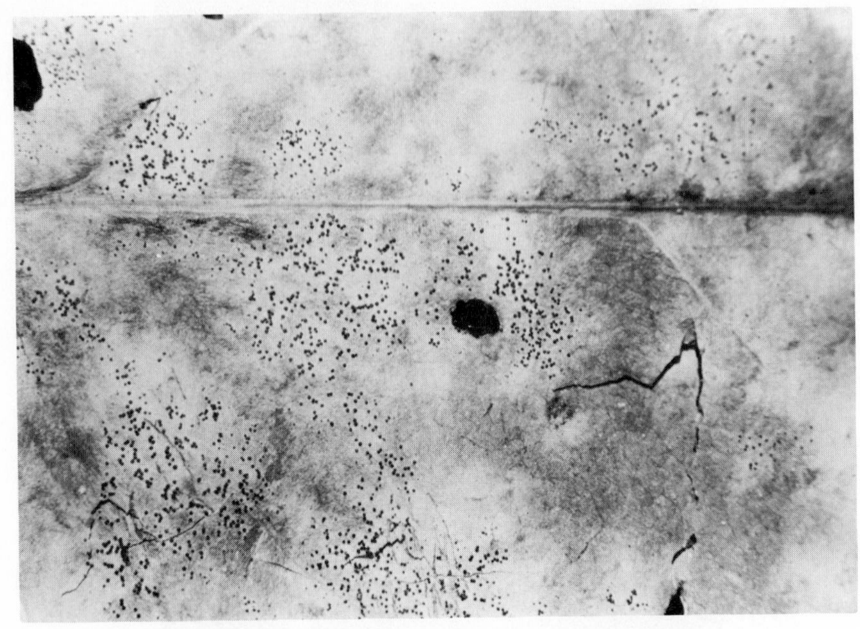

FIGURE 10-6 Mildew cleistothecia on a leaf surface.

FIGURE 10-7 Closeup of single *Microsphaera* sp. cleistothecium.

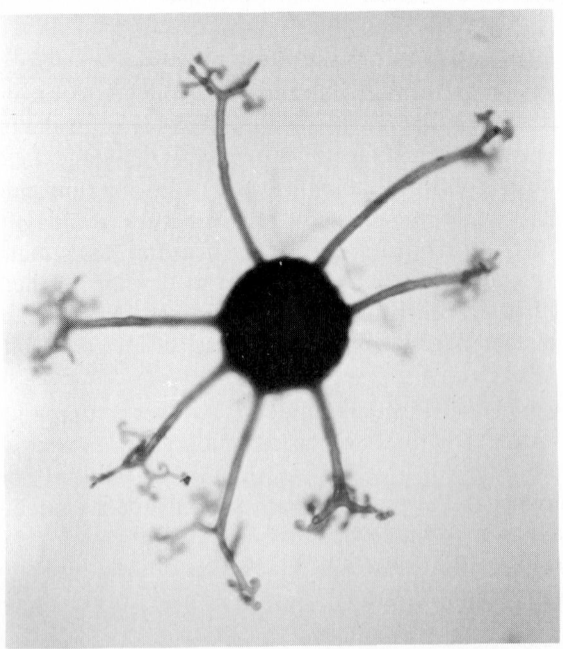

The powdery mildew fungi produce distinctive appendages on the cleistothecia. This makes specific identification of the fungus very easy, as can be seen from the following key to the genera of fungi causing powdery mildews.

1. Cleistothecial appendages simple hyphae
   2. One ascus per cleistothecia — *Sphaerotheca* spp.
   2. Several asci per cleistothecia — *Erysiphe* spp.
1. Cleistothecial appendages branched or bushlike at tips
   3. One ascus — *Podosphaera* spp.
   3. Several asci — *Microsphaera* spp.
1. Cleistothecial appendages curled at tip, several asci — *Uncinula* spp.
1. Cleistothecial appendages with swollen bases, several asci — *Phyllactinia* spp.

The powdery mildew fungi are unique in that the mycelium does not penetrate the host. Specialized nutrient-absorbing structures called haustoria are inserted into the leaf cells but do not break through the plasma membrane, so that the cell remains functional. The haustoria absorb nutrients from the living cell through the plasma membrane to supply the mycelium and sporulation on the surface of the leaf.

Because mildew generally develops late in the year, control is not warranted. If a problem occurs earlier in the season, fungicide sprays can be effectively utilized on high-value trees. The mycelium on the plant surface is readily killed with fungicides.

Leaf blister Leaf blister, common on oaks (Fig. 10-8), is caused by a species of *Taphrina*. *Taphrina* is a primitive Ascomycete that produces its asci on the surface of the host in no specialized ascocarp structure. The eight ascospores often undergo additional mitotic divisions, so that the ascus may have many ascospores (Fig. 10-9). This fungus differs from the typical foliage pathogen in that the ascospores are shot in the fall. The spores overwinter on the bark of the tree and are washed onto the newly emerging foliage in the spring. A single infection period occurs each year. Therefore, a strong fungicide applied during the dormant season will eliminate much of the potential disease inoculum for the next season.

The *Taphrina* fungus produces disruption of cell division and enlargement, so that a distorted blistered leaf is produced. Some of the *Taphrina* spp. infect flower parts and petioles of trees and produce enlargement and distortion of these parts.

FIGURE 10-8 Leaf blister disease of red oak.

FIGURE 10-9 Asci and ascospores of *Taphrina* sp. found on the surface of the blistered leaf as seen in the microscopic section.

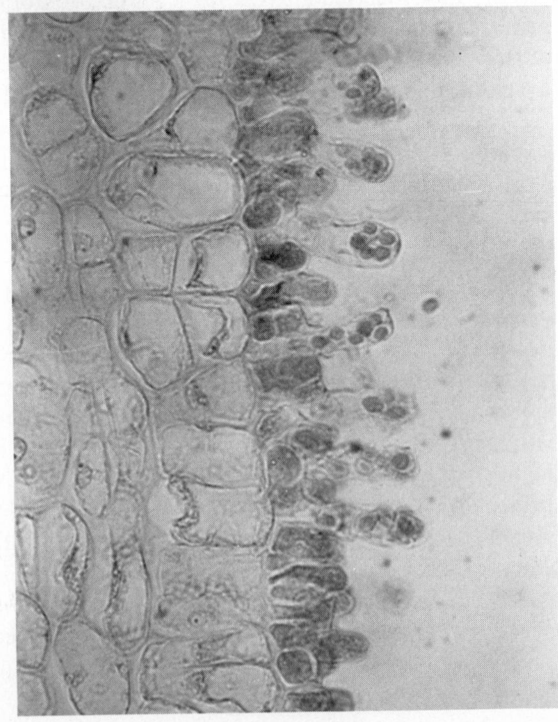

FIGURE 10-10 Tar spot disease of red maple. (Photograph compliments of Dr. Wayne Sinclair.)

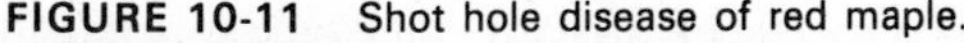

FIGURE 10-11 Shot hole disease of red maple.

Tar spot and shot hole Tar spot (Fig. 10-10) caused by species of *Rhytisma* and shot hole (Fig. 10-11) caused by species of *Phyllosticta* are two other common foliage diseases. These are characterized by black tar-like stroma on the leaf surface in the tar spot and circular necrotic spots in the shot hole disease.

There are many additional foliage diseases which can be readily diagnosed based on host, symptoms, and specific signs of the pathogen.

Conifer Foliage Diseases

Foliage diseases of conifers are important because, when severe, they reduce growth and even cause death. Conifers do not have the ability to refoliate, and therefore reduced photosynthetic surface causes the tree to become less and less vigorous.

Some of the most serious diseases of exotic conifers planted in various locations are foliage diseases. These diseases are often caused by fungi that cause minor damage in the natural range of the species but, when introduced into a new area, develop very rapidly to limit the potential of the exotic tree species.

Dothistroma blight *Dothistroma pini* is an example of a damaging foliage disease to exotic Monterey pine (*Pinus radiata*) grown in the southern hemisphere. The fungus can infect over 30 pine species as well as Douglas fir and European larch. In North America, the disease occurs on a number of pines in the West but is confined to young plantations and ornamental plantings of Austrian and ponderosa pines in the East.

Dothistroma is the asexual stage of an Ascomycete fungus *Scirrhia pini* which is reported only from the West.

The story behind the *Dothistroma* disease is most interesting because it demonstrates the type of problem encountered in attempting to maximize forest production.

Plantings of Monterey pine were first initiated near the end of the nineteenth century in Chile, Africa, New Zealand, and Australia. Excellent growth and form in these locations encouraged additional plantings and the development of lumber and paper industries, particularly in Chile and New Zealand.

No major diseases occurred on these plantings until 1957, when the *Dothistroma* blight became an important threat. *Dothistroma* was known from native Monterey pine in California but only as a minor pest. Examination of historical material documents the presence of the fungus as early as 1940 in Chile, parts of Africa, and New Zealand, but the extent of the problem at that time apparently did not warrant serious concern.

No one knows what triggered the explosive epidemic of *Dothistroma* in

Chile, New Zealand, and East Africa, but not in Australia and South Africa. It could be associated with a new virulent race of the fungus or particular weather conditions. In any case, the extensive plantings were ideal for the rapid development of a major problem.

An example of the explosive potential for this disease is shown in a study in California. Foliage infection of Monterey pine increased from 2.7% to 94% in 2 years, and mortality increased from 0.2% to 67.6%.

Disease increase is directly related to climatic conditions. Cool, moist conditions favor dispersal and infection by the pathogen. Seedlings and saplings are most susceptible to the disease and are often killed. Larger trees experience growth reduction in relation to the degree of foliage infected.

Excellent control is obtained with the use of copper fungicides. In New Zealand, fungicides are applied to large areas with aircraft on young plantations in three or four operations over a period of 15 years. After 15 years the trees have sufficient natural resistance to maintain themselves.

One might use this disease epidemic as an example of the difficulties encountered in maximizing forest production. Problems of this type might exemplify the need for managing natural forest systems. Actually, there is more to be learned from the *Dothistroma* problem than "don't fool with Mother Nature."

It is generally accepted that natural forest management systems are less likely to be upset by epidemic pest problems. The real question is whether the natural forest can supply the wood needed by our industrial civilization. Just as with agricultural production, it will be necessary to intensify wood production to keep up with demand. Just as with agriculture, the intensification generates new sets of problems. If the potential for problems is recognized and anticipated, disaster can be avoided or minimized.

In New Zealand, pathologists were conscious of the potential for this type of disease. Survey and research efforts by pathologists anticipated the problem. They have made spray recommendations to accommodate this disease in the short run. Over the long run, they are also evaluating resistance and other management practices.

The *Dothistroma* problem is not a disaster; it is an example of the need to integrate pathology into intensive forest management.

Elytroderma needle blight Elytroderma needle blight of Ponderosa pine is caused by *Elytroderma deformans*. This disease produces a conspicuous reddened foliage in the spring. On the needles are seen elongated black fruiting structures of the fungus. Repeated or systemic infection of trees causes a clustering of branches called witches' brooming.

This native pathogen appears to intensify due to specific environmental conditions. During wet climatic cycles, the disease intensifies. In areas where high moisture occurs because of local topography, localized outbreaks are common. Disease centers persist during dry periods because the fungus is able to parasitize and survive in twigs and buds. Therefore, the fungus can persist and expand its influence whenever environmental conditions are favorable.

Brown spot needle blight Another foliage disease of a native conifer caused by a native pathogen is brown spot needle blight of longleaf pine. *Scirrhia acicola*, the pathogen, infects needles of longleaf pine seedlings, reduces the growth of the seedlings, and prolongs the grass stage of the tree for 4 to 10 years longer than normal (Figs. 10-12 and 10-13). Once the tree starts height growth, which will not occur until the seedling has developed at least 1 inch of diameter growth at the ground line, the disease has very little effect on the tree.

This disease is typical of a disruption of natural balances caused by human attempts to utilize a plant species. Longleaf pine is very tolerant of fire even in the seedling stage, so natural fires and mixed ages of trees prevented serious buildup of the pathogen. Trees grown in nurseries are highly susceptible to the rapid spread of infection via the asexual stage. Even though chemical controls are applied in the nursery, some of the stock planted out is infected. This infected stock provides the initial source of inoculum for most plantations, which spreads rapidly through the remainder of the plantation of uniform infectable-sized trees.

Controlled burns of seedlings in the grass stage have provided some measure of control of the disease, but it is difficult to control the intensity of the fire necessary to reduce inoculum; it must not be so hot as to kill the terminal bud. One often gets some reduction in stocking because of fire kill, particularly of the resistant trees that have started height growth. The amount lost to fire may or may not be about as great as would be lost due to the disease.

Future control for this disease may lie in the resistant varieties presently being developed. Until they are available, there will be a minimum usage of longleaf pine.

Another form of control for this disease may involve silvicultural stand manipulations. A shelterwood overstory of medium-density longleaf pine prevents seedling infection for up to 8 years. The disease is more serious on poor sites and on sites with minimum ground-cover density.

Therefore, it is possible to regenerate longleaf pine on good sites through a shelterwood cut. Removal of overstory within 8 years and a subsequent burn during the winter of the second season following the cut should regenerate a reasonable stand. The burn is merely to reduce infection buildup. It occurs at a time when needle litter and logging debris have deteriorated sufficiently to produce a moderate intensity burn.

FIGURE 10-12 Grass stage of longleaf pine infected by *Scirrhia acicola.* (Photograph compliments of Dr. Savel Silverborg.)

FIGURE 10-13 Closeup of the brown spot disease on longleaf pine needles. (Photograph compliments of Dr. Savel Silverborg)

Christmas tree foliage diseases Foliage diseases are a serious problem in eastern Christmas tree plantations. The Christmas tree plantation diseases sometimes begin in the nursery stock, and quickly spread throughout the uniform plantation. Another feature of plantations is the limited range of genetic variability of the stock. This occurs inadvertently because the seed for most nurseries is collected from just a few trees or from a limited area. Narrowing of the gene pool by unconscious selection of a few seed trees is just asking for problems with pathogens that are capable of rapid spread by asexually produced spores. If a pathogen can infect one tree in the plantation, it can infect any tree, because they are genetically very similar.

A foliage disease currently of concern to eastern Christmas tree plantations is Lophodermium needle cast (*Lophodermium pinastri*) of Scotch pine and red pine (Fig. 10-14). Brown spot needle blight (*Scirrhia acicola*) is also a problem on Scotch and red pines. On Douglas fir, Rhabdocline needle cast (*Rhabdocline pseudotsugae*) is characterized by mottling of needles and

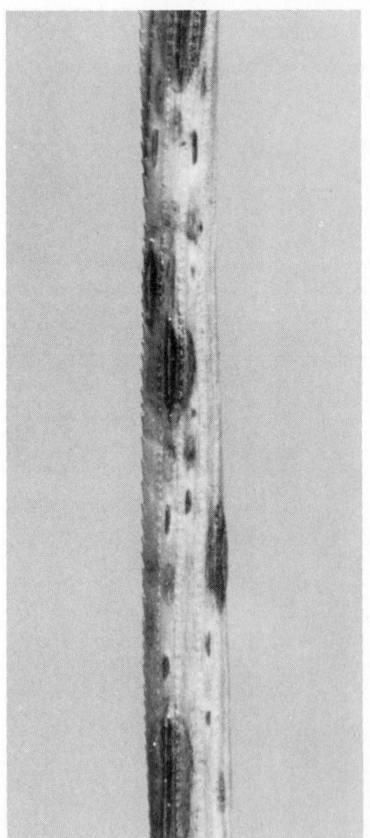

FIGURE 10-14 *Lophodermium pinastri* hysterothecia on pine needles. (Photograph compliments of Dr. Wayne Sinclair.)

FIGURE 10-15 *Rhabdocline* hysterothecia on lower surface of Douglas fir needles. (Photograph compliments of Dr. Savel Silverborg.)

FIGURE 10-16 *Phaeocryptopus gaumanni* perithecia on lower surface of Douglas fir needles. (Photograph compliments of Dr. Savel Silverborg.)

elongate apothecia produced on current-year needles (Fig. 10-15), and Swiss needle cast (*Phaeocryptopus gaumanni*) is characterized by the production of multiple rows of black perithecia-like locules on the underside of 2-year-old needles (Fig. 10-16). Fungicide sprays are the most widely recommended controls. Reduction of weeds in the plantation by moving or by use of herbicides reduces the humidity necessary for infection of leaves on lower branches. Therefore, weed control is advisable.

Diplodia tip blight Although shoot blights could represent another class of disease, they have many features common to foliage diseases. The *Diplodia pinea* fungus, for example, infects through elongating shoots or needles by direct penetration, very much like a foliage pathogen.

The *Diplodia* disease is reported for at least 11 species of hard and soft pine throughout the eastern and midwestern states. It is most serious on

FIGURE 10-17 Diplodia-killed branch of Austrian pine.

Austrian, Scotch, mugo, ponderosa, and red pines, particularly where the species are used for ornamental plantings.

At about 20 years of age, the infection of shoots begins. Killing back of the current year's growth detracts from the ornamental value of the trees (Fig. 10-17). Growth stunting and ragged appearance of the tree eventually reduce the ornamental value. Death of the trees is uncommon because they are usually removed before death occurs.

Fungicidal sprays during the time of shoot emergence can be used to protect trees, but it is important to protect the elongating tip by more than one spray.

CONTROL OF FOLIAGE DISEASES CAUSED BY FUNGI

A limited array of control procedures are known for foliage fungi of trees. In agriculture, some are controlled by fungicides and others by breeding disease-resistant varieties. Foliage diseases in Christmas trees and ornamentals can be reduced by the application of fungicides. Most treatments prevent new infections rather than eradicate older ones. Therefore, it is important to anticipate the problem and spray before infection has a chance to develop. With the present concern over pesticide pollution of the environment, it is appropriate to evaluate the benefits of application more closely rather than just spray. An example of evaluation procedures is discussed in Chapter 19.

Prevention by making proper species selection for specific sites is a much more appropriate control. Foresters and nurserymen are often more impressed by super growth or super form of an exotic than they are concerned regarding potential future problems. Based on past experiences on many continents, one can predict with reasonable accuracy that large-scale plantings of exotic trees will be seriously plagued, particularly by foliage diseases. Decisions must be made at the beginning of such an introduction program as to whether the desirability of the introduction, based on growth or form, is worth the increased risk and future costs associated with inevitable pest problems. (I say that decisions must be made, but I recognize that they are usually ignored even today by federal, state and private agencies developing and supplying new materials and methods.)

Control of foliage diseases by selection and breeding is possible but is not without problems. The subject of disease-resistance breeding is expanded in Chapter 20.

Control through site selection and manipulation will be possible choices once we learn how to quantify the qualitative elements of the environment that affect disease intensification. Quantification of qualitative elements affecting disease intensification in the late blight of potato and scab of apples shows that one can predict disease-intensification periods and therefore advise growers when to spray. The best example of disease control in trees through site selection is based on the studies of Van Arsdel on white pine blister rust. This topic is expanded upon in Chapter 11.

REFERENCES

BERRY, F. H., and W. LAUTZ. 1972. Anthracnose of eastern hardwoods. USDA For. Ser. For. Pest Leafl. 133. 6 pp.

BOWERSOX, T. W., and W. MERRILL. 1976. Stand density and height increment affect incidence of Septoria canker in hybrid poplar. Plant Dis. Rep. *60*:835–837.

BOYER, W. D. 1975. Brown spot infection on released and unreleased longleaf pine seedlings. USDA For. Serv. Res. Pap. 50–108. 9 pp.

CHILDS, T. W. 1968. Elytroderma disease of ponderosa pine in the Pacific Northwest. USDA For. Res. Pap. PNW–69. 45 pp.

CHILDS, T. W., K. R. SHEA, and J. L. STEWART. 1971. Elytroderma disease of ponderosa pine. USDA For. Serv. For. Pest Leafl. 42. 6 pp.

COBB, F. W. JR., B. UHRENHOLDT, and R. F. KROHN. 1969. Epidemiology of *Dothistroma pini* needle blight on *Pinus radiata*. Phytopathology *59*:1021–1022.

GIBSON, I. A. S. 1972. Dothistroma blight of *Pinus radiata*. Annu. Rev. Phytopathol. *10*:51–72.

HOCKING, D., and D. E. ETHERIDGE. 1967. Dothistroma needle blight of pines. I. Effect and etiology. Annu. Appl. Biol. *59*:133–141.

LIGHTLE, P. C. 1954. The pathology of *Elytroderma deformans* on ponderosa pine. Phytopathology *44*:557–569.

LIGHTLE, P. C. 1960. Brown-spot needle blight of longleaf pine. USDA For. Serv. For. Pest Leafl. 44. 7 pp.

MORTON, H. L., and R. E. MILLER. 1977. Rhabdocline needle cast in the Lake States. Plant Dis. Rep. *61*:801–802.

MORTON, H. L., and R. F. PATTON. 1970. Swiss needle cast of Douglas-fir in the Lake States. Plant Dis. Rep. *54*:612–616.

NICHOLLS, T. H., and D. D. SKILLING. 1974. Control of lophodermium needlecast diseases in nurseries and Christmas tree plantations. USDA For. Serv. Res. Pap. NC–110. 11 pp.

PETERSON, G. W. 1977. Infection, epidemiology, and control of Diplodia blight of Austrian, ponderosa, and Scots pine. Phytopathology *67*:511–514.

PETERSON, G. W., and D. A. GRAHAM. 1974. Dothistroma needle blight of pines. USDA For. Serv. For. Pest Leafl. 143. 5 pp.

SCHWEITZER, D. J., and W. A. SINCLAIR. 1976. Diplodia tip blight on Austrian pine controlled by benomyl. Plant Dis. Rep. *60*:269–270.

SINCLAIR, W. A., W. T. JOHNSON, and G. W. HUDLER. Anthracnose diseases of trees and shrubs. N.Y. State Coll. Agr. Cornell Tree Pest Leafl. A-2. 7 pp.

SKILLING, D. D., and T. H. NICHOLLS. 1974. Brown-spot needle disease biology and control in Scots pine plantations. USDA For. Serv. Res. Pap. NC–109. 19 pp.

WAKELEY, P. C. 1970. Thirty-year effects of uncontrolled brown-spot on planted longleaf pine. For. Sci. *16*:197–202.

11

FUNGI AS AGENTS OF TREE DISEASES: RUST DISEASES

- Types of rusts
- Mode of action of rusts
- Rust disease cycle
- Symptoms of rust diseases
- Diagnosis of rust diseases
- Examples of rusts and their control

Rust fungi produce major diseases of leaves and stems of agricultural as well as forest crops. Some aspects of the foliage infection by rust fungi are very comparable to other foliage diseases. Other aspects of the stem canker phase of rusts are very comparable to canker diseases. Why, then, do we discuss rusts as a separate group of diseases rather than include them in the foliage and canker chapters? Rust diseases are caused by a group of fungi all of which are obligately parasitic. Obligate parasites require living hosts for normal development. The fungi of the order Uredinales of the class Basidiomycetes have evolved a complex life cycle not found in any other group of fungi. The obligately parasitic complex life cycle, usually involving two very different plant hosts, results in a unique disease cycle worthy of separate discussion from other fungus diseases.

It is difficult to understand why a fungus should evolve such a complex life cycle as the rusts. Such complexity would require long-time mutual association of the two hosts with the pathogen, during which time both the hosts and the pathogen would experiment with numerous mutations and gene combinations for betterment of resistance or virulence positions. Increased resistance in the host would put a negative selection pressure on the pathogen, leading eventually toward extinction. On the other hand, the pathogen with genes for increased virulence (ability to cause disease) would be at a selective advantage and, therefore, would increase in the pathogen population, but this would eventually result in a reduced host population. This evolution, involving the interrelationship of hosts and pathogens, eventually results in a standoff. Neither can gain a strong advantage over the other for any length of time. The result is a mutual tolerance of the host for a certain amount of parasitism and of the pathogen for a certain amount of inhibition in its invasion of the host. The trend is toward mutualistic symbiosis.

If rusts and their hosts have evolved a tolerance of each other, why are rusts a problem? There are three types of conditions under which rusts and other balanced disease relationships become a problem. One results from the *introduction of the fungus into a new area* where host plants, similar to those in the natural range, have not evolved genetic tolerance because they have never been exposed to the fungus. A plant that does not have the genetically evolved capacity to tolerate the fungus generally is not able to produce the correct defense mechanisms to prevent the fungus from rapidly invading. This would be similar to a small child finding a large jar of candy for the first time. The candy (the host) will become consumed and the small child (the pathogen) very sick in a short period of time. Another possibility resulting from this first introduction is the overshock on the part of the plant to this new invader. The defense reaction mechanism may be so great that the plant kills itself. In either case, we have a rapid destruction of the host plant as a result of a serious rust disease and also an eventual decrease in the pathogen.

Many of our most serious disease epidemics are associated with introduced pathogens. White pine blister rust, Dutch elm disease, and chestnut blight are classic examples.

The second way a balanced relationship can be upset is through our agricultural and forest practices (our eagerness to produce the maximum food or fiber per acre). Overenthusiasm with maximum production or unconscious selection of better-looking disease-resistant individuals for future crops eventually *reduces the genetic variability within the crop population.* The narrowing of genetic variability in relation to disease resistance places a selection pressure on the pathogen population for a mutant or gene combination with the capacity to overcome the resistance. It is just evolution speeded up. By the time the fungus comes forth with its champion there is often extensive acreage of the "resistant" variety. The variant strain, called a race, wastes no time in destruction of the crop, and we have a serious rust disease. In wheat, where races of *Puccinia graminis* have been looked for, researchers have found more than 200 different races of the fungus. Disease-resistance breeding is more complex than just selecting the resistant individuals. We discuss this topic in more detail in Chapter 20.

Extensive planting of limited genetic variability are not as common in forest practice as they are in agriculture. But present activities with poplar, sycamore, and other species could lead to reduced genetic variability and susceptability to serious disease. The use of clones in urban trees is increasing and can be expected to cause problems in the future.

The third way a rust can become a serious disease problem results from our utilization of new plants for the products we desire. We generally put the kiss-of-death on ecological balances as soon as we decide to utilize a particular plant. Changing timber and fiber species will bring to light a whole new array of problems, some of which will be rust diseases.

Fusiform rust is a good example of a disease problem caused by *changing ecological balances through forest management practices.*

TYPES OF RUSTS

The common name "rusts" was derived from the rust appearance of the uredial stage in the cereal rusts. The common name in current usage is applied to any fungus in the Uredinales order of the class Basidiomycetes.

Many rust fungi of trees initially infect and cause diseases of the foliage or succulent shoots. But some eventually invade the cambium and phloem regions of stems and branches. The latter group are referred to as stem rusts. Some rusts occur on cones.

Diseases of trees caused by rusts can result from infection by basidiospores, as in the case of the *Cronartium* stem rusts of pines, or from infection by aeciospores and urediospores, as in the case of *Melampsora* leaf rust of poplars.

Alternate hosts of tree rust diseases can be herbaceous plants, woody shrubs, or other trees.

MODE OF ACTION OF RUSTS

Rust fungus hyphae penetrate between the cells of the host and derive nutrients through haustoria, as described in Chapter 10. The rust fungus is an obligate parasite dependent upon the living cells of the host for growth and development.

It was assumed for many years that obligate parasites such as rusts would not grow on culture media in the laboratory because they needed some unknown vital requirement which they got from living hosts. But plant pathologists have learned to grow rusts, within the last 10 years, on general culture media enriched with peptone or yeast extract.

If rust fungi can be grown on media like other fungi, a necessity for an essential specialized nutrient does not account for host specificity or the need for living cells. The rust fungi, like the mycorrhizal fungi, probably have adapted themselves to elicit a minimal host chemical defense response which they can tolerate. The rust fungi avoid interference and competition for nutrients from other microorganisms that are held in check by the host chemical defense response.

RUST DISEASE CYCLE

Figure 11-1 depicts the stages in a typical heteroecious (two-host) long-cycled rust. Although rust fungi have a unique complex life cycle, the functions of the various stages are basically the same as those found in most fungi. The urediospore stage (Fig. 11-2) of rust fungi is comparable to conidia produced by the Imperfect Fungi. Each spore has two nuclei (n & n). We shall see that two compatible nuclei become associated in a common thallus during the pycnial phase. These asexual spores can reinfect the same plant or other plants of the same species to intensify the infection. Urediospores are generally capable of withstanding the variable temperature and moisture conditions of long-range dispersal. The rust fungus can survive indefinitely in some instances, with no other spore stage. If we were to eliminate white pine from North America, we would still have a rust disease of gooseberries and currants because of urediospores.

In the teliospore stage (Fig. 11-3), the two compatible nuclei fuse to form a diploid ($2n$) nucleus. The teliospores in the *Puccinia* genus also provide a dormant stage capable of overwintering in dry or cold climates.

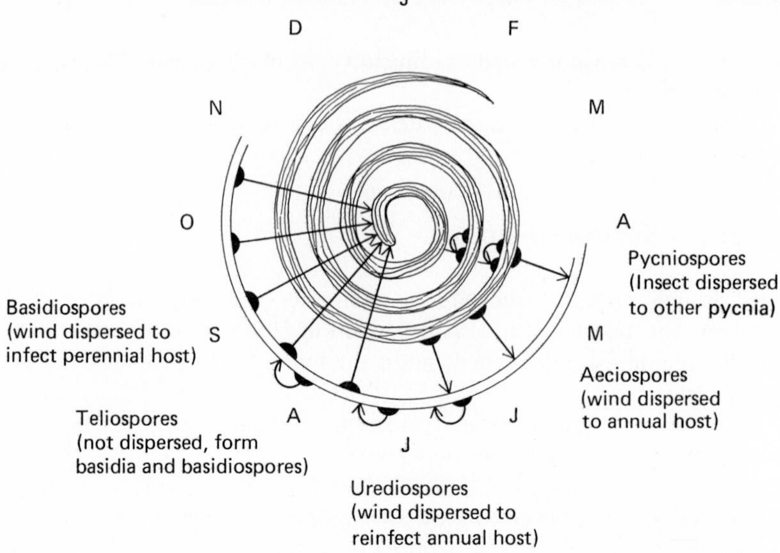

FIGURE 11-1 Life-cycle diagram of a *Cronartium* rust on a monthly calendar. Two hosts (═══ uredial host, ≋≋≋ aecial host) and the periods and direction of spore transfer (↑) are depicted. Note the 3 years of development in the perennial host before production of inoculum to infect the annual host. (Modification of a life cycle diagram by Ziller, 1974.)

FIGURE 11-2 Photomicrograph of a section through a uredium of *Cronartium ribicola* on a *Ribes* leaf.

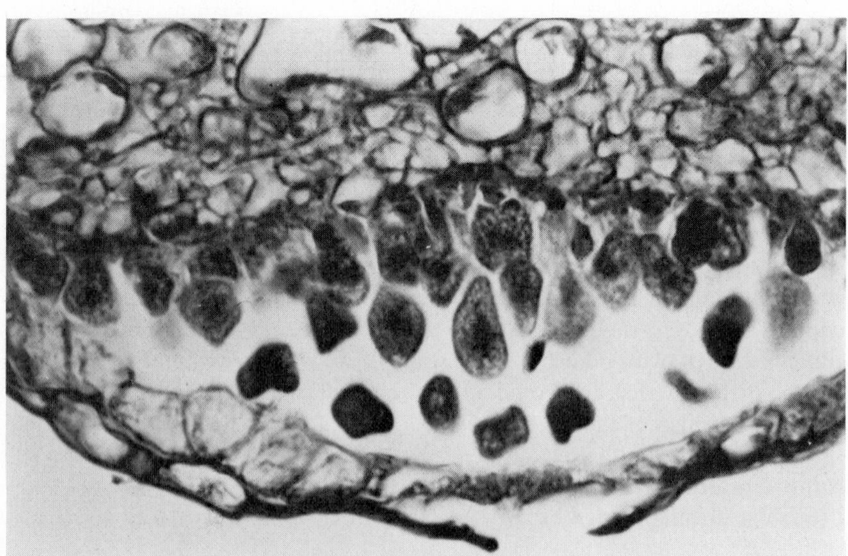

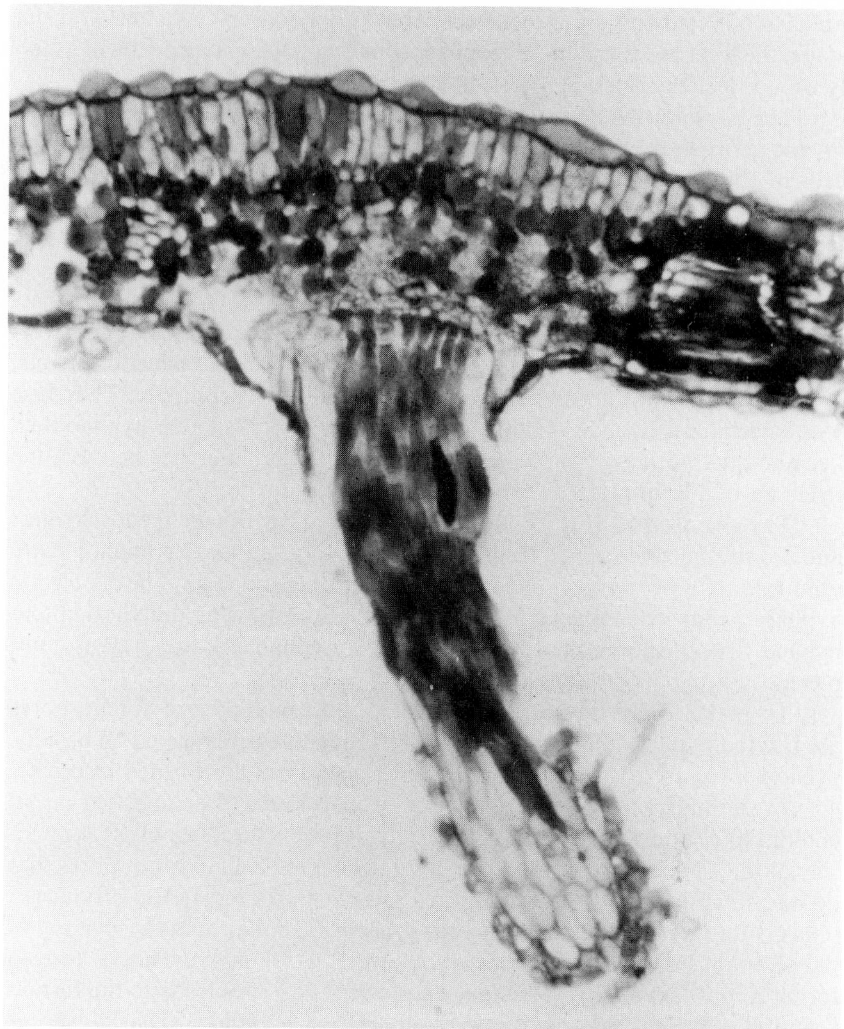

FIGURE 11-3 Photomicrograph of a section through a telium of *Cronartium ribicola* on a *Ribes* leaf.

At the time of nuclear fusion, the teliospore germinates to form a basidium. The diploid nucleus undergoes meiosis to produce four haploid ($1n$) nuclei. Each nucleus becomes separated from the others by the production of septa which segment the basidium into four cells. Each cell produces a basidiospore that contains a single haploid nucleus.

The basidiospores of rusts are very temperature- and moisture-sensitive. As such, they are produced only under specific environmental conditions and are disseminated very short distances. They generally serve as a host-transfer stage. Basidiospores produced from germinating telia on one host will usually only infect the alternate host.

The haploid ($1n$) hyphae, produced upon germination of basidiospores, invade between the cells of the new host. Nutrients are obtained from living cells of the host without invading the cytoplasm. The metabolically active host cells allow a certain amount of nutrient to diffuse out to specialized feeder hyphae called haustoria, which penetrate the cell wall but not the plasma membrane.

Pycnia containing pycniospores are produced from haploid hyphae. Pycniospores serve as spermatia. Insects are attracted to pycnia by the sweet smell and taste of the pycnium. Pycniospores, passively attached to insects, are transferred to other pycnia. Pycniospores do not germinate. They fuse with specialized hyphae within the compatible pycnia. These hyphae then become dikaryotic (contain $n + n$ nuclei). (Note the difference between the pycnidia of the Imperfect Fungi and the pycnia of the rusts.)

The aeciospores (Fig. 11-4) are produced from the dikaryotic hyphae and are specialized for host transfer. They are very similar to conidia but are unique in that they will not infect the host they are formed on. The dikaryotic hyphae, produced upon germination, invade the host as did the haploid hyphae. Urediospores are formed from this hyphae and the cycle is completed.

To make things simple, urediospores are formed in uredia, teliospores are formed in telia or in telial columns, basidiospores are formed on basidia, pycniospores are formed in pycnia, and aeciospores are formed in aecia.

In summary, the typical rust produces two kinds of dikaryotic ($n + n$) conidia-like spores, aeciospores for host transfer and urediospores for intensification of infection on one of the hosts. The rust produces two spores that do not function like spores, teliospores and pycniospores. It also produces a sexual spore type, basidiospores. Teliospores are similar to the hymenium of the Homobasidiomycetes. In dikaryotic ($n + n$) teliospores, nuclei fuse to form diploid ($2n$) nuclei. Germination of the teliospores leads to the formation of basidia in which meiosis takes place. Four haploid ($1n$) basidiospores are formed on the septate basidium. Basidiospores produced from telia on one host generally infect the alternate host. The haploid ($1n$) mycelium resulting from infection by basidiospores produces haploid pycniospores. Pycniospores function as spermatia to transfer nuclei to hyphae of compatible pycnia. Nuclear division and migration of nuclei results in a dikaryotic ($n + n$) mycelium from which aeciospores are formed.

Some rust fungi shorten the five-stage life cycle by not producing one spore stage, as in the case of *Gymnosporangium* spp. (cedar apple rust). Others shorten and change the life cycle by using the host-transfer aeciospore stage for reinfection of the pine host, as in the case of *Endocronartium* stem rust of pines.

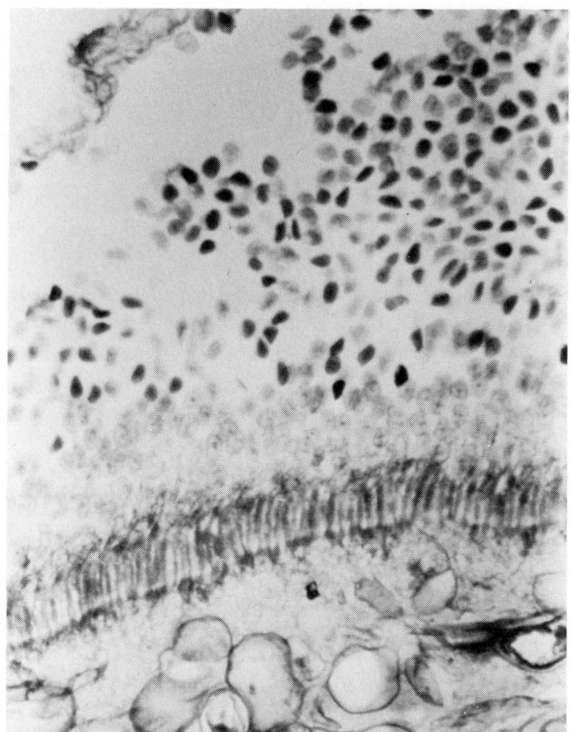

FIGURE 11-4 Photomicrograph of a section through an aecium of *Cronartium ribicola* on a white pine branch.

SYMPTOMS OF RUST DISEASES

Rust diseases produce very distinctive aecia, uredia, or telia signs on the infection lesion. In the absence of signs, rusts may look like many other foliage or stem disorders.

Rust specialists can recognize infection prior to sporulation. Color, shape, and size of the infection lesion when inoculated on specific host varieties are used as fingerprints for specific races of rust.

Symptoms of stem rusts are more characteristic. The shapes of galls or cankers are somewhat characteristic. Remnants of pycnial and aecial pustules around the margin of the canker are also specific signs for stem cankers.

Rusts reduce growth, yield, and sometimes cause death of parts or whole trees. Death of branches and tops from white pine blister rust can be easily recognized from a distance. The distinctive dead branches are called flags.

DIAGNOSIS OF RUST DISEASES

Diagnosis of rust diseases is based on identification of the causal organism. Arthur's *Manual of the rusts in the United States and Canada* (1934) is the most comprehensive treatment of rusts for diagnostic purposes. The book *The tree rusts of western Canada* (Ziller, 1974) is a more recent and very useful reference for most tree rusts.

Rusts are obligate parasites, so one generally assumes that a rust sporulating on a diseased host is the cause of the problem. Proof of pathogenicity for rusts requires a modification of Koch's postulates. It may be necessary to find the alternate host to get basidiospore inoculum for testing pathogenicity. Although methods for growing rusts in culture are known, inoculation with a pure culture of a rust is not commonly done.

EXAMPLES OF RUSTS AND THEIR CONTROL

White Pine Blister Rust

White pine blister rust is caused by the fungus *Cronartium ribicola* (Figs. 11-5 and 11-6). The pathogen, native to Asia, was introduced into Europe by plant collectors before the nineteenth century. It was introduced early in the twentieth century into eastern and western North America from Europe on infected nursery stock of eastern and western white pines.

It may seem peculiar that nursery stock for planting in North America with species native to North America should come from Europe. At the turn of the century, Europeans supplied both the skills and the materials for forest practice to the North American continent, where people were just beginning to recognize that exploitation could not continue indefinitely without some management.

The disease cycle of blister rust is shown in Fig. 11-7. Infection of pine needles occurs via basidiospores produced on *Ribes* bushes during late summer or fall (Figs. 11-8 and 11-9). A small yellow or orange fleck develops at the infection site on the needles within a few weeks. By the following spring the fungus hyphae have grown down the needle to the twig. It will persist in the cambium region until the branch or the tree dies. A year later pycnia are produced on the infected twig. It is not known if spermatization by pycnia is necessary in this fungus. Aeciospores are formed on the same tissue during the spring of the next year (Fig. 11-10). Aeciospores infect the *Ribes* leaves. Urediospores produced on the *Ribes* leaves increase the number of infections on *Ribes*. During the late summer and fall telial columns are found rather than uredia. The teliospores of *Cronartium* fungi are cemented together in a column. Germination of the teliospore results in haploid basidiospores for infection of pines.

FIGURE 11-5 Death of upper crown branches (flagging) of eastern white pine resulting from *Cronartium ribicola* infection of upper crown branches.

FIGURE 11-6 Blister rust stem canker on eastern white pine.

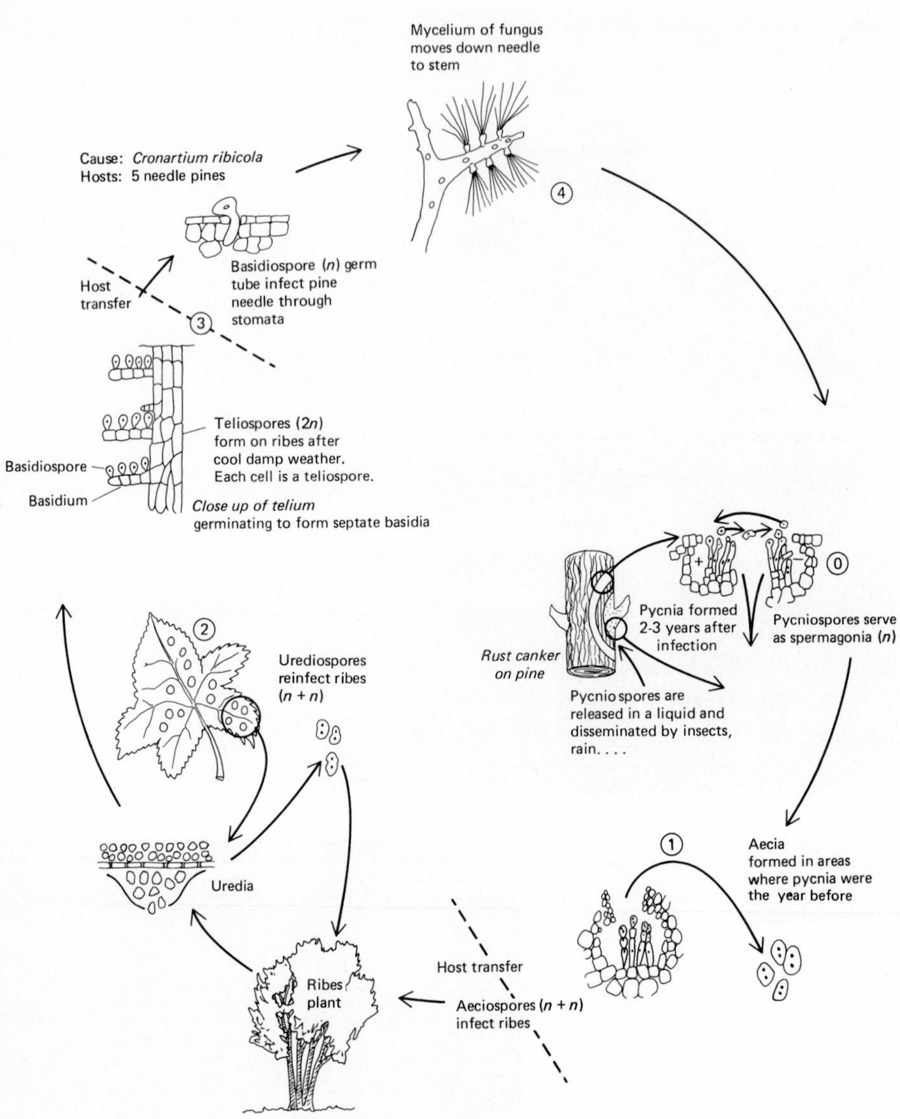

FIGURE 11-7 Disease-cycle diagram of white pine blister rust.

FIGURE 11-8 *Ribes* bush.

FIGURE 11-9 Closeup of undersurface of *Ribes* leaf, showing uredial and telial spore stages of *Cronartium ribicola.*

FIGURE 11-10 Young blister rust stem canker showing oval-shaped remnants of pycnial scars.

As with any introduced pathogen, it takes a number of years to recognize the new pathogen and assess the potential problem. By the middle of this century the disease had spread throughout the ranges of the eastern and western white pine and sugar pine.

Major efforts were carried out by the U.S. Forest Service, state, and private concerns to control the disease by eradication of *Ribes.* The Forest Service began its control program to save the western white pine of the northern Rocky Mountains in 1930. By 1934, 11,000 workers were employed in the blister rust control program. Phasing out of the program began 30 years after it began and was ended completely by 1965.

Ribes were not eradicated and the disease was not controlled. From virtual armies of people involved in control of this disease, we have moved to a point where almost no effort is applied to direct control today. In the range of western white pine, management practices presently recommend favoring other species in the area and total abandonment of white pine planting.

In the east, large numbers of people were employed by the various states in blister rust control programs up until 1970. Federal monies supported this effort. By 1970, rust control funds were restricted to control efforts on only the highest-quality white pine stands.

Hindsight shows us that the control efforts were doomed to failure and causes us to ask why so much effort was spent so poorly. But before we criticize our predecessors for their lack of vision, we should recognize that human beings do not have perfect foresight. Rather than criticize, it would be appropriate for us to review the programs, to see both the successes and the failures, so as to enhance our own chances of success in combatting other diseases in the future.

A general principle widely accepted by pathologists is that if we understand the life cycle of a disease agent, we can pick out the weak link as a point to apply direct control. Many plant pathologists engaged in basic research on specific pathogens justify their work on this premise. The major blister rust control effort was directed at the weak link in the rust life cycle, the need for two hosts. Eradication of the alternate host *Ribes* spp. should eliminate the blister rust problem, since infection of pines occurs only from basidiospores produced by the fungus growing on *Ribes* plants. The principle, in this case applied to *Cronartium ribicola,* is correct, but in practice it did not work for much of the white pine range. Why did eradication of *Ribes* work in only a limited number of cases? Initially, *Ribes* bushes were eliminated by digging and pulling. If any of the root was left, it would sprout to produce a new plant, so eradication often took two or three attempts. Later use of herbicides improved the number of plants killed with one application. Another difficulty in *Ribes* eradication involves the recolonization of a site through seeds. The seeds of *Ribes* plants can remain dormant for many years. Sites that were thought to be free of *Ribes* bushes can be recolonized from dormant seeds, especially following site disturbances caused by logging. *Ribes* eradication works only if it is thoroughly and repeatedly applied to stands of limited size. In the East, control of blister rust is effective if the size of the stand is small enough for thorough *Ribes* eradication.

From the *Ribes* eradication programs we can conclude that another widely accepted principle may not be entirely applicable. It is assumed that, in the absence of total eradication of the pathogen, a partial reduction in the pathogen is better than nothing. Theoretically, disease is directly related to the amount of inoculum, so that a reduction in inoculum should directly affect the amount of disease. If one reduces the number of *Ribes* bushes in an area, the amount of inoculum or spores should be reduced and therefore the amount of disease. It is true that numbers of spores are reduced, but the real question concerns how many spores are necessary to produce an unacceptable level of disease. With the blister rust in the West, it was found that only one infected bush per acre was sufficient to supply all the necessary inoculum to cause a serious level of disease. In this instance, the theory does not apply, because the fungus produces such an abundance of inoculum from a single infected bush. Disease epidemics are discussed more thoroughly in Chapter 19, but we can recognize now that reduction in the inoculum of a disease with a high potential for spread may not significantly affect the final amount of disease in a long-term crop.

In the late 1950s and early 1960s a program of blister rust control using antibiotic sprays gained momentum. Two materials, cycloheximide (Actidione) and, later, phytoactin were used. The latter chemical was applied in a mist from an aircraft, much as insecticides are applied to forest areas.

The aircraft method of application is about the only method that can be used for the application of chemicals to forest areas. Even this least expensive method is not entirely satisfactory, because it is difficult to get complete coverage of the trees. Streams and nontarget trees in the treatment area may also get sprayed.

Blister rust control through aerial application of chemicals was not abandoned because of problems associated with aerial application. Chemical control programs were abandoned because the chemicals did not control the disease. Why, then, were they used? Improperly interpreted results of experimental tests with the chemicals showed some control of blister rust. The Forest Service was painfully aware that the *Ribes* eradication programs were costly and not entirely satisfactory. The time was right for any new idea, so the promising chemicals were accepted with open arms. It was very difficult for the people involved to recognize the mistake and reverse the momentum of the control effort. Hopefully, in the future we will avoid premature acceptance of experimental results. [I am concerned at present that the enthusiastic acceptance of Benomyl to control Dutch elm disease is possibly just another replay of the blister rust episode (see Chapter 13). I hope that 10 years from now I will be the one to apologize rather than the large number of highly competent forest pathologists who are presently working on chemical control of Dutch elm disease.]

In the early 1960s, another method for the control of blister rust was developed, based upon an understanding of environmental factors affecting production, spread, and infection of pines by basidiospores. The studies of Van Arsdel at the Lake States Forest Experiment Station are a classic example of how an understanding of the epidemiology of forest diseases can be used in making management decisions. Beginning with the general knowledge that climatic factors directly affect the development of blister rust, it was found that for successful infection of white pine, a period of 2 weeks of relatively cool weather in late summer was needed to cause the rust to produce telia rather than urediospores on the *Ribes* leaves. For the production of basidiospores from the telia, dissemination of viable spores to pine needles, germination of basidiospores, and infection of needles, 48 hours of continuous 100% relative humidity and temperatures below 68°F are required. One can separate geographical areas into infection hazard zones based on the frequency of the these exacting climatic conditions. In the

Lake States, the region surrounding the lakes frequently experience these conditions. It is almost impossible in this area to prevent infection by means of *Ribes* eradication control procedures. Lesser levels of blister rust hazard are found in areas where the influence of the lakes is less.

Unfortunately for western foresters, fall rains and fog are common in mountainous areas, so most of the western white pine and sugar pine grow in high-hazard areas.

Local climatic conditions vary with topography and aspect. Therefore, region wide hazard ratings must be modified with local features. Prevention of rust disease can be accomplished in medium- to low-hazard zones by avoiding planting in small openings, pruning lower branches, keeping the canopy closed, or using a thin-crowned overstory nurse crop. All of these practices are aimed at reducing microclimatic high humidity caused by dew.

Application of epidemiological principles provides the resource manager with information to help avoid disease problems where possible and to appropriately apply control procedures where they will do the most good. This concept of learning how to live with disease and manipulating disease through environmental awareness is one of the most important contributions that forest pathologists can make to forest management.

Another approach to control involves the use of disease resistance. In the early 1950s, a selection program was initiated to look for disease-resistant white pines in both eastern and western white pines. The fruits of the initial selection programs are now available to western foresters on a limited scale. Seed orchards of the original selections are producing seed with more resistance than can be expected from a random seed collection. The western program has continued research on understanding and manipulation of the genetic mechanisms of control, so that further improvements can be expected in years to come.

A very nice summary of disease resistance breeding in forest trees is provided in the published proceedings of an advanced study institute meeting held in 1969 in Moscow, Idaho (Bingham et al., 1972).

Benefits of disease resistance can be obtained through regeneration practices as well as resistance breedings. Superior-form, nondiseased seed trees can generally be expected to produce better-form, more-disease-resistant seedlings than can nonselected trees. Careful selection of seed trees can therefore lead to genetic improvement.

Disease control in forest trees through resistance is a second very important contribution that pathologists can make to forest management. Disease resistance, like epidemiology, is an approach that will require a high degree of sophistication in management. It is not a cure-all for all our problems. Selectively utilized resistance together with epidemiological understanding can be used to reduce future losses for tree disease problems. Concepts, procedures, and expected results of disease resistance breeding of trees are discussed in Chapter 20.

FIGURE 11-11 Fusiform-rust-infected slash pine.

FIGURE 11-12 Fusiform-rust-infected water oak.

Fusiform Rust of Southern Pines

The fungus *Cronartium fusiforme,* unlike *C. ribicola,* is a pathogen native to the southern United States and represents a serious rust disease because of our utilization of loblolly and slash pine (Figs. 11-11 and 11-12). Prior to 1900 the disease was rare, but because of fire control, planting of infected seedlings, and expansion of the range of the susceptible species, the disease has intensified to the point that it is now a serious problem.

The life cycle of *C. fusiforme* involves alternation between pines and oaks, particularly members of the red oak group. In the spring, aeciospores produced on the pines infect newly emerging oak leaves. The uredial increase stage on oaks becomes intermixed with telia, which are produced within a few weeks of infection of the oaks. The telia germinate to produce basidia and basidiospores during periods of high moisture. Basidiospores infect the new and expanding growth on the pines throughout spring and summer. The pathogen survives from year to year as mycelium within the pine host. Galls are formed on twigs and stems in reaction to the fungus presence in the cambium-phloem region. The growth disruption in the stem disrupts the overall growth rates of the trees and produces a weak point in the stem where breakage often occurs.

Why has man's activity caused this pathogen to develop into one of the principal tree-related problems in the South? Southern pines are fire subclimax types that are maintained only by repeated fires. In the absence of fire, the sites revert to oak/hickory. Our attempts to preserve the forest by preventing fires has caused oaks to increase in the pine stands, thereby producing a more intimate association of the two hosts for *C. fusiforme.*

Fusiform rust is very common in nurseries because it is almost impossible to eliminate oaks from around the nurseries. Fungicide sprays are used but are not totally effective. A certain amount of diseased stock is sent out with each shipment. If oaks are present in areas where the stock is planted, the rest of the young trees are in an excellent position to become infected at an early age.

The use of fungicide in the nursery may actually increase the incidence of rust in the field. Disease-susceptible seedlings, artificially maintained with fungicides in the nursery, quickly become infected when planted in the field, where fungicidal control is too costly.

Loblolly and slash pine have been intensively planted on sites that were formerly occupied by the rust-resistant longleaf pine. A foliage disease of longleaf pine, brown spot needle blight (discussed in Chapter 10) has made regeneration to longleaf very difficult, and therefore there has been a shift to other species in some longleaf pine areas. Shortleaf pine has problems with littleleaf disease on sites, so it cannot be used widely, even though it is highly resistant to the rust. Littleleaf disease is discussed in Chapter 18.

Foresters' attempts to grow trees faster, by site modifications and silvicultural practices, only enhance the spread of the rust. Obligate parasites such as rusts are generally most destructive on vigorously growing plants. Fertilization or attempts to improve vigor and growth only make the plants more susceptible to an obligate parasite. We find that facultative parasites, on the other hand, are most destructive on weakened or slow-growing plants.

Control of fusiform rust has become a serious problem. Eradication of oaks to control the rust is even more difficult than eradication of *Ribes* for control of blister rust. Prevention of infection in the nursery by fungicide spray is somewhat successful and might be applied to young plantations with some added success, but the costs are a bit high. Epidemiological studies have shown that for infection to occur, certain weather conditions are required. But, unfortunately, such conditions are relatively common in spring and summer throughout the range of the disease.

So far, disease resistance has been the most successful means of control. Disease-resistance breeding has developed some resistant lines by crossing susceptible species with resistant shortleaf pine. Hybrids with shortleaf pine are resistant to fusiform rust and susceptible to pine–oak rust (eastern gall rust), a very similar rust that forms round rather than elongate galls on the pines. There seems to be some genetic control of resistance or susceptibility within the species, since consistent variation between different seed sources and clones is seen in the field and in test nurseries.

A very effective method for detecting resistance in seed collections is currently being utilized. Young seedlings are uniformly inoculated in the greenhouse. Gall development on test seedlings is compared to gall development on a similar batch of standard seedlings to determine a relative fusiform rust susceptibility rating. Seedlings with low rust susceptibility in the greenhouse test have generally proven to be less diseased in the field.

Pine-Oak Rust (Eastern Gall Rust)

Pine-oak rust caused by *Cronartium quercuum,* mentioned above as causing a disease of shortleaf pine, is a compounding problem when looking for resistance to fusiform rust. It also represents a problem on southern pines and on northern jack and Scotch pine. The life cycle is very similar to *C. fusiforme,* so much so that some workers would not separate these two fungi into species but would call them races of the same species.

Pine-Pine Rust (Western Gall Rust)

To make things more confusing, another gall rust, caused by *Endocronartium harknessii,* produces round galls similar to pine-oak rust on jack and Scotch pine. This disease, called pine-pine rust, western gall rust, or, in the northeast, Woodgate rust, was introduced into the eastern United States from the West, where it occurs on ponderosa and lodgepole pine. Pine–pine rust causes extensive galling on some Scotch pine Christmas tree plantings in New York and Pennsylvania (Fig. 11-13). A unique feature of this fungus which aids in the rapid spread of the pathogen is the elimination of an alternate host. The aeciospores produced on pines are capable of reinfecting the pine.

To complete our discussion of the *Cronartium* rusts of pine, we should indicate that there are yet other stem and limb and cone rusts caused by *Cronartium* spp. We have only scratched the surface. We have seen that introduction of a pathogen such as *C. ribicola* to a new area has resulted in serious disease. Also, the modification of cultural practices can cause a pathogen such as *C. fusiforme* to develop into a serious problem. If the *Endocronartium harknessii* ever gets into southern pines, we may see another disease significantly negate all present fusiform-rust-resistance work. There is a good possibility that a transition from the northeast to the south will take place because of extensive Christmas tree plantings, which provide a corridor of pines from the northeast through the oak–hickory forests of Pennsylvania to the southern pines.

FIGURE 11-13 Pine-pine rust on Scotch pine.

FIGURE 11-14 Cedar apple rust galls on juniper. (Photograph compliments of Dr. Josiah Lowe.)

Cedar Apple Rust

Cedar apple rust, caused by various species of *Gymnosporangium,* is the last rust disease that we will discuss. This is really more of a fruit production than a forestry problem, but the disease is common on ornamental apples and junipers (Fig. 11-14).

The telial stage (Fig. 11-15) of the fungus causes very little damage to *Juniperus* species. One should note that the disease should really be called juniper apple rust, because *Juniperus* spp., not *Thuja* spp., are the teliospore hosts.

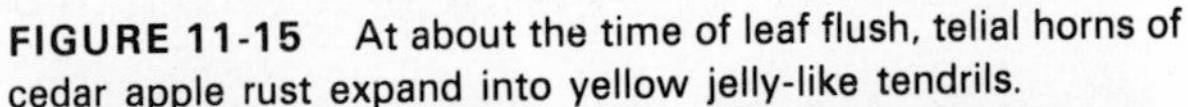

FIGURE 11-15 At about the time of leaf flush, telial horns of cedar apple rust expand into yellow jelly-like tendrils.

FIGURE 11-16 Cedar apple rust on the upper surface of hawthorn, showing black pycnia in the center of yellow infection lesions. (Photograph compliments of Dr. Josiah Lowe.)

FIGURE 11-17 Closeup of elongated aecial cups on the lower surface of hawthorn leaves. (Photograph compliments of Dr. Josiah Lowe.)

The urediospore or increase spore stage has been eliminated in this fungus, so that there is an absolute dependence upon the intimate presence of both hosts for survival of the fungus.

Pycnia and aecia form on apple leaves and fruit, reducing the yield and quality of the fruit (Figs. 11-16 and 11-17).

FIGURE 11-18 The common practice of mixing junipers and hawthorn in ornamental plantings often results in serious disease in one or both plants.

Cedar apple rust is one of the more common problems of ornamental plantings. Hawthorns and flowering crabs regularly used in the Northeast are good hosts for the pycnial and aecial stages of the fungus. These are often associated with upright and prostrate junipers in ornamental plantings. The 2- to 3-inch slimy orange tentacled galls, formed during wet periods in the spring, cause a great deal of apprehension on the part of homeowners who do not understand the disease. Later in the season, unsightly reddish brown spots on flowering crabs and hawthorns show the next stage of the disease. The best control measure is not to use the two hosts in the same or nearby plantings (Fig. 11-18).

REFERENCES

ANDERSON, N. A. 1973. Eastern gall rust. USDA For. Serv. For. Pest Leafl. 80. 4 pp.

ANDERSON, R. L. 1973. A summary of white pine blister rust research in the lake states. USDA For. Serv. Gen. Tech. Rep. NC-6. 12 pp.

ARTHUR, J. C. 1934. Manual of the rusts in the United States and Canada. Purdue Research Foundation, Lafayette. 438 pp.

BINGHAM, R. T., R. J. HOFF, and G. I. MCDONALD. 1972. Biology of rust resistance in forest trees. Proceedings of a NATO–IUFRO advanced study institute. USDA For. Serv. Misc. Publ. 1221. 681 pp.

BLISS, D. E. 1933. The pathogenicity and seasonal development of *Gymnosporangium* in Iowa. Iowa State Coll. Agr. Mech. Arts Res. Bull. 166, pp. 340–392.

BOYCE, J. S. 1957. The fungus causing western gall rust and Woodgate rust of pines. For. Sci. *3*:225–234.

CHARLTON, J. W. 1963. Relating climate to eastern white pine blister rust infection hazard. USDA For. Serv. Northeast. For. Exp. Sta. 38 pp.

CZABATOR, F. J. 1971. Fusiform rust of southern pines—a critical review. USDA For. Serv. Res. Pap. SO-65. 39 pp.

DINUS, R. J. 1974. Knowledge about natural ecosystems as a guide to disease control in managed forests. Proc. Am. Phytopathol. Soc. *1*:184-190.

HIRT, R. R. 1964. *Cronartium ribicola:* its growth and reproduction in tissues of eastern white pine. N.Y. State Univ. Coll. For. Syracuse. 30 pp.

HOFF, R. J., and G. I. MCDONALD. 1972. Stem rusts of conifers and the balance of nature. In Biology of rust resistance in forest trees. USDA For. Serv. Misc. Publ. 1221, 525–536 pp.

HOFF, R. J., and G. I. MCDONALD. 1977. Selecting western white pine leaves-trees. USDA For. Serv. Res. Note INT-218-FR 13.

HOFF, R. J., G. I. MCDONALD, and R. T. BINGHAM. 1976. Mass selection for blister rust resistance: a method for natural regeneration of western white pine. USDA For. Serv. Res. Note INT-202-FR 13.

KETCHAM, D. E., C. A. WELLNER, and S. S. EVANS, JR. 1968. Western white pine management programs realigned on northern Rocky Mountain national forests. J. For. *66*:329–332.

LAIRD, P. P., and W. R. PHELPS. 1975. A rapid method for mass screening of loblolly and slash pine seedlings for resistance to fusiform rust. Plant Dis. Rep. *59*:238–242.

PETERSON, R. S. and F. F. JEWELL. 1968. Status of American stem rusts of pine. Annu. Rev. Phytopathol. *6*:23–40.

POWERS, H. R., JR., J. P. MCCLURE, H. A. KNIGHT, and G. F. DUTROW. 1974. Incidence and financial impact of fusiform rust in the South. J. For. *72*:398–401.

SAVILE, D. B. O. 1971. Coevolution of the rust fungi and their hosts. Quart. Rev. Biol. *46*:211–218.

SIGGERS, P. V. and R. M. LINDGREN. 1947. An old disease—a new problem. South. Lumberman *175*(2201):172–175.

VAN ARSDEL, E.P. 1961. Growing white pine in the lake states to avoid blister rust. USDA For. Serv. Lake States For. Exp. Stn. Pap. 92. 11 pp.

VAN ARSDEL, E. P. 1965. Micrometeorology and plant disease epidemiology. Phytopathology *55*:945–950.

VAN ARSDEL, E. P. 1967. The nocturnal diffusion and transport of spores. Phytopathology *57*:1221–1229.

ZILLER, W. G. 1974. The tree rusts of western Canada. Can. For. Serv. Publ. 1329. (Dept. Environ.) 272 pp.

12

FUNGI AS AGENTS OF TREE DISEASES: CANKER DISEASES

- Types of cankers
- Mode of action of canker diseases
- Canker disease cycle
- Symptoms of canker diseases
- Diagnosis of canker diseases
- Examples of canker diseases
- Control of cankers

Fungi that cause canker diseases are generally adapted for parasitic existence in the living cells of the phloem, cambium, and outer xylem region of the tree, as well as for saprophytic existence upon dead phloem and xylem cells. This type of organism is called a facultative parasite.

Canker fungi cause extremely diverse effects on host trees. They range from rapid killing of the host to contributing saprophytes in deterioration of branches weakened by other factors.

Canker diseases emphasized in this chapter are all caused by Ascomycete fungi. There are cankers caused by Phycomycetes, such as *Phytophthora cactorum*, which causes bleeding canker of maples and other tree species. Basidiomycete fungi also may cause cankers. Rust cankers are discussed in Chapter 11. Some Basidiomycete decay fungi, such as *Inonotus obliquus (Poria obliqua)* and *Phellinus igniarius (Fomes igniarius),* also cause cankers on northern hardwoods. *Inonotus hispidus (Polyporus hispidus), Phellinus spiculosa (Poria spiculosa),* and *Spongipellis pachyodon (Irpex mollis)* are canker rots of southern hardwoods. Basidiospores of these fungi infect through wounds. Mycelial invasion of sapwood results in decay up to 2 meters above and below the canker infection point and cambium death. See Chapter 14 for a more complete discussion of decay.

TYPES OF CANKERS

Canker fungi can be grouped into (a) saprophytes contributing to declines, (b) annual cankers, (c) facultative parasites which produce perennial target cankers, and (d) diffuse cankers. Each of these four types of cankers will be discussed.

MODE OF ACTION OF CANKER DISEASES

Invasion of living tissue of the cambium, phloem, and outer xylem provides canker fungi access to readily utilizable transport and storage nutrients of trees. These fungi may produce toxins to overcome host defensive mechanisms. The host produces callus wound tissue around the invaded area in an attempt to wall off the infection. Some canker fungi have been shown to utilize cellulase enzymes for the invasion and decomposition of wood.

Slight differences in mode of action separate the four types of canker fungi. Many saprophytic canker fungi are associated with weakened trees. They cannot penetrate the morphological defense barriers of healthy trees, nor can they tolerate chemical defense mechanisms of normal healthy trees. Weakened trees are unable to produce sufficient morphological and chemical barriers to prevent invasion.

Saprophytic canker fungi may overcome normal host defenses by simultaneous massive invasion through large numbers of insect wounds. The insects may also weaken the host, making infection and invasion easier.

An annual canker fungus is an opportunist that infects soon after wounding and invades quickly before the host has a chance to respond. Eventually, the host's response catches up with invasion to halt further expansion of the canker. A large number of fungi can induce annual cankers.

A perennial canker fungus is much like an annual canker fungus, in that it invades when the host's defenses are slow to respond. Invasion of host tissues during late summer, fall, and winter induces less host response. Perennial canker fungi differ from annual canker fungi in that the former are good saprophytic competitors on dead cankered tissues. They also must tolerate some host defense responses better than annual canker fungi, because they appear to break through the host defenses each year.

The interaction of host and fungus in perennial cankers involves more than just what one is led to believe from published research reports. I have personally made thousands of inoculations with many different cultures of *Nectria galligena* and *Ceratocystis fimbriata*. Most of the inoculations produced cankers the first season, a few expanded the second season, but none were still active after 5 years. Some type of balanced relationship must be established between host and pathogen to develop a typical perennial canker. We have little understanding of this relationship.

The fungal pathogen dominates in the diffuse canker system. Host response is minimal or ineffectual. Toxins produced by the fungus and rapid mycelial invasion are characteristics of diffuse cankers.

If one looks back over the described mode of action for the four types of canker pathogens, a gradient should be obvious. At one end of the spectrum the host dominates. At the other the pathogen dominates. In the middle are the opportunistic annual canker fungi which invade through wounds before the host is able to respond. Once the host response occurs, they are held in check. A step further up the pathogenicity ladder are the perennial canker fungi. Perennial canker fungi persist on a host over extended periods of time by limiting invasion to periods when the host response is minimal.

CANKER DISEASE CYCLE

It is generally assumed that canker fungi infect trees through wounds in the stem or through broken branch stubs. These assumptions result from observations of dead branch stubs in the center of many cankers and the development of cankers following artificial inoculation of wounds. Most artificial inoculation techniques involve insertion of massive amounts of mycelial inoculum into wounds. It is erroneously assumed that spores will infect similar wounds.

Those cankers that have been studied in detail show that infection by spores is a different problem, which we do not yet understand. Infection may also be through the foliage, as in the case of Septoria canker of poplars. Infection may occur by means of ascospores or conidia. Information on the importance of each is rather limited.

Invasion involves varying degrees of interaction of pathogen and host as discussed earlier.

Sporulation usually occurs initially as conidia. Ascospores are produced sometime later, presumably after fertilization with a second compatible strain of fungus. The need for fertilization by a compatible strain is also poorly documented.

Long-term survival of perennial and diffuse canker fungi as saprophytes on dead tissue of the canker is common. Some of the fungi survive as saprophytes in the soil. Others survive as saprophytes on dead branches of trees.

Dissemination of spores to infection sites is usually by wind, although splashing rain, insects, rodents, and human beings may also play a role in dissemination.

SYMPTOMS OF CANKER DISEASES

Symptoms of canker diseases range from small irregularities of the stem caused by the callusing of annual cankers, to massive areas of dead bark and the death of trees. Discoloration of the bark of smooth-barked trees is common. Callusing of cankers may close the canker wound or may produce excessive enlargement of the stem around the canker. On thick-barked tree species, cankers may enlarge under normal-appearing bark and be evident only after removal of some of the bark.

DIAGNOSIS OF CANKER DISEASES

The diagnosis of canker diseases is usually straightforward. Characteristic symptoms are often associated with specific fungus signs. Positive diagnosis depends upon identification of the fungus.

The canker-causing Ascomycetes have both a sexual and an asexual spore stage. The two names for the canker-causing fungi are sometimes initially confusing to students but are very convenient to the pathologist, who may encounter either spore stage when diagnosing the cause of a particular canker.

In the absence of signs, it may be necessary to isolate the suspected pathogen in pure culture and follow Koch's postulates to prove pathogenicity. Anyone who tries to isolate from cankers will find a large number of fungi present. It is usually easier to isolate the causal pathogen from the margin of the canker at the interface of diseased and nondiseased tissue. A large number of saprophytic fungi can compete and survive in the dead substrate in the center of the canker, but only the pathogen should be able to compete and survive against the host defense mechanisms around the margin of the canker.

EXAMPLES OF CANKER DISEASES

Saprophytic Canker Fungi Contributing to Declines

See Chapter 18 for a more detailed discussion of decline.

Cytospora canker One of the best examples of a saprophytic canker fungus contributing to decline is the Cytospora canker of spruce, particularly Colorado blue spruce (Fig. 12-1). The Cytospora canker problem is characterized by the death of low branches of ornamental blue spruce, usually trees that are 20 years of age or older. The death of branches continues slowly from the bottom upward until the tree is no longer acceptable as an ornamental. Pycnidia of the Imperfect fungus *Cytospora kunzei* are commonly found on dying branches.

The commonly accepted method of "control" is to fertilize and water the trees and prune out the dead branches. The spraying of trees with fungicides is recommended by some workers, but there is no experimental evidence to indicate that spraying does more than enrich those being paid to apply the spray.

Why do we refer to this problem as Cytospora canker and not just call it senescence decline of ornamental spruce? The common presence of pycnidia on dead branches and the constant isolation of *Cytospora* from dead branches fulfills the first step in Koch's postulates for proof of pathogenicity. Inoculation with pure cultures produces cankers and the death of branches. Reisolation recovers *Cytospora* and completes Koch's postulates. So it is obvious that *Cytospora* is a pathogen. Not necessarily!

One has to look carefully at the experimental procedures used to prove Koch's postulates. First, the common association as shown by isolation from diseased bark is to be expected, because one can readily isolate the fungus from healthy bark. The fungus is a very common associate of both healthy and diseased bark. Secondly, inoculation of wounds in branches with pure cultures does not mean that only a single organism is present in the wound. It is almost impossible to prevent contamination of the inoculation site by other organisms, so we are working with a mass of different organisms in the wounds. When the noninoculated control wounds commonly result in symptoms similar to those of inoculated wounds and *Cytospora* is recovered from

FIGURE 12-1 Mortality of lower branches of Colorado blue spruce. This disease is called Cytospora canker.

inoculated as well as noninoculated wounds, there is no reason to assume that the pure culture introduced into the tree had a unique effect on the host. Third, if a large mass of fungus hyphae is introduced into a wound and the wound expands, one has to be cautious in interpreting pathogenicity. This is a very unnatural mass of fungus coming in contact with a host so rapidly that the host does not have time to distinguish pathogen from nonpathogen. Using this method, one can produce wound expansion with a large number of saprophytic fungi. Therefore, the significance of a single fungus as a possible pathogen is questionable.

The foregoing discussion is intended to demonstrate for this disease that even though *Cytospora* is a common associate of dying branches of spruce and Koch's postulates can be used to demonstrate proof of pathogenicity, there is reason to question the interpretation of the causal relationship of *Cytospora* with the death of branches in spruce. Therefore, I prefer to refer to this type of canker as a contributing factor in the decline and separate it from other canker types. The name for this disease is too widely accepted to be rejected, so the "senescence decline of ornamental spruces" will remain Cytospora canker. The subject of declines is developed more extensively in Chapter 18. Consider the predisposing, inciting, and contributing factors described there as they relate to Cytospora canker.

Beech scale Nectria canker The most widely accepted explanation for beech bark disease or beech scale Nectria canker (Fig. 12-2) involves the interaction of the scale insect *Cryptococcus fagi* and the canker fungus *Nectria coccinea* var. *faginata.* Even though the disease complex causes extensive destruction of commercial beech, very little work was done on this disease between 1934, when the early work was done, and the mid-1970s. This disease, like Cytospora canker, can best be understood as a decline involving predisposing, inciting, and contributing factors.

The scale insect, *C. fagi,* was introduced into North America from Europe around the turn of the century and has spread, from the introduction site in Nova Scotia, westward throughout New England and into New York and northern Pennsylvania. The beech stands of Ohio, West Virginia, and central Pennsylvania have not yet come under attack by this insect.

The wingless young scale insects are attracted to rough places on the smooth beech bark. They feed on the phloem by inserting their stylets through the bark to the phloem. Once they begin to feed, they no longer move.

FIGURE 12-2 Crown death of beech tree associated with beech scale Nectria canker of the main stem.

FIGURE 12-3 Cottony masses of the scale insect *Cryptococcus fagi.*

Female scale insects produce young parthenogenetically (in the absence of fertilization), causing the size of the population to increase rapidly. The bark becomes covered with a cottony scale (Figs. 12-3 and 12-4), produced by the sedentary insects to protect themselves. What effect the feeding of the scale insect has on the tree is unknown. It does not, as a rule, result in the death of the trees.

FIGURE 12-4 Closeup of *Cryptococcus fagi* scale. The cottony wax has been melted away to show some of the small oval scale insects.

FIGURE 12-5 Oval patches of *Nectria coccinea* var. *faginata* fruiting on beech.

About 10 years after invasion by scale insects, the fungus, *N. coccinea* var. *faginata*, is often found fruiting on the portions of the stem formerly occupied by scale insects (Figs. 12-5 and 12-6). Associated with the appearance of the fungus is the death of the trees. Eventually, almost all mature trees are killed. The mortality of beech in the Northeast has been about 1 to 2% per year between 1959 and 1977. The disease appears to progress faster in dense, pure, mature beech stands.

Control recommendations are to salvage the merchantable beech before the disease destroys them. The wounds produced by both insects and canker fungi are rapidly invaded by saprophytic decay fungi, which quickly destroy the saw-timber value of the stand. Therefore, salvage cuts should be made as soon as the scale insect appears.

The death of beech was at one time thought to ideally suit the management objectives of northern hardwood stands. The death of beech should enhance the development of birch and maple, the more economically valuable associates. In fact, selective cutting of maple and birch had produced stands stagnated by overmature beech throughout much of the Northeast. Changing markets and better utilization of beech have changed the outlook for beech bark disease.

Another consequence of beech bark disease that changed thinking was the fact that in some instances the loss of many beeches produced an abundance of combustible residue and consequently an increase in fire problems. The ecological effects of the destruction of any major component of the complex northern hardwood type cannot be taken lightly.

Today very little is known about this disease. We know almost nothing about the fungus that supposedly plays a prominent role. Is it a saprophytic native fungus or an introduced fungus? How does the beech bark disease fungus, *N. coccinea* var. *faginata,* differ from *N. galligena,* a fungus that causes target canker of many hardwoods? What, if any, activity of the fungus causes death of the trees? How does the scale insect affect the fungus? No one has artificially produced beech bark disease, so we do not even know that we are looking at the real cause of the death of the trees. The development of the symptoms could just as easily result from a virus or mycoplasma introduced into the tree by the sucking insect. The *Nectria* fungus could be just a saprophyte on the dead portions of the tree. The root rot fungus *Armillariella mellea (Armillaria mellea)*, commonly associated with decline disease, is found in beech stands dying from beech bark disease. What role does root rot play in the problem? The fact that these basic questions, and many others on the influence of environmental and site factors on the rate of development of the problem are yet unanswered, places serious limitations on our ability to make significant recommendations regarding control of this disease.

FIGURE 12-6 Closeup of clusters of perithecia of *Nectria coccinea* var. *faginata* on beech.

Annual Cankers

Fusarium canker Canker fungus-host combinations, in which the fungus is checked by callus development after the first year's invasion, are called annual cankers. Annual cankers are very common but, because of the minimum effect on the host, are often overlooked. Irregularly roughened bark caused by callusing of the wounds is a typical symptom. In species of trees used for lumber, annual canker infections often show up as dark streaks in the wood.

A typical example of an annual canker is Fusarium canker of sugar maple. *Fusarium solani* and a number of other fungi were isolated from annual cankers on sugar maple in Pennsylvania. Inoculation with pure cultures of the fungi demonstrated the pathogenicity of *F. solani*.

Annual cankers have generally received no more than cursory diagnostic study. An exception was an epidemiological study of the problem in Pennsylvania. It was found that infection occurred only over a period of a few years. There appears to be a relatively short period of time in the development of the tree when it is susceptible. Therefore, we see this type of disease develop only once during the life of a stand of trees.

Perennial Target Cankers

Nectria canker Nectria canker, caused by *Nectria galligena,* is a classic perennial target (Figs. 12-7 and 12-8). The fungus, after infection, destroys a minimal amount of the cambium each year. In the spring, the tree produces a callus ridge to wall off the fungus. During the spring and early summer, the fungus must survive as a saprophyte on dead xylem and phloem. During late summer, fall, and winter, the fungus moves out and over the callus and invades another minimal portion of the cambium beyond the previous year's development. This alternation of fungus growth and tree growth results in a target type of canker.

If the tree produces sufficient callus, the canker development may be checked. This arresting of fungus growth may occur on part of the canker, resulting in an irregular-looking target.

Perennial target cankers are the result of a balanced interaction of host and fungus and seldom cause death of the tree.

The importance of perennial cankers on the development of the host is really quite minimal. A survey of Nectria canker of aspen showed that cankers were more common on the dominant trees of the stand and that growth reduction of cankered trees did not occur until the trees were at least 90% girdled by one or more cankers.

The importance of perennial cankers on the commercial value of hardwood timber is very significant. Cankers on the main stem may totally

FIGURE 12-7 Nectria canker on trembling aspen.

FIGURE 12-8 Cross section of a Nectria canker on trembling aspen.

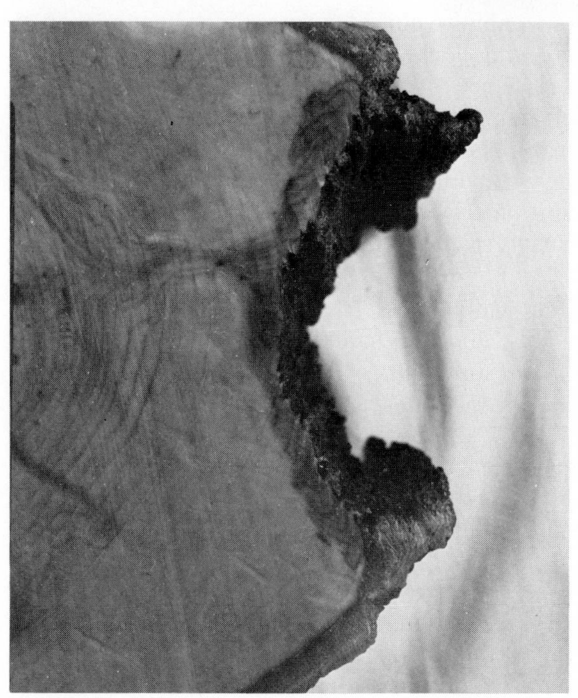

FIGURE 12-9 Perithecia of *Nectria galligena* on bigtooth aspen.

or significantly destroy the economic value of the most important portion of the tree. The discoloration and decay of xylem behind the canker makes the canker more than a surface defect.

Nectria canker is a very common disease of many hardwoods, including aspens, maples, birches, elm, basswood, and apple. The red perithecia of *Nectria galligena* that may fruit on the margin of the canker positively identify it (Fig. 12-9). The *Cylindrocarpon* imperfect stage that sometimes fruits on the margin of young cankers is the stage of the fungus most often seen in culture.

The fungus can invade healthy trees through frost cracks and other wounds. Small injuries that would occur in a branch axil because of heavy snow pushing the branch down are points of infection in the Northeast. In apples grown in the western United States, infection occurs through conidia infecting leaf scars in the fall of the year. Whether leaf scar infection occurs in forest trees is not known. There is no known control for the disease. Canker-infected trees will not produce high-quality wood products and therefore should be cut during thinning operations.

Ceratocystis canker Ceratocystis canker is sometimes diagnosed based on the target-like series of concentric rings. Nectria canker is sometimes diagnosed based on a similar appearance, so confusion and error sometimes result. There are very few reports of perithecia of *Nectria* fruiting on target cankers of trembling aspen, but perithecia of *Ceratocystis fimbriata* are commonly seen fruiting on these cankers in Colorado and Minnesota. Controlled inoculation experiments with *C. fimbriata* and *N. galligena* show that both are capable of producing cankers in some instances.

Why would it be important to be able to distinguish between Nectria and Ceratocystic cankers? We do not know how to differentiate the two cankers now. It is assumed that Nectria cankers on one hardwood species potentially represent a threat to the other hardwood species in the stand. Therefore, if one species is heavily cankered, one might consider selective removal of that species from the stand to prevent infection of the others. If *Ceratocystis* is the cause of the canker on aspen, there is no value in selectively removing aspens from hardwood stands to prevent infection of the more valuable sugar maple and birch.

Research in forest pathology attempts to accumulate new information that will be helpful in preventing disease losses as well as in understanding or disproving the effectiveness of already accepted practices. The latter is more difficult but is very important to the maintenance of credibility. The student should recognize that forest pathology is a developing profession, with new as well as modified concepts being developed all the time. Discrepancies in our concepts are not presented here to confuse, but rather to present the real world.

Ceratocystis fimbriata is a canker pathogen of a number of other tree species. London plane utilized as an ornamental in many eastern cities is cankered by *C. fimbriata.* The disease is called canker stain because of the deep penetration of stain in the infected stem. The fungus was, at one time, readily spread from tree to tree in wound-dressing paint. Addition of fungicide to the wound dressing avoids disease spread through wound treatments. Natural dissemination of the fungus through insects and splashing rain provides sufficient inoculum for infecting wounds in healthy trees. Caution should be exercised during pruning to avoid spreading the fungus on pruning tools. A chlorox dip of the tools between cuts will disinfect them and avoid this common method of disease spread.

Eutypella canker In the Northeast another perennial target canker occurring only on maples and sometimes confused with Nectria canker is Eutypella canker, caused by *Eutypella parasitica.* Eutypella canker is easily distinguished from Nectria canker by the characteristic signs of the fungus always found associated with it (Figs. 12-10 and 12-11). Black perithecia of *Eutypella parasitica* are embedded in the bark on the canker face. The necks of the perithecia extend above the bark as short hair-like projections. Cutting into the bark readily reveals the perithecial cavities. Another sign of Eutypella canker is the presence of a mycelial fan in the cambium at the upper and lower margins of the canker, easily seen by removing the bark.

FIGURE 12-10 Eutypella canker of sugar maple.

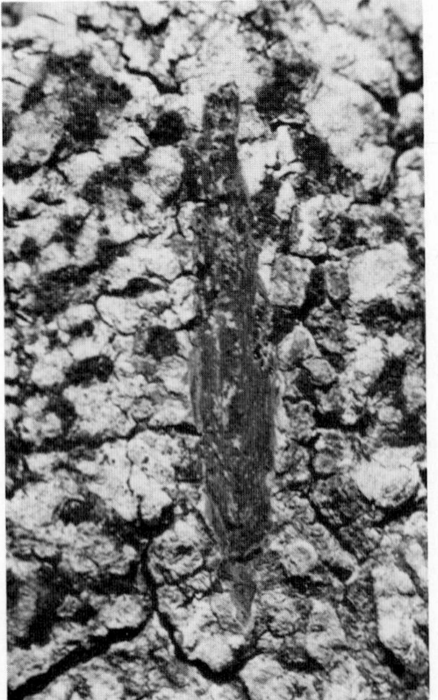

FIGURE 12-11 Closeup of Eutypella canker, showing clusters of black projecting perithecial necks and perithecia embedded in the bark as revealed in the knife cut.

Diffuse Cankers

Chestnut blight Another type of canker fungus-host combination results when the fungus is so aggressive that the host is unable to produce a callus defense wall. These are called diffuse cankers. Chestnut blight is a classic example of a diffuse canker. The fungus destroys the cambium and girdles the tree in a very few years.

Endothia parasitica, the cause of chestnut blight, is a king among fungus pathogens. Introduced into the United States early in this century, within 50 years it had spread so rapidly that the American chestnut was largely destroyed throughout its natural range from Maine to Mississippi. An estimated equivalent of 9 million acres (3.6 million ha.) of pure chestnut stands were destroyed by this fungus (Figs. 12-12 to 12-14). A few sprouts and tangled debris of fallen giants remain of what was once the most important hardwood timber species in the United States.

What makes *E. parasitica* the king of pathogens? A look at the disease cycle of chestnut blight will show all the features of a super pathogen.

A super pathogen should produce an abundance of spores on a wide variety of substrates over an extended period of time. Masses of conidia are extruded from pycnidia produced over the entire diseased chestnut bark.

FIGURE 12-12 Chestnut blight killed the large chestnuts of this stand. Sprouts produced over a long period of time from the stumps result in a brush field.

FIGURE 12-13 Chestnut-blight-infected sprout.

FIGURE 12-14 Closeup of Chestnut blight canker, showing pustules on the bark. These are necks of the pycnidia or perithecia of the fungus which are embedded below in the bark.

A super pathogen should not be host-specific, but should increase on a number of hosts or dead substrates. The ability to survive on more than one host enhances the survival ability in the absence of the primary host. A number of oaks, chinkapin, shagbark hickory, red maple, and sumac serve as saprophytic substrates for *E. parasitica*.

A super pathogen should utilize a number of means to disseminate its spores. *E. parasitica* conidia are produced in sticky tendrils that are ideally suited for dissemination by any insect that comes in contact with the bark of diseased chestnut. Birds that feed on the insects or light on the diseased bark can pick up and disseminate the spores. Almost anything that moves can carry spores from the surface of diseased bark to wounds on healthy trees. Splashing water can disseminate the spores short distances. Ascospores produced in perithecia on diseased bark are forcibly discharged and wind-disseminated to infect other hosts. To summarize dissemination, wind, water, and many vectors are utilized to spread the fungus from diseased to healthy trees.

A super pathogen should be capable of infecting through commonly produced infection courts over a variety of environmental conditions. Those pathogens with very exacting environmental requirements and specific infection courts are less capable of expansion throughout a wide environmental range. The climatic conditions from Maine to Mississippi are not at all uniform, yet this fungus expanded over this range. Infection courts are any type of wound produced by insects, birds, man, and wind breakage, to name a few.

A super pathogen should invade the host rapidly enough to avoid induction of resistance mechanisms in the host. One of the distinguishing characteristics of the diffuse canker is the rapid invasion of the host, thereby avoiding the callus defense reaction characteristic of annual and perennial cankers.

To wipe out a species of plant with a pathogen, the virulence, or capacity to cause disease, must exceed all resistance accumulated in the individuals of the host population. The pathogen must kill the host fast enough, and repeatedly, so that the host does not have time to recombine and select individual gene combinations for resistance to the pathogen. The chestnut blight disease, by killing the original population rapidly and killing most sprouts before they have a chance to produce seed, prevents the host from genetic improvement through natural selection. Normally, natural selection for increased resistance would occur in a reproducing population of plants subjected to the pressures of an introduced pathogen.

A super pathogen should be able to survive during extended periods of cold or dry conditions. *E. parasitica* can survive years on dried herbarium specimens as well as a wide range of fluctuating environmental conditions in the field.

Many other very serious disease-causing fungi have some of the characteristics of the ideal super pathogen described but none have combined them all. One can understand differences between different disease-causing fungi by examining or comparing their capacity with the super pathogen.

Much work was done on the chestnut blight disease, particularly during the 1950s, but most of the effort was abandoned, so that by the late 1960s and early 1970s very little was being done. Today, there is a new flurry of activity centered around hypovirulent (nonvirulent) strains of the fungus. In the laboratory, the hypovirulent strain has been shown to induce hypovirulence in other strains when grown in the same culture dish. If the hypovirulent strain of the fungus is introduced into wounds around cankers on field-grown chestnuts, sometimes the cankers stop enlarging and the host calluses over the wound. The lack of virulence appears to be associated with a piece of double-stranded RNA which presumably can be transferred through hyphal fusions of virulent and nonvirulent lines of the fungus.

What does all this mean for the future of the American chestnut? At this point it is difficult to say, but the approach is the first glimmer of hope for the species. It is obviously not going to be practical to physically introduce the hypovirulent strain into every canker that occurs in a chestnut stand. The ideal situation would be to allow natural spread of the hypovirulent strain to infiltrate a stand. Natural spread of the hypovirulence is limited due to reduced sporulation of these strains. Another limitation involves the inability of various strains of the fungus to fuse and transfer the hypovirulence factor. Some unknown incompatibility factor is involved. Obviously, much research is needed before we can even determine whether the approach is potentially useful. If something develops from this type of work, it will have major significance not only to chestnut blight but to many other diseases.

Hypoxylon canker A discussion of diffuse cankers should include a few that are not as devastating as chestnut blight, to show that it is possible to manage tree species with diffuse cankers. This is not to minimize the overall importance of diffuse cankers, since they are very destructive, but to show that we can learn to live with most of them.

Hypoxylon canker caused by *Hypoxylon mammatum*, is a disease of bigtooth and trembling aspens that was first reported in 1921 (Figs. 12-15 and 12-16). In contrast to *E. parasitica*, this is a pathogen native to North America, which has just recently been recognized in Europe. European forest pathologists are concerned with the potential for this disease, as it affects native *Populus tremula* and hybrids with *P. tremula, P. tremuloides,* and *P. grandidentata* parentage.

In the Lake states, where aspen is the most important commercial wood

FIGURE 12-15 Hypoxylon canker of aspen.

FIGURE 12-16 Hypoxylon canker often results in the stem breaking off at the canker.

FIGURE 12-17 Asexual spores of the *Hypoxylon mammatum* fungus develop on peg-like pillars below the outer bark.

species, up to 10% of the trees are affected. In individual stands up to 90% may be infected. An annual loss equivalent to one-fourth of the annual increment for the species is reported.

The disease is readily diagnosed by the characteristic black and white mottle it produces in infected bark. Bark of young cankers and margins of older cankers become initially yellow to orange rather than the normal green color. With time, diseased bark and wood tissue becomes blackened and cracked in a checkerboard-like fashion. The asexual stage of the fungus appears, the year following infection, on peg-like pillars that cause the outer layer of bark to blister and peel back (Fig. 12-17). The sexual stage appears 2 to 3 years after infection. Ascospores are produced in perithecia embedded in fungus stroma (Fig. 12-18). The stroma and perithecia are initially white but become black with age.

A long and eminent list of forest pathologists have contributed to our understanding of Hypoxylon canker, yet we still have many questionable and unanswered steps in the life cycle to unravel. One notable gap in our understanding is the exact site and conditions under which infection takes place. No reproducible demonstration of ascospore infection of trees in the field has yet been made. Many workers have attempted indirectly to determine the site and conditions of infection, so we have a number of reports on this topic. Insect wounds, branch stubs, small twigs, bark saprophytes, bark fungal-inhibitory toxins, fungus host-killing toxins, bark moisture, and branch axils are all implicated by different groups of researchers.

Another area that has produced a number of concepts involves the role of environmental conditions on disease development. Both poor and good sites are implicated with disease. Reduced amounts of disease in dense stands implies a stand-density relationship.

A model accounting for 92.5% of the variation in Hypoxylon canker incidence was developed through multivariate statistical analyses of environmental factors. This recent study in New York found Hypoxylon canker incidence correlated to reduced aspen cover density, soil moisture, soil mottles, soil consistence, and exchangeable soil Mn, Ca, and Na. Hypoxylon canker incidence was positively related to soil temperature, bulk density, soil fraction over 2 and 10 mm, soil depth, and rooting depth.

FIGURE 12-18 Closeup of Hypoxylon canker, showing clusters of *Hypoxylon mammatum* perithecia embedded in fungus stoma.

The relationship between site and Hypoxylon incidence could be predicted based on a vegetation site index in which the presence of blueberries (*Vaccinium* spp.) and raspberries (*Rubus* spp.) was related to a high canker incidence. Ferns (*Osmonda* spp.) and Sphagnum (*Sphagnum* spp.) were related to sites with little or no canker incidence. Maples (*Acer* spp.) were correlated to a moderate canker incidence.

Another advancement in our understanding of this disease comes from recent work on genetics of resistance in aspen to Hypoxylon canker. Three mechanisms of resistance were found. Resistance due to callus formation appears to be controlled by a few major genes, but pathogen variation may overcome this type of resistance. Resistance due to branch death is low and the heritability for resistance is also low. Retardation of canker enlargement is one of the best forms of resistance but is difficult to evaluate in short-term experiments. A more complete discussion of the genetics of resistance is included in Chapter 20.

A third contribution to our understanding of this disease involves understanding how the disease increases within stands of aspen. This topic is developed more fully in Chapter 19, but the important point to recognize with Hypoxylon canker is that intensification within an even-aged stand is not dependent upon the amount of disease in the stand. Diseased trees within even-aged stands do not contribute inoculum for infection of other trees. Trees are most susceptible to infection by Hypoxylon canker when they are seedlings. Three years between infection and sporulation allow most of the trees surrounding an infected tree to grow out of the susceptible stage before sporulation of the fungus occurs.

Our present understanding of site factors, genetics of resistance, and the epidemiology of disease intensification provides the forest manager with the capacity to control this disease. First, he or she must be concerned with the selection of sites for management of aspen. Dry sites should be avoided. Wet sites are generally more valuable for other species, so medium sites with moderate predisposition to Hypoxylon canker are the most common sites used.

As we move to more controlled management systems, the selection of genetic resistance in aspen can be used to reduce losses due to Hypoxylon. Today, we generally use natural reproduction to generate aspen stands.

Silvicultural manipulation of sites can be effectively utilized to reduce losses. The most effective treatment is to manage large blocks of even-aged stands through cutting and/or burning practices. In mixed-aged stands or with small blocks, inoculum from infected trees is concentrated near the young susceptible sprouts and seedlings. Infected trees will be concentrated around the inoculum sources, resulting in areas of the stand with high mortality. With large blocks of even-aged trees, single infected trees will occur randomly throughout the stand. These random infections will not intensify into centers because the trees will develop beyond the susceptible stage before secondary inoculum can be produced. The limited number of randomly infected trees will not seriously affect the total stand development. They will contribute to necessary natural thinning.

Scleroderris canker The last canker disease to be discussed is a diffuse canker of conifers. Scleroderris canker, caused by *Gremmeniella abietina* (formerly known as *Scleroderris lagerbergii*), is an excellent example to show that even highly sophisticated mid-twentieth-century forest pathologists can have a new disease introduced under their noses twice and not recognize its importance until it is well established.

The first detailed description of this most important disease of red and jack pines was presented in 1966. Failure of plantings to survive had been noticed for at least 10 years, but the cause could not be determined until a visiting European forest pathologist provided the idea that the disease was the same as a well-known disease in Europe. Death of lower branches and trees is recognized as Scleroderris canker by culturing or recognition of pycnidia of *Brunchorstia pinea,* the imperfect stage, or the presence of apothecia of *G. abietina.* A characteristic symptom of the disease is the yellowing of needle bases on infected branches and the yellowish-green color of cambium and wood tissues (Figs. 12-19 and 12-20).

FIGURE 12-19 Necrosis of needle base associating with Scleroderris canker in the branch.

FIGURE 12-20 Red pine branch tips killed by Scleroderris canker as seen in early summer.

Today, Scleroderris canker occurs in plantations in the states of Michigan, Wisconsin, Minnesota, Vermont, New Hampshire, Maine and New York. In Canada, work in the provinces of Ontario, Quebec, and New Brunswick has shown the wide distribution and importance of the disease. When people know what to look for, the disease is shown to be present and accounts for plantation failures formerly thought to be due to some attribute of the site.

Usually, infected plantations occur in areas where cold air accumulates, such as topographic depressions or openings in the forest canopy. A strong dependence on cool, damp, environments is seen for this disease. The fungus also appears to grow and compete better at lower temperatures.

The disease is initially distributed to the plantations on infected nursery stock. Recognition of nurseries that produce infected stock is an important step toward prevention of additional problems. It is very difficult to detect infection in nursery stock until it is planted out, so one has to use records of sources of stock for known infected plantations to determine which nurseries are the source of the problem.

Thorough distribution of inoculum throughout vast areas of natural regeneration of jack pine in northern Ontario was also observed. In this case, inoculum, came from a distant source, because no remnants of the former infected stand were apparent. The spores were probably carried to the area on a weather front from the south and deposited in the young stand by rain.

Epidemic disease development following long-range spore dispersal of a massive amount of inoculum is rarely observed for tree diseases. It may be more common than we think, though, because long-range dispersal is well documented for agricultural crops.

After about 10 years of intensive study in the United States and Canada, Scleroderris canker was reasonably well understood and controlled through fungicide sprays in the nursery. Of course, it was also important to avoid replanting where the disease was well established.

In 1975, a new Scleroderris canker disease was reported from New York. This disease occurred on large trees up to 18 m in height. Mortality was extensive on red and Scotch pine (Fig. 12-21). It was determined that the *Gremmeniella* fungus causing death of large pine was different from the *Gremmeniella* fungus occurring on small trees in the Lake states. The disease in New York was caused by a strain of the fungus that closely resembles some European strains of the fungus. By 1977, approximately 14,000 hectares were known to be infected.

So our research efforts have begun all over again. It is necessary to determine the environmental factors affecting the new disease. We must quickly determine the potential for spread within the red and Scotch pine of

FIGURE 12-21 Mortality of large red pine caused by Scleroderris canker.

the region and also the potential for spread to the Lake states region, the West, and the South. Introduced pathogens have, in the past, caused some of our most serious diseases. Only time will tell whether our current capabilities to understand and control introduced diseases are any better than those possessed by our predecessors.

CONTROL OF CANKERS

Control measures, such as keeping trees healthy by fertilizing and watering, avoiding wounding, and selectively removing diseased trees, are sometimes recommended, but no evidence is available for measurable reduction in disease infection using these methods. Sound control recommendations are based on epidemiological understanding of environmental, host, and pathogen interaction. Application of this understanding to the control of cankers awaits changes in forest management practices. The forest manager must recognize the potential to manipulate the disease before it develops. At the present time, pathological understanding is most often requested after the problem develops, and then the pathologist is able to do little more than name the disease and predict the loss.

REFERENCES

ANONYMOUS. 1954. Chestnut blight and resistant chestnuts. USDA Farm. Bull. 2068. 21 pp.

BEATTIE, R. K. and J. D. DILLER. 1954. Fifty years of chestnut blight in America. J. For. *52:*323-329.

BIER, J. E. 1940. Studies in forest pathology III. Hyoxylon canker of poplar. Can. Dept. Agr. Pub. 691 Tech. Bull. 27. 40 pp.

BRUCK, R. I. and P. D. MANION. 1980. Interacting environmental factors associated with incidence of Hypoxylon canker on trembling aspen. Can. For. Res. *10:*17-24.

BRUCK, R. I. and P. D. MANION. 1981. An indicator species spectrum for the determination of environmental predisposition of trembling aspen to Hypoxylon canker. Phytopathology

CROWDY, S. H. 1952. Observations on apple canker IV. The infection of leaf scars. Ann. Appl. Biol. *39:*569-587.

DORWORTH, C. E. 1970. *Scleroderris lagerbergii* Gremmen and the pine replant problem in central Ontario. Can. For. Serv. Ont. Reg. For. Res. Lab., Sault Ste. Marie, Ont. Inf. Rep. O-X-139. 12 pp.

EHRLICH, J. 1934. The beech bark disease, a *Nectria* disease of *Fagus*, following *Cryptococcus fagi* (Baer). Can. J. Res. *10:*593-692.

FILIP, S. M. 1978. Impact of beech bark disease on even-age management of a northern hardwood forest (1952 to 1976) USDA For. Serv. Gen. Tech. Rep. NE-45. 7 pp.

FRENCH, W. J. 1969. Eutypella canker on *Acer* in New York. N.Y. State Univ. Coll. For. Syracuse Tech. Pub. 94. 56 pp.

HOUSTON, D. R., E. J. PARKER, R. PERRIN, and K. J. LANG. 1979. Beech bark disease: a comparison of the disease in North America, Great Britain, France, and Germany. Europ. J. For. Pathol. *9:*199-211.

HUBBES, M. 1964. New facts on host-parasite relationships in Hypoxylon canker of aspen. Can. J. Bot. *42:*1489-1494.

JORGENSEN, E., and J. D. CUTLEY. 1961. Branch and stem cankers of white and Norway spruce in Ontario. For. Chron. *37*:394–404.

MACDONALD, W. L., F. C. CECH, J. LUCHOK, and C. SMITH, eds. 1978. Proc. Am. Chestnut Symp. W.V. Univ. Morgantown, W.Va. 122 pp.

MANION, P. D. 1975. Two infection sites of *Hypoxylon mammatum* in trembling aspen *(Populus tremuloides).* Can. Bot. *53*:2621–2624.

MANION, P. D., and M. BLUME. 1975. Epidemiology of Hyoxylon canker of aspen. Proc. Am. Phytopathol. Soc. *2:*101.

MANION, P. D., and D. W. FRENCH. 1967. *Nectria galligena* and *Ceratocystis fimbriata* cankers of aspen in Minnesota. For. Sci. *13:*23–28.

MCCRACKEN, F.I., and E.R. TOOLE. 1974. Canker-rots in southern hardwoods. USDA For. Serv. For. Pest Leaf. 33. 4 pp.

MILLER-WEEKS, M., and J. T. O'BRIEN. 1978. Beech bark evaluation survey. USDA For. Serv. Northeast. Area, State Private For. Eval. Rep P-78-2–4.

SHIGO, A. L. 1972. The beech bark disease today in the Northeastern U.S. J. For. *70:*286–289.

SKILLING, D. D. 1977. The development of a more virulent strain of *Scleroderris lagerbergii* in New York State. Eur. J. For. Pathol. *7:*297-302.

SPAULDING, P., T. J. GRANT, and T. T. AYERS. 1936. Investigations of Nectria diseases in hardwoods of New England. J. For. *34:*169-179.

VALENTINE, F. A., P. D. MANION, and K. E. MOORE. 1976. Genetic control of resistance to Hypoxylon infection and canker development in *Populus tremuloides.* Proc. 12th Lake States For. Tree Improv. Conf. USDA For. Serv. Gen. Tech. Rep. NC-26, pp. 132–146.

VAN ALFEN, N. K., R. A. JAYNES, S. L. ANAGNOSTAKIS, and P. R. DAY. 1975. Chestnut blight: biological control by transmissible hypovirulence in *Endothia parasitica.* Science *189:*890–891.

WATERMAN, A. M. 1955. The relation of *Valsa kunzei* to cankers on conifers. Phytopathology *45:*686–692.

WEIDENSAUL, T. C., and F. A. WOOD. 1974. Analysis of a maple canker epidemic in Pennsylvania. Phytopathology *64:*1024–1027.

WOOD, F. A. and J. M. SKELLY. 1964. The etiology of an annual canker on maple. Phytopathology *54:*269–272.

13

FUNGI AS AGENTS OF TREE DISEASES: VASCULAR WILT DISEASES

- Types of wilts
- Mode of action of wilt diseases
- Wilt disease cycle
- Symptoms of wilt diseases
- Diagnosis of wilt diseases
- Examples of wilt diseases and controls
- Origin of wilt diseases

Photosynthesis and transpiration by leaves require large amounts of water. In vascular plants, water is taken up from the soil by the roots and transported via the vessels or tracheids of the xylem to the leaves.

A continuous unbroken column of water from the roots to the leaves allows water to be lifted hundreds of feet. Root pressure is part of the upward pumping mechanism. Some energy for upward pumping may be supplied by living parenchyma cells around the vessels. However, the basic mechanism of water transport relies on the cohesive tension of water molecules. Molecule-by-molecule replacement of transpired water in the leaves with molecules of water picked up in the roots occurs to keep a closed water system. Any breaks in the closed system immediately induce an interruption or void space in the column of water. A vacuum can enlarge the void space but cannot lift water above 9.75 m, because this is the maximum vertical displacement of a column of water under atmospheric pressure.

Transpiration of individual trees varies from zero on rainy nights to 200 to 400 liters per day on sunny days. Any disruption in the transport of water therefore quickly induces wilting and death.

Rapid wilting and death of "normal"-looking trees during hot, dry periods is a dramatic thing to see, and therefore wilt disease epidemics produce a high degree of emotional concern by individuals and communities.

TYPES OF WILTS

Wilting of leaves can be caused by bacterial or fungal activity in the vessels of plants. It can be caused by bacterial or other microorganisms that disrupt or destroy uptake of water by feeder roots. It can be caused by drought. It can also be caused by canker fungi girdling branches.

The types of wilts discussed in this chapter are caused by the fungal invasion of xylem vessels.

MODE OF ACTION OF WILT DISEASES

Wilt-disease-causing organisms invade the vessels of plants, disrupt water movement, and subsequently cause wilting.

Wilt diseases can occur in all plants with vessels, but are especially serious in ring-porous trees. Upward water movement in ring-porous trees takes place in the current year's vessels. Therefore, plugging of the current year's vessels causes rapid wilting. Diffuse porous trees transport water in vessels of a number of annual rings, so many small vessels need to be plugged before wilting occurs.

The mechanisms of disruption are varied and not completely understood. One basic mechanism involves the production of tyloses by the plant. Tyloses are a defense reaction plugging of vessels produced by disruption of the plasma membrane of parenchyma cells adjoining the vessels. Membranes and the contents of parenchyma cells expand through the connecting pits into the vessels. Another mechanism of wilting involves plugging of vessel endwall perforation plates by the fungus spores and hyphae caught on the endwall plate. A third mechansim involves production of toxins by wilt-causing organisms that disrupt the water-pumping mechanisms either at the leaves or along the vessels. A fourth mechanism involves changes in viscosity of vessel fluids as the result of enzymatic degradation and solubilization of cell-wall material, thereby slowing down the flow rate of water to the leaves. A fifth mechanism involves parasitism of parenchyma cells, with dysfunction of the vessels as a secondary reaction. A sixth mechanism involves embolism or introduction of air bubbles into the vessels through fungus disruption of cell walls.

The effect of any of these six mechanisms is to damage or slow down water transport within the tree. Air is introduced into the vessels and the affected part of the system stops functioning.

WILT DISEASE CYCLE

The inoculation point for wilt pathogens is a wound in either the stem or the root system, depending upon the disease. The intense vacuum of the vessel system often draws spores well within the host during wounding.

Germination and subsequent mycelial or budding-type growth occurs. Plugging of vessels and disruption of vessel integrity by the fungus causes additional breakage and suction columns. The fungus may be moved about rapidly. Invasion of parenchyma cells and movement between vessels occurs through pits, expanding the sphere of influence of the fungus.

Sporulation and disseminaiton of wilt pathogens occur after the tree dies. Sporulation is affected by how fast the tree dries out. Different types of relationships have evolved between the wilt pathogens and insect vectors. These are more fully developed with specific examples.

SYMPTOMS OF WILT DISEASES

Initially, limp or drooping normal-colored leaves may be observed. Within a few days, the wilted leaves change color to yellow brown and die. They may be shed or remain on the tree.

Vascular discoloration is usually evident. Cross or diagonal sections of infected branches will show discoloration of the current year's vessels.

DIAGNOSIS OF WILT DISEASES

The diagnosis of wilt diseases is initially based on symptoms. Verifiction of a specific causal agent is accomplished by isolation and identification of the fungus or agar media. Branches collected from recently dead areas are surface-sterilized, and chips of outer xylem are aseptically transferred to culture media. About 1 week later, the fungi are identified based on characteristic asexual fruiting structures.

Proof of pathogenicity of a new suspected pathogen is accomplished using Koch's postulates.

EXAMPLES OF WILT DISEASES AND CONTROLS

Dutch Elm Disease

Infection by *Ceratocystis ulmi* (Figs. 13-1 to 13-5) takes place as a result of feeding wounds in small branch crotches by *Scolytus multistriatus* (smaller European elm bark beetle). The infection takes place in the spring of the

FIGURE 13-1 Dutch-elm-diseased American elm, showing branches with wilting leaves in the lower crown and dead as well as normal-looking branches in the upper crown.

FIGURE 13-2 Closeup of wilting leaves of American elm.

year, when large springwood vessels are near the surface. Elm trees that do not produce large springwood vessels have been shown to be somewhat resistant to the disease. Infection can also take place in large stems as a result of feeding activity of the native elm bark beetle (*Hylurgopinus rufipes*). In both cases, the beetle feeds on the phloem cambium region. Some chewing into the xylem exposes recently formed vessels. Spores of *C. ulmi* on the surface of the feeding beetle may be dislodged from the beetle and introduced into injured vessels. The fungus moves passively down the tree as suction columns develop in disrupted vessels. It moves around the circumference of the tree in the root region, and eventually plugs vessels around the whole tree. The fungus gains access to vessels of adjacent trees through root grafts between trees.

Fungus sporulation in beetle pupal chambers and galleries is ideally coordinated with the activity of the two types of bark beetles. Spores are passively carried as contaminants on the surface of beetles which emerge from infected dead trees and fly to healthy trees to feed.

Control of Dutch elm disease (DED) is more a political problem than a biological problem. The importance of the elimination of insect-breeding places in dead logs and the severing of root grafts has been known for a long time. Spraying with insecticides has never been completely satisfactory and is definitely not sufficient as a single means of control. Extensive losses of street elms result because political bodies are unwilling to properly care for the

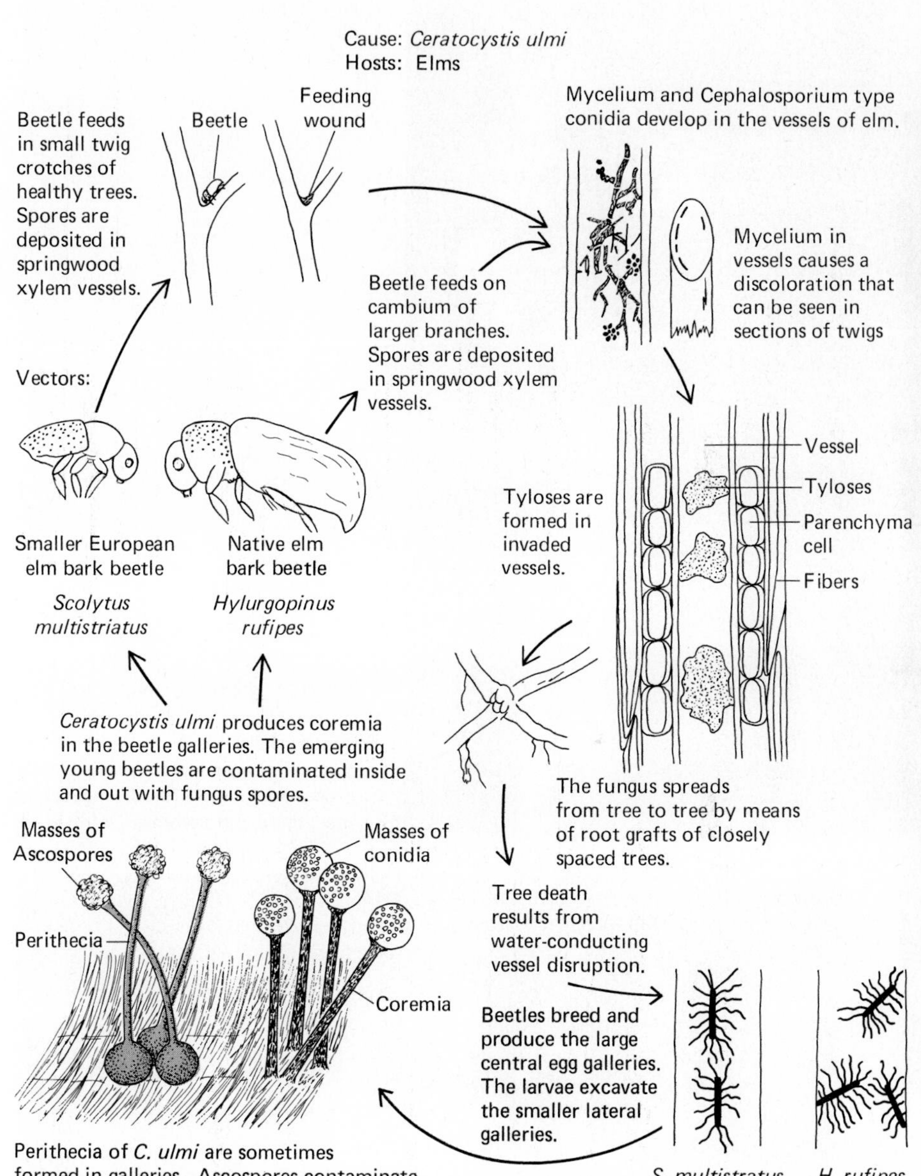

FIGURE 13-3 Disease cycle of Dutch elm disease.

FIGURE 13-4 Smaller European elm bark beetle galleries in American elm.

FIGURE 13-5 Coremia of *Ceratocystis ulmi* are produced in bark beetle galleries. Elm bark beetles become contaminated by the conidia produced in white study masses at the tips of the coremia.

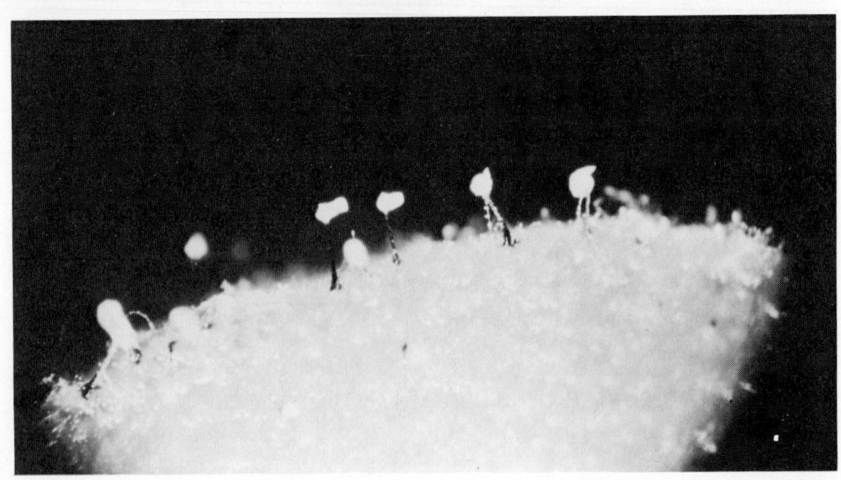

trees. The cost to remove an elm that has been dead for 2 years is similar to the cost to remove a recently killed tree. Removal of a few living trees to improve spacing is probably less expensive than removal after they die, because of root grafts to DED-infected trees. The cost is the same, but early removal of dead trees and improvement of spacing or severing root grafts reduces subsequent losses in the remaining population of elms.

Control of DED must be centered around a good sanitation program to remove all dead elm material. Dead branches in standing live trees as well as dead elm trees must be removed from the control area and burned or buried. In the absence of careful elimination of all dead elm material, it is almost impossible to effectively manipulate DED.

Can Dutch elm disease be controlled? Many communities try. Many fail. Very few are satisfied. One has to consider the economics of the system. There is no question that Dutch elm disease control programs cost money. But costs to the community are 37 to 76% less with control programs than without. A community can expect to maintain most of its elm population only with a sustained high-performance control program. Any suspension in effort cannot be reversed. Losses will occur. But costs and losses will occur in any mature tree population. Therefore, the answer is yes—Dutch elm disease can be controlled. Elms can be maintained as cost effectively as can any other species.

Overenthusiasm today regarding the use of systemic fungicides for the control of Dutch elm disease is somewhat misplaced. At a minimum cost of $40 per tree per treatment, how many trees can one afford to treat? How much more effectively could the money be utilized for the destruction of root grafts and beetle breeding places? The questionably effective "control" by the fungicide is only good for 1 year, so the high cost of application must be continued each year. The method of application involves pressure injection through wounds made in the stem.

Wounds produced for injection are extensively invaded by stain and decay fungi. All of these drawbacks make fungicidal injections practical for a limited number of very important trees.

My note of sarcastic pessism in relation to DED is intended. Hopefully, people like you, who have an appreciation of the biologists' ways of controlling nature, will provide a political base to help avoid such problems in the future. There will be many new problems involving ornamental trees in the future no matter what species is planted.

A point to keep in mind when evaluating or considering control is what level of disease incidence you are willing to accept, because absolute control is not possible. One must also recognize that trees are not immortal and that disease prevention or control is more difficult with older trees. Expectations for saving the original overmature population of elms in our eastern cities were overly optimistic. Many of these trees were dying in the absence of

Dutch elm disease. A young, vigorously growing, well-cared-for population of elms, if planted today, could be expected to do better than most of the other tree species presently being planted in our cities. The role of elms in urban tree management is discussed further in Chapter 22.

Oak Wilt

Oak wilt caused by *Ceratocystis fagacearum* is a serious disease, particularly of red and black oaks. The disease is distributed from Minnesota to Pennsylvania south to North Carolina and Arkansas. Pathologists have been con-

FIGURE 13-6 Oak wilt disease cycle.

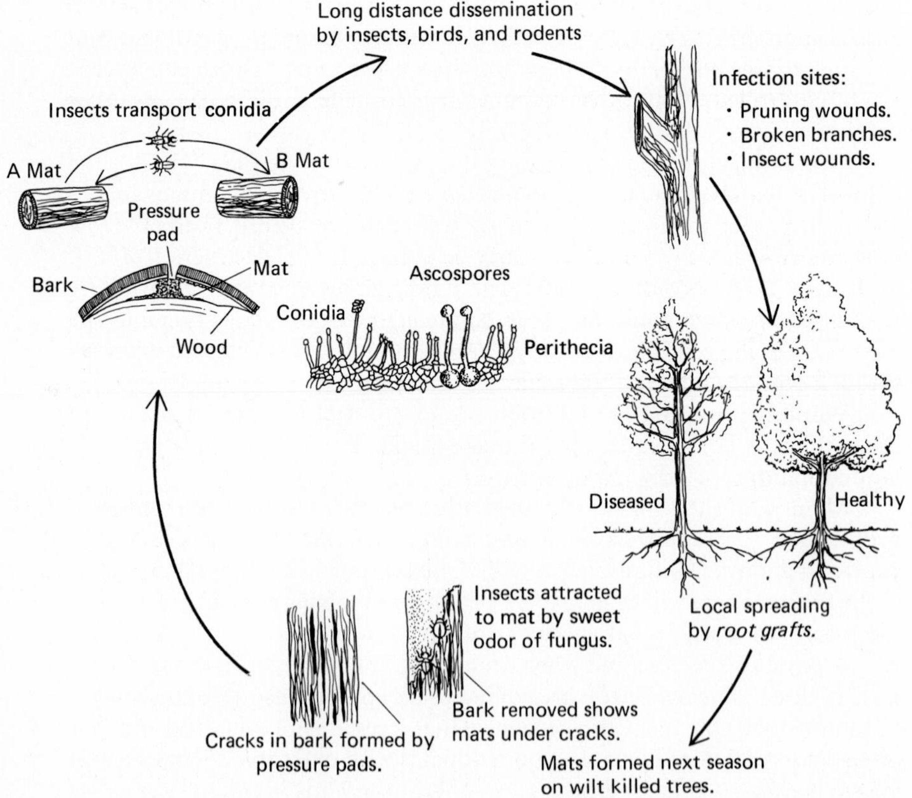

FIGURE 13-7 Oak wilt infection center in Wisconsin.

cerned over the effect this disease would have on oaks of the deep South, where it has not been reported until recently. Inoculations of oaks in southern Arkansas caused death of the inoculated trees, but the pathogen did not sporulate and spread to additional trees. A recent report of *C. fagacearum* associated with the decline of live oak in Texas may change our concepts of this disease.

The pathogen does not occur on the European continent, and there is a strong effort to prevent its introduction. For some unknown reason, the pathogen also does not occur in New York or New Jersey.

The disease cycle of the oak wilt pathogen is similar to Dutch elm disease in some ways but is very different in others (Figs. 13-6 to 13-8). The first difference is in the means of dissemination and infection. The means of dissemination and specific site of infection are still not fully understood for oak wilt. There are good indications that insects are involved, but researchers cannot agree upon which insects.

FIGURE 13-8 Closeup of wilted leaves.

FIGURE 13-9 Ceratocystis fagacearum pressure pads produced below the bark. The sweet odor of the fungus mats attracts insects, birds, and rodents, which become vectors of the fungus spores produced on the fungus mat. (Photograph compliments of Dr. Savel Silverborg.)

Within vessels, the *C. fagacearum* pathogen is similar to the *C. ulmi* pathogen. Root grafts are also a major means of spread from one tree to another.

In some areas, sporulation occurs on pressure pads which develop under the bark of recently killed trees (Fig. 13-9). Expansion of the fungus pressure pad causes the bark to crack, allowing insects access to the fungus. Many insects, as well as birds and rodents, are attracted by the sweet smell of the pads. These are some of the probable vectors.

Control of oak wilt has traditionally involved sanitation similar to Dutch elm disease, but the reason for the removal of the dead trees was to eliminate material on which fungus mats were formed, rather than to eliminate beetle breeding areas. Methods of deep girdling to cause rapid death and drying of infected trees, as well as cutting of infected trees and surrounding trees, were used in major control efforts in West Virginia and Pennsylvania, respectively. Evaluation of the effectiveness of control efforts compared to noncontrolled areas demonstrated the ineffectiveness of both treatments in preventing spread of the disease.

At the present time, one can recommend that pruning be avoided during the growing season in areas where oak wilt occurs. This is because it appears that wounds attract a number of insects that may be vectors of the pathogen. Root grafts should be severed between infected and healthy trees. Diseased trees should be cut and the wood disposed of to prevent the production of pressure pads.

Whereas at one time, many research programs were producing much new information on oak wilt, now we have reached a plateau and are getting little new information. Yet we still do not have the disease in check.

Verticillium Wilt

Verticillium wilt, caused by *Verticillium dahliae* or *V. albo-atrum* is the last wilt disease that will be discussed in detail. There are a number of other wilts of local importance, and there may be additional wilts that will become important in the future.

The *Verticillium* fungus causes wilts on a number of agricultural crops as well as on trees. The most seriously affected trees are maples and elms, but the symptoms are slow to develop and not as striking as the last two wilts. If Dutch elm disease is present in an area, we very seldom recognize a Verticillium-wilt-affected elm tree. In maples, the death of branches and the tree is slow enough that the condition is often diagnosed as part of the maple decline problem (Figs. 13-10 and 13-11).

Verticillium wilt causes a gradual wilting and death of branches, and eventually the whole tree. One characteristic symptom of the disease is a

FIGURE 13-10 Verticillium-wilt-infected branch on Norway maple.

FIGURE 13-11 Closeup of Verticillium wilt, showing necrotic blotch symptoms on Norway maple leaves and green discoloration in the xylem of the infected branch.

green or brown streaking in the xylem of affected branches, stems, and roots. Positive identification is possible only by culturing the fungus.

The disease cycle is not thoroughly understood, but work with agricultural crops and trees has demonstrated that infection takes place through the roots from inoculum in the soil. The *Verticillium* fungus produces sclerotia for survival in the soil. Recall from Chapter 8 that sclerotia are compact masses of hyphae which function as a dormancy or survival stage, particularly for soil-borne fungi. The sclerotia can remain dormant for extended periods. Replanting trees in infested soil results in a high probability of infection of the new plantings. Therefore, the only recommendation that can be made with regard to this disease is to cut the diseased trees and not to replant with another susceptible species.

Verticillium wilt may become a more serious problem in the future for urban populations of trees, because of the present lack of concern for the problem. Many shade-tree nurseries today are established on former agricultural land. If the former agricultural crop included tomatoes or potatoes, the level of *Verticillium* sclerotia in the soil will be very high, because these crops and agricultural practices are ideally suited for the development of the pathogen. At the present time, trees showing symptoms of Verticillium wilt are not discriminated against. The nursery operator cuts off the branches showing the symptoms and sells these trees along with the healthy ones. They have a good chance of surviving the guarantee period, after which the nurseryman is off the hook. We do not know the long-term effect of this practice or the conditions that cause the disease to flare up again in infected trees. We can predict that there will be future problems due to this practice.

We do not fully understand the possible consequences of many of our actions today. Future generations will have to absorb the costs of our mistakes with Verticillium wilt of maples, just as this generation had to absorb the cost of the mistake involved in planting too many elms.

ORIGIN OF WILT DISEASES

It is interesting to consider the origin of the wilt pathogens we are presently plagued with. The *Verticillium* fungus has a wide host range and distribution. It is well adapted for existence in the soil and appears to gain access to plants through root wounds. The disease increase depends upon the amount of inoculum in infested soil, which in turn is dependent upon the previous crops maintained on the site. We can speculate that this fungus has become a serious disease agent of plants only because of cultivation of the same plants on the same site year after year.

The origin of the oak wilt pathogen is a mystery. Some have suggested that it was introduced, based on a high rate of disease increase, but further data have not substantiated the high rates of disease increase. Oak wilt does not occur on any other continent, so it must be native to North America. We might speculate that *C. fagacearum* originated from a saprophytic stain fungus.

The hypothesis of evolution from a saprophytic stain fungus is not unique to oak wilt. The idea has been proposed for the Dutch elm disease pathogen and is backed up by some observations. The *C. ulmi* pathogen as a serious disease agent originated on the European continent around the turn of the century. All introductions to other continents can be traced to this center of origin. The pathogen spreads like an introduced pathogen in Europe as well as anywhere it becomes established.

An interesting finding with the *C. ulmi* fungus is that some strains or cultures of the fungus are not pathogenic. These survive as saprophytic stain fungi on dead elm logs.

The association of bark beetles and species of *Ceratocystis* is more than coincidence. The long-necked perithecia with an accumulation of sticky ascospores at the top of the perithecia neck, as well as the long coremia, with sticky spores in a mass at the top, appear to be an evolutionary adaptation for insect dissemination and association. Both spore stages are found in insect galleries. The genus *Ceratocystis* is well known for its association with insects and wood stain.

To make the transition from a saprophytic existence in association with bark beetles to a parasite of elm trees would require a mutation that allows the fungus to survive within the vessels of living trees. It must be able to tolerate the defense mechanisms or bypass the activation of defense mechanisms by the tree to survive in the living tree. We do not know enough about the mechanism of interaction between the tree and the fungus to speculate as to what type of mutation occurred to allow survival within the tree. Once such a mutant occurred, it would by chance be deposited in the vessels during normal feeding by beetles. The fungus that could survive in the tree and cause its death would probably predominate in the staining of the wood and invasion of beetle galleries, so a strong selection pressure for the pathogenic strain would develop in the fungus population.

A postscript to the foregoing speculation regarding the origin of *C. ulmi* should indicate that additional mutations and selection for pathogenic capacity can occur. Pathologists on the European continent are now finding that the elms they developed and planted widely because of resistance to the Dutch elm disease are now dying of a wilt disease they would like to call American elm disease. They speculate that we exported a more virulent strain of the fungus *C. ulmi* back to Europe on elm logs presently being exported to Europe for manufacturing into veneers.

There is one feature common to these three wilt diseases. They all probably are of recent serious pathological origin. If this is the case, we can expect wilt diseases of other species to develop along similar lines—we have not seen the last of the wilt disease epidemics. Hopefully, we can learn from the ones we now have, so as to avoid extensive losses from future wilt epidemics.

REFERENCES

CAMPANA, R. J. 1974. DED controls: Will systemics work? Weed, Trees and Turf, May(74):16–17.

CANNON, W. N., JR., J. H. BARGER, and D. P. WORLEY. 1977. Dutch elm disease control: intensive sanitation and survey economics. USDA For. Serv. Res. Pap. NE 387. 9 pp.

CANNON, W. N., JR., and D. P. WORLEY. 1976. Dutch elm disease control: performance and costs. USDA For. Serv. Res. Pap. NE 345. 7 pp.

DIMOND, A. E. 1970. Biophysics and biochemistry of the vascular wilt syndrome. Annu. Rev. Phytopathol. *8*:301–322.

ELGERSMA, D. M. 1969. Resistance mechanisms of elms to *Ceratocystis ulmi*. Phytopathol. Lab. "Willie Commelin Scholten" 77, Baarn, The Netherlands. 84 pp.

HIMELICK, E. B. 1969. Tree and shrub hosts of *Verticillium albo-atrum*. Ill. Nat. Hist. Surv. Biol. Notes 66. 8 pp.

JEWELL, F. F. 1956. Insect transmission of oak wilt. Phytopathology *46*:244–257.

JONES, T. W. 1971. An appraisal of oak wilt control programs in Pennsylvania and West Virginia. USDA For. Serv. Res. Pap. NE-204. 15 pp.

JONES, T. W. 1972. Oak wilt. USDA For. Serv. For. Pest Leafl. 29. 7 pp.

KRAMER, P. J., and T. T. KOZLOWSKI. 1960. Physiology of trees. McGraw-Hill Book Company, New York. 642 pp.

MERRILL, W. 1967. The oak wilt epidemics in Pennsylvania and West Virginia: an analysis. Phytopathology *57*:1206–1210.

MERRILL, W. 1968. Effect of control programs on development of epidemics of Dutch elm disease. Phytopathology *58*:1060.

MILLER, H. C., S. B. SILVERBORG, and R. J. CAMPANA. 1969. Dutch elm disease: relation of spread and intensification to control by sanitation in Syracuse, New York. Plant Dis. Rep. *53*:551–555.

SINCLAIR, W. A., J. L. SAUNDERS, and E. J. BRAUN. 1975. Dutch elm disease and phloem necrosis. Cornell Tree Pest Leafl. A-9. 20 pp.

SINCLAIR, W. A., and R. J. CAMPANA, eds. 1978. Dutch elm disease perspectives after 60 years. Northeast. Reg. Res. Publ. Search. Vol 8, No. 5. 52 pp.

SMITH, L. D., and D. NEELY. 1979. Relative susceptibility of tree species to *Verticillium dahliae*. Plant Dis. Rep. *63*:328-32.

WILSON, C. L. 1965. *Ceratocystis ulmi* in elm wood. Phytopathology *55*:447.

ZIMMERMAN, M. M. 1963. How sap moves in trees. Sci. Am. *208(3)*:132–142.

14

FUNGI AS AGENTS OF TREE DISEASES: WOOD DECAY

- Types of wood decay
- Mode of action of wood decay fungi
- Wood decay "disease" cycle
- Symptoms and effects of decay
- Methods for recognition of decay
- Identification of species of decay fungi based on fruit bodies
- Examples of heart rot and sap rot decays
- Role of heart rot in plant successions
- Control of heart rot decay
- Examples of wood decay in the home
- Control of wood decay in products

Wood decay is a very broad subject involving at least two major subdivisions. One aspect of wood decay deals with decay of living trees. Another deals with decay of wood products. Decay of living trees, heart rot, accounts for more loss in saw timber than fire, insects, weather, or any other disease agent. About one-third of all losses in saw timber are caused by heart rot. Decay of wood products is also significant. Ten percent of the annual cut of timber is utilized to replace wood that has deteriorated because of decay fungi.

TYPES OF WOOD DECAY

Wood decay in its broadest sense, wood deterioration, is caused by a wide variety of insects, marine animals, fungi, bacteria, and physical factors of the environment. This chapter deals primarily with decay caused by fungi that enzymatically digest woody cell walls.

Even the limited area of fungi that enzymatically digest woody cell walls is too broad, because fungi that cause cankers and fungi that cause wilts can enzymatically digest cell-wall material. In an attempt to narrow the subject, we limit our discussion to those fungi that live and derive nutrients primarily by digestion of woody-cell-wall materials. Those fungi that utilize living cell contents as major source of nutrients but also digest wood are discussed in other chapters.

Heart Rot and Sap Rot

A basic separation of wood decay can be made on the basis of whether the fungus survives primarily in living trees or in dead trees and wood products. The decay fungi of living trees are heart rots. The decay fungi of dead trees and wood products are sap rots. We like to categorize the decay fungi into these two groups, but they may not recognize our system. For example, *Ganoderma applanatum* (*Fomes applanatus*) operates very much like a sap rot fungus in the colonization and decomposition of American elm stumps, but operates like a heart rot fungus by causing a root and butt rot of living trembling aspen. One cannot, therefore, totally categorize the fungi into these two groups without considering the host.

As with almost any system of categorizing natural phenomenon, there are some fungi that straddle the fence. *Fomitopsis pinicola* (*Fomes pinicola*) is a decay fungus of dead trees in the East, but in the Pacific Northwest it also causes heart rot.

Top Rot and Root and Butt Rot

The heart rots are further separated into top rots and root or butt rots, depending upon which portion of the tree is affected. Top rot decay fungi are found in the wood of upper portions of trees. The fungi seldom progress very far into the roots and therefore do not spread from one tree to another via roots or from the stumps of a harvested crop to the next generation of sprouts or suckers.

Root and butt rot fungi colonize the lower stem and roots of trees. They may parasitize the cambium of roots or may remain in the central xylem tissues. Those that parasitize cambium are discussed more fully in Chapter 16.

Root and butt rot fungi can be serious problems with stands regenerated from stump sprouts and root suckers. Decay in the parent trees and infection of cut stumps may inoculate the new stand with root rot fungi. Fire and logging scars are also common points of infection by root and butt rotters.

Slash Rot and Products Decay

The sap rot fungi are separated into slash rotters and products decay fungi. Slash rotters decompose branches, stems, and roots of dead trees to recycle the carbon, minerals, and other basic elements tied up in the complex organic structure of wood. The products decay fungi are a group of slash rotters commonly associated with the deterioration of wood products.

FIGURE 14-1 Appearance of decayed wood.

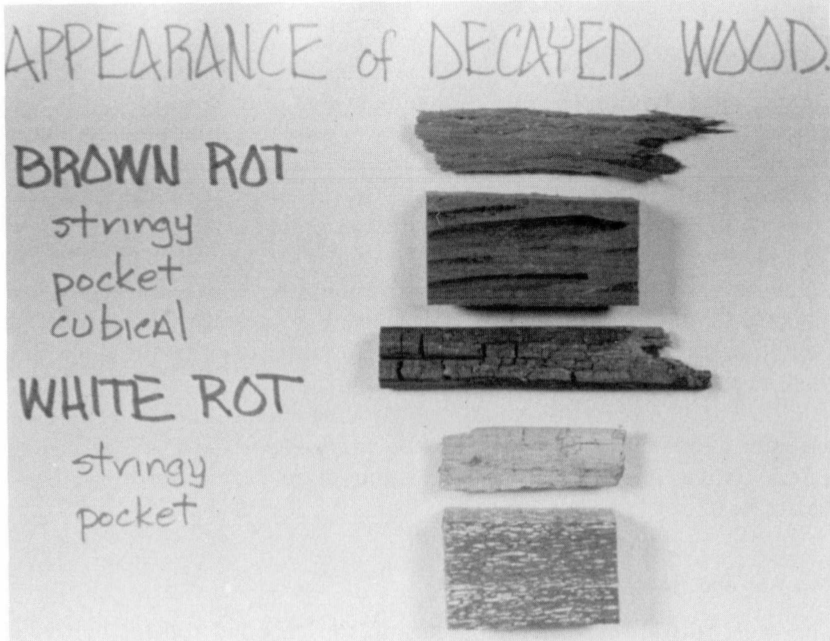

White Rot, Brown Rot, Simultaneous Rot, and Soft Rot

Another method of separating wood decay fungi is based on the types of enzymatic activity utilized for decomposition of the wood. All decay fungi and many other fungi produce cellulase enzymes for digestion of cellulose, the main structural component of wood. Some fungi, called white rotters, produce enzymes for digestion of lignin, the second most abundant component of wood. Those wood decay fungi that enzymatically digest only cellulose are called brown rotters. Those that enzymatically digest both cellulose and lignin are called white rotters. The brown rot fungi produce a brown wood residue following decay, and the white rot fungi produce a white wood residue following decay (Fig. 14-1).

Some people recognize a third group of decay fungi, called simultaneous rotters, which decay wood very rapidly and utilize cellulose and lignin at about the same rate.

A fourth type of decay fungus, called a soft rotter, decomposes cellulose in localized pockets in the secondary wall. Soft rot occurs only in wood that is saturated with water. Soft rots are caused by fungi in the class Ascomycetes and Fungi Imperfecti, while the other decays are usually caused by Basidiomycete fungi.

The white rots are further subdivided into those that produce a stringy appearance in the wood and those that produce pockets of rot.

Brown rots can be stringy, produce pockets, or more commonly produce across-the-grain fractures in the wood which break the wood up into cubes. The latter category is called brown cubical rot.

An interesting, but not totally consistent, feature of the white and brown rots is the association of white rots with hardwoods and brown rots with conifers.

MODE OF ACTION OF WOOD DECAY FUNGI

The separation of wood decay fungi into white, brown, simultaneous, and soft rot is better understood once one develops an understanding of the basic chemistry and structure of wood and the basic enzymatic reactions of decay fungi. A wood decay model will tie together the chemistry of wood structure and enzymatic activity of fungi.

Wood Chemistry and Structure

Wood is a unique, strong, durable material that is decomposed by a limited number of microorganisms. This may sound like a contradiction to the main subject of this chapter, but it is important to recognize that wood is a rather durable material. Durability and strength are the reasons woody plants are able to survive and compete for long periods of time. Durability, strength, economy, and widespread occurrence are reasons wood is one of our most important structural materials.

The durability and strength of wood are associated with specialized structural cells of the xylem, tracheids in conifers, and fibers in hardwoods. The xylem is made up of other specialized cells for transport and storage, but it is the tracheids and fibers that represent the main structural elements.

Structural woody cells have thick walls composed primarily of cellulose and lignin. The holocellulose or total cellulose fraction of wood is about 70 to 75% of the extractive-free weight of softwoods and 75 to 82% of the extractive-free weight of hardwoods. Lignin represents 18 to 30% of the extractive-free weight or most of the noncellulose weight.

The term "extractive" refers to a diverse group of chemical materials in wood that are readily extracted from wood with water or organic solvents. The extractives are not part of the cell wall. They are storage products within living cells, metabolic end products of normal cell death, or metabolic reaction products of wound-response mechanisms. Storage products are sugars and starches. The other compounds are a complex of phenolic materials.

Cellulose is a long-chain polymer of 10,000 glucose units held together with β-1, 4 glycosidic bonds (Fig. 14-2). Cellulose chains may aggregate into microfibrils in which a number of cellulose chains are cross-linked by hydrogen bonding into crystalline cellulose. Other portions of the cellulose chains are more loosely associated. There are a number of theories on the structure of microfibrils, but all have crystalline, highly organized cellulose regions and noncrystalline, amorphous, cellulose regions.

① β-1,4 glucose linkage to form cellulose

FIGURE 14-2 Cellulose chemical structure.

Lignin is impregnated within the noncrystalline, amorphous portions of the cellulose matrix and is bound by some means to the cellulose. Lignin is a complex polymer with many cross linkages between the basic building blocks, phenylpropane (Fig. 14-3).

Part of the holocellulose fraction of wood is made up of something other than β-linked glucose polymers. This other fraction, hemicellulose, is made up of glucose, galactose, mannose, glucuronic acid, arabinose, and xylose

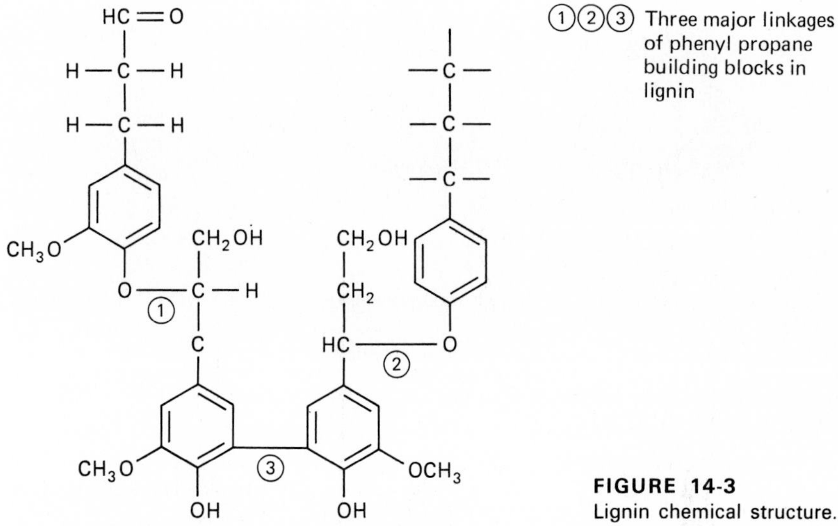

①②③ Three major linkages of phenyl propane building blocks in lignin

FIGURE 14-3
Lignin chemical structure.

units linked by glycosidic bonds between sugars. The hemicellulose molecules are intermixed with the lignin polymer in the amorphous cellulose, around the crystalline cellulose regions.

True cellulose, lignin, and hemicellulose account for about 95% of the chemical constituents of the woody cell. The remaining fraction consists of extractives such as starch, phenols, pectins, and minerals. Pectins are glycosidically linked polymers such as hemicellulose, made up primarily of glucuronic and galacturonic acids. Starch is very much like cellulose in that it is made up of glucose units, but the glycosidic linkages between the glucose molecules in starch are α -1,4 and α -1,6 linkages. What seems like a small difference between the cellulose beta linkages and the starch alpha linkages is singificant enough to prevent most animals, including man, from utilizing cellulose for food. Phenols are hydroxylated aromatic compounds of numerous configurations. The odors, colors, and natural decay resistance of wood are the result of specific phenols.

The basic chemical building blocks of wood just described are organized into the woody cell (Fig. 14-4). Cellulose, lignin, and hemicellulose are aggregated into substructures, which are further organized into fibrils with specific orientation within the various layers of the cell wall. A thin inner portion of the secondary wall (S_3) is made up of microfibrils with an oblique angular direction in relation to the long axis of the cell. The middle portion of the secondary wall (S_2), where most of the structure and mass of the cell wall is found, is made up of microfibrils with an orientation very much along the long axis of the cell. The outer layer of the secondary wall (S_1) has microfibrils oriented at almost right angles to the main axis of the cell.

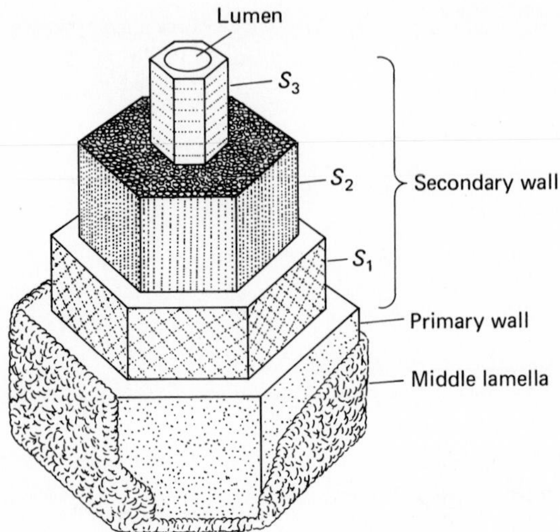

FIGURE 14-4
Woody-cell-wall structure.

The secondary cell wall is surrounded on the outside by a primary wall (P), composed of hemicellulose and pectin. A small fraction of the primary wall is made up of protein. The protein is bound within the wall and is not readily available to fungi, which find the wood substrate rather deficient in nitrogen.

The inner portion of the secondary wall is covered by a warty layer composed of debris left over when the cell died, as well as deposition products like phenols impregnated into the cells during the transition from sapwood to heartwood.

The middle lamella, composed primarily of lignin and pectin, cements the cells together into a woody structural material. During chemical pulping of wood, the middle lamella is chemically degraded, thereby releasing the fibers, which are then regrouped and pressed together into paper.

Fiber and tracheid cells are formed from lateral meristem cambium cells. An elastic primary cell wall is formed first. The wall stretches to allow the cell to expand. Secondary-cell-wall material is synthesized within the cell and laid down to give the cell rigidity. Once the cell walls are complete, the protoplasm of the living cell disintegrates to form a warty layer on the inner wall.

The sapwood or outer portion of the stem is composed of (a) dead fibers and vessels in hardwoods, or dead tracheids in conifers; and (b) living parenchyma and ray cells. In some species, such as oak or walnut, the parenchyma and ray cells eventually die, so that the central portion of the stem is entirely composed of dead cells. During the process of dying, the ray and parenchyma cells synthesize the phenolic compounds that make up the distinctive color, odor, and decay resistance of the heartwood. Other species, such as aspen, maple, and birch, do not form true heartwood. A few living parenchyma cells can be found well within the central xylem. Any color changes in these species are due to the activities of microorganisms.

Enzymes Produced by Wood Decay Fungi

The woody material just described is remarkably resistant to degradation by all but a very few microorganisms. The capacity to degrade and utilize woody material for energy and fungus synthesis is limited to those fungi that produce extracellular cellulase enzymes which are capable of permeating the lignin/cellulose complex of the cell wall. Many fungi, representing all classes of fungi, produce cellulases that can degrade materials such as cotton fibers, which are composed of cellulose not impregnated with lignin. Therefore, the unique characteristic of the Basidiomycete wood decay fungi and a few Ascomycete wood decay fungi is the ability to degrade cellulose in the presence of lignin.

Cellulase enzymes are of two principal types. Endocellulase hydrolytically splits the long cellulose chain randomly, thereby producing many short cellulose chain fragments of varying length. Another type, exocellulase, systematically degrades cellulose by splitting off two glucose units at a time from one end of the chain. The resulting two-glucose-unit disaccharide (cellobiose) is water-soluble. Cellobiose is taken up by the fungus by transport through the fungus cell wall, and plasma membrane and is utilized within the fungus hyphae.

Another group of enzymes produced by some wood decay fungi are lignase oxidative enzymes capable of splitting up the lignin polymer and oxidizing the phenyl propane molecule. None of the lignin degrading enzymes have been isolated and characterized, but their mode of action can be postulated based on the chemical analysis of partially decayed wood. Oxygenases are probably involved in demethylation and ring cleavage. Dehydrogenases are probably involved in the oxidation of alcohol groups.

The enzymatic degradation of the remaining cell components—hemicellulose, pectins, and starch—have received very little attention by people interested in wood decay. Hemicellulases are thought to play a role in the preference of white rot fungi for hardwoods and brown rot fungi for conifers. It is contended that hemicellulose is removed before the crystalline cellulose regions of the microfibril are accessible to the cellulase enzymes.

Wood Decay Model

Free moisture is necessary for decay. Recall from the discussion of the requirements for growth of fungi in Chapter 8 that free moisture is necessary for most fungi. Fungi decay wood only if the moisture content is above fiber saturation.

If one allows a dry piece of wood to absorb moisture in an atmosphere of 100% relative humidity, the first moisture that enters the wood will be absorbed and bind, by hydrogen bonds, to openings in the amorphous regions of the cellulose fibrils. Once all the potential bonds are fulfilled, the wood is said to be at fiber saturation. The actual percentage of water at this point is variable, depending upon the density of the wood, but an average figure for fiber saturation is around 28% moisture content. Theoretically, wood maintained just at fiber saturation does not have any free water within the lumens of the cells, but slight temperature fluctuations can cause moisture to condense and form a film of water in the lumen of the cell, allowing decay to commence.

The wood of living trees, containing in excess of 100% moisture based on oven-dry weight, can be decayed by fungi. Wood maintained in water will accumulate water until all of the cell lumens are filled with water. At or near this total saturation point, the white and brown rot decay fungi do not operate very well because of lack of oxygen. Therefore, for rapid decay to take place, the wood must be above fiber saturation but not totally saturated.

Soft rot fungi can degrade wood saturated with water. Soft rot fungi are often from the Ascomycete or Fungi Imperfecti class. These fungi degrade the surface layers of submerged or continually wetted wood of greenhouse benches, boats, dock posts, and other structures. Only the surface layers of the wood are degraded, because of lack of oxygen in the interior of the wood. Oxygen dissolved in the water supplies the fungus with its oxygen requirements. Microscopic cavities in the secondary wall of conifer woods are produced by soft rot fungi. On hardwoods, the soft rot produces major erosions or trenches in the secondary wall.

Cellulase and lignin oxidative enzymes released from the growing fungus hyphae are diffused to the woody cell-wall components in the free water within the lumen of the cell. Cellobiose and other degradation products diffuse back to the fungus and are taken up and utilized within the fungus hyphae.

At the chemical level, the postulated wood decay model involves a number of enzymes (Fig. 14-5). An X_1 enzyme is postulated to break the ligno-cellulose bonds. Another group of enzymes, the hemicellulases, degrade the hemicellulose matrix surrounding the crystalline cellulose region of the microfibril.

FIGURE 14-5 Organization of cellulose, hemicellulose, and lignin into wood and the enzymes involved in its decomposition.

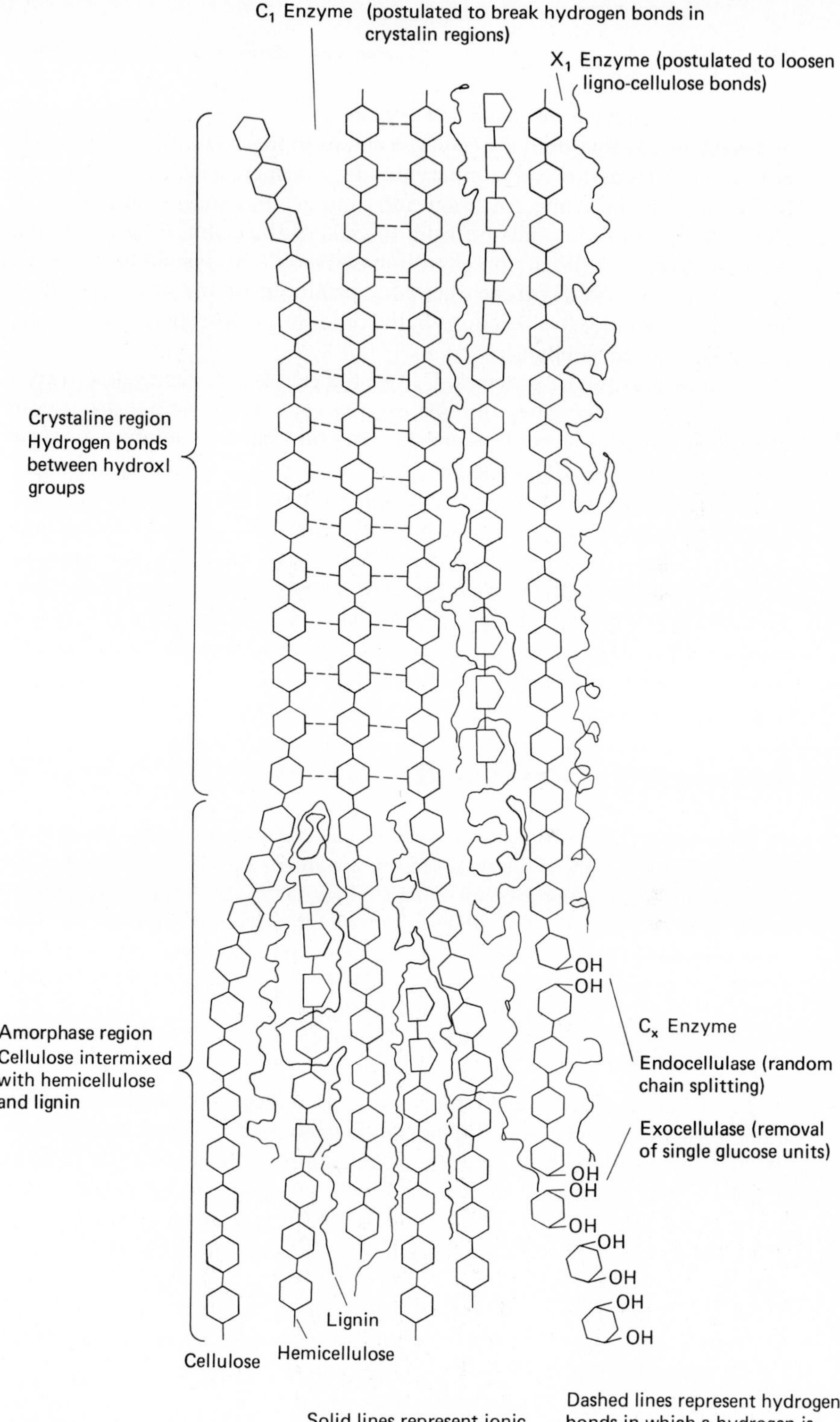

Solid lines represent ionic C—C and C—O linkages

Dashed lines represent hydrogen bonds in which a hydrogen is shared between two oxygens.

Another enzyme or group of enzymes, C_1, are postulated to break the hydrogen bonds that bind the cellulose chains in the crystalline regions. The crystalline regions are very impervious to enzymes. The C_1 enzymes are found only in the white rot fungi and may act as exocellulases, splitting glucose and cellobiose units off from the end of the chain. Brown rot fungi are postulated to utilize a non-enzyme-based oxidation system to loosen the crystalline cellulose. Hydrogen peroxide produced by the brown rot fungi and iron cause oxidative reactions in the cellulose chains, thereby enhancing the activity of endocellulases.

A fourth group of enzymes, C_X, are the cellulase enzymes described in the previous section. They may be exo- or endocellulases but are usually considered to be endocellulases which randomly split the cellulose molecule into various shorter fragments.

The fifth group of enzymes, lignin-oxidizing enzymes, are also postulated enzymes. They have not been identified, but analysis of partially decayed material indicates the probable activity of extracellular oxygenases in demethylation and ring cleavage and dehydrogenases in alcohol oxidation. White and brown rot fungi both have lignin-degrading capacity, but the brown rot fungi appear to lack the ring-cleavage enzymes necessary for complete degradation of lignin to CO_2 and H_2O.

WOOD DECAY "DISEASE" CYCLE

It is somewhat inappropriate to refer to all decays as disease, because we generally expect that, to be diseased, an organism should be living. Sap rot and product decay are problems of nonliving tree products. Heart rot, on the other hand, is a problem of living trees involving a dynamic interaction of the host and the pathogen. Therefore, in some instances we use the qualification "disease" when describing a disease cycle.

Heart Rot Disease Cycle

The disease cycle of a heart rot fungus (Fig. 14-6) begins when a basidiospore randomly disseminated by the wind comes in contact with a wound in a tree. If the temperature is right and the wound supplies the proper condition of moisture, nutrients, and absence of inhibitors, the spore germinates. The spore germinates to form a germ tube which further expands into a hypha (Fig. 14-7). The hypha releases enzymes for the enzymatic digestion of cell-wall material. The hypha branches and grows in the lumens of fiber and vessel cells. Movement from one cell to another is generally through pits in cell walls but, as decay progresses, bore holes are formed by enzymatic digestion of the cell wall (Fig. 14-8).

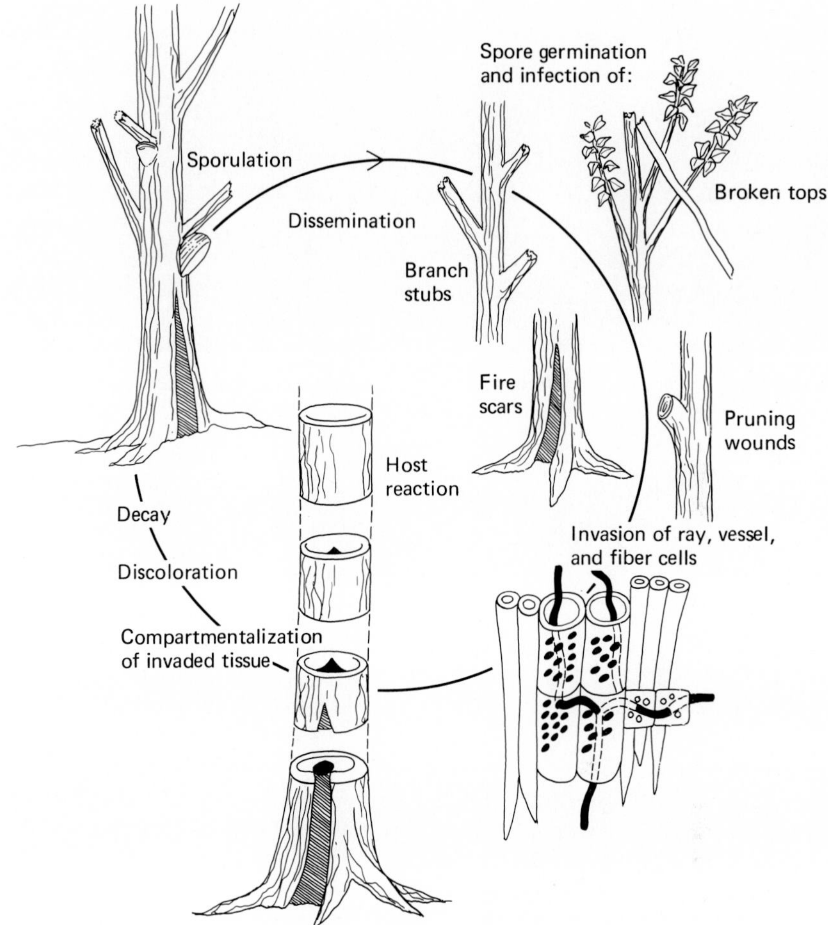

FIGURE 14-6 Heart rot disease cycle.

Trees respond to wounding and invasion. A tree is made up of a number of subcompartments. The most obvious is the new cone laid down each year over the last in the form of an annual ring. Vital functions of trees, in temperate climates, take place in the recent cone. The older cones are compartmentalized from the new cone by a series of small-diameter, thick-walled, highly lignified cells produced during the summer.

The stem is further compartmentalized loosely into wedge-shaped segments by rays, which provide living cell communication between the older inner xylem and the recent outer xylem and phloem.

The rays form a screen of small groups of living cells capable of a short-term response to invasion. Because they maintain contact with the living

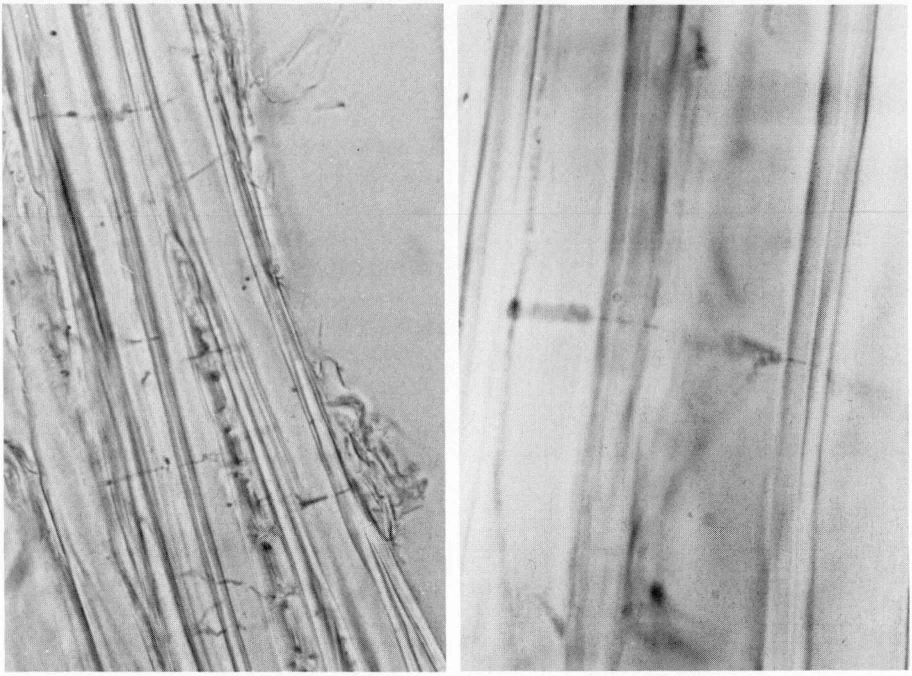

(a)

(b)

FIGURE 14-7 (a) Photomicrograph of *Phellinus tremulae* hyphae in aspen fibers. (b) Note the reduction in the size of the hyphae as it penetrates the cell wall in this enlargement.

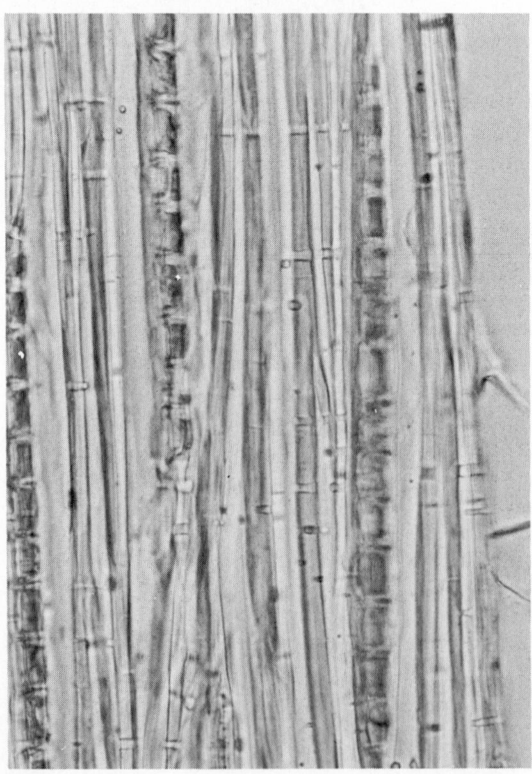

FIGURE 14-8 Photomicrograph of bore holes in aspen fibers caused by *Phellinus tremulae* decay.

cambium of a vigorous tree, the rays may be continually resupplied with nutrients for extended response and containment of invasion by the decay fungus.

The upper and lower ends of the wood response compartment are formed by tyloses and extractives plugging vessels and tracheids. The tyloses of hardwoods are discussed in Chapter 13. Tyloses are also produced in vessels in response to wounding and invasion by decay fungi. Conifers and hardwoods utilize a massive accumulation of extractives to fill the lumens of tracheid cells. Resinous accumulation of extractives are readily seen in the dead branches of both hardwoods and conifers.

The outer surface of the response compartment is the strongest side of the defense-response mechanism. Living cells of the cambium produce a wall of defense which generally prevents expansion of the invading fungus into wood laid down after the initial wound. But some fungi interact parasitically with the cambium wound response to produce a canker which overcomes the outer response wall.

Wounding a tree induces a series of oxidative reactions. Further chemical reactions and morphological responses follow to compartmentalize the invading microorganisms and heal over the break in structural integrity of the tree stem or branch (Fig. 14-9). Tyloses form in vessels above and below the wound to seal the upper and lower ends of the compartment. Parenchyma cells of the rays produce toxic phenolic compounds to impregnate cells around the sides of the wound. The summer wood cells of each annual ring slow down the penetration of fungi into the interior of the stem. The cambium on the margin of the wound produces callus cells to seal off the injury, so that it will be separated from wood produced by the tree in future years.

Callus cells are the best defense wall produced by the tree. Evidence from artificial inoculations of wounds shows that discoloration and decay fungi do not progress into wood laid down after the wound was healed over. The upper and lower ends of the compartment are eventually penetrated by decay fungi as the vertical extent of discoloration and decay progresses up to 30 cm per year. Chemical barriers on the side are very slowly broken down or may become future zone lines, marking the limits of decay activity. The rings of summer wood cells provide a passive barrier which is eventually penetrated. Because the number of parenchyma cells able to respond diminishes with depth into the tree, the interior side of the wound is poorly defended by the tree. Discoloration and decay organisms may break down this poorly formed chemical barrier to utilize the central portion of the tree. The center or heart of the tree is least well defended, hence the name heart rot for decay of this portion of the tree.

If the tree is successful in its compartmentalization, the wound is quickly closed and the modification within the compartment is restricted to chemical changes.

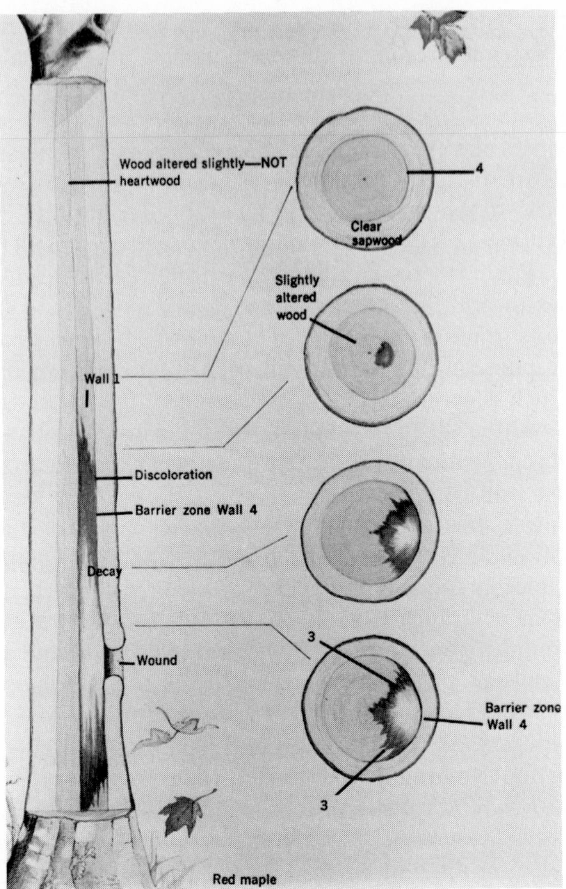

FIGURE 14-9 Compartmentalization of a decay column following a wound in red maple. (From a drawing by Dave Carrol, compliments of Dr. Alex Shigo, U.S. Forest Service.)

If fungi become established on the open wound, they may be able to expand the affected compartment to additional compartments by rapid growth of the mycelium. The chemical responses of the tree are specifically intended to inhibit the growth of the invading fungi, so that a tree in good vigor may be able to produce enough chemical response soon enough to stop the invading fungi. The compartment with entrapped fungi is then walled off from the rest of the tree.

Although the tree compartmentalizes the decay fungus away from the most functionally active and structurally important outer portions of the stem, additional wounds may break down the outer compartment callus and chemical walls to initiate a new series of reactions, leading to a new compartment outside the first. A long branch stub that cannot be totally callused over is an ideal point of invasion by decay fungi. The wound is continually opened and reopened for expansion of decay outward beyond original compartment boundaries.

Some fungi are specifically adapted to tolerate the chemical-response compounds of the tree. These fungi would represent the limited number of heart rot decay fungi, at least one per tree species. These fungi utilize the chemically modified ecological niche to their advantage. By tolerating the chemicals produced by the tree, they do not have to compete with other microorganisms which cannot tolerate the toxic chemicals.

The infecting basidiospore has a single genome ($1n$), so it is necessary in the life cycle of some of the decay fungi for a second basidiospore to infect the same substrate. The second spore initiates an infection colony. The two genomes are brought together by anastamosis of the hyphae of the two colonies.

The dikaryotic mycelium ($n + n$) continues to develop within the stem until some triggering mechanism, which is still not known, causes the fungus to move out of the stem along paths of least resistance to fruit. Dead branch stubs commonly provide ideal paths for the fungus to progress out to the surface to fruit, as well as expansion beyond the original compartment as just described. Fruit bodies are produced under or on the broken dead branches. Nutrients generated by the enzymatic activity of the hyphae within the wood supply the fruiting body with large amounts of energy, nutrients, and water necessary for sporulation.

Recall the nuclear fusion necessary to form a diploid nucleus and meiotic division within the basidium, discussed in Chapter 8. This stage completes the disease cycle.

It is interesting to consider the number of spores that are produced by a fruit body. I was amazed to find that a *Phellinus tremulae* (*Fomes igniarius* var. *populinus*) fruit body could produce in excess of 100,000 spores per square millimeter of pore surface (Fig. 14-10). If you recall this example from Chapter 8, you will remember that this number is a daily estimate of the spore-producing capabilities of the fungus.

If one measures the amount of nutrients the fungus must derive from wood for synthesis of the vast number of spores, a serious inconsistency is evident. The digestion of wood does not supply enough nutrients. The total nitrogen in the spores produced is in excess of the total nitrogen available in the decay column area. We do not know where the fungus derives the nitrogen it needs. Some investigators are developing a hypothesis of nitrogen fixation by bacteria in association with fungus hyphae in wood or in the fungus fruit body. Others would speculate that the fungus actually parasitizes living cells and transport cells around the continually festering branch stub wound.

FIGURE 14-10 Note the white spore casts on the slides. The naturally occurring *Phellinus tremulae* fruit body inside the plastic box produced 100,000 spores per square millimeter of slide surface in less than 12 hours.

Concepts of Infection and Invasion by Heart Rot Fungi

The science of forest pathology began in 1874 when Robert Hartig described the association of fungus hyphae decaying wood with the fruit bodies found on the outsides of trees. Hartig observed the causal relationship between the Basidiomycete fungus fruiting on the surface and the decay of wood. The observations expanded into the classic concepts of decay. Infection of wounds by basidiospores eventually results in decay. Wounds, dead branches, broken tops, and fire scars are commonly seen as points where heart rot decay is continuous with the outside of the stem, so they are obviously points of infection. The concept was expanded by others to assume that for infection to take place, the wound must expose heartwood.

In 1938, Haddow published a comprehensive study of heart rot in eastern white pine caused by *Phellinus pini* (*Fomes pini*). His observations indicated that the fungus infected small twigs as small as a few millimeters in diameter. Dead twigs of very young trees were associated with infection. White pine weevil-killed leaders were also points of infection. His work emphasized the infection by heart rot fungi of very small branches. The Haddow concept of infection was later further developed by Etheridge, who worked with the Indian paint fungus, *Echinodontium tinctorium,* on western firs. He found numerous points of infection associated with small dead branches. Etheridge claims that after initial invasion of the small twig, the fungus becomes dormant until some point in the future when the tree is stressed or injured. Wounds made during logging do not represent points of infection but rather points where host response to the wound causes activation and invasion of already existing infections.

A third concept of decay infection and invasion, developed and popularized recently, is the Shigo concept of succession leading to decay. The concept involves initally a host response to wounding, next invasion of the modified wood by pioneer organisms such as bacteria and nonhymenomycete fungi, and finally the infection and invasion by the hymenomycete fungi of the wood that has been modified by the pioneer organisms. Nonhymenomycete fungi are primarily Fungi Imperfecti. Hymenomycete fungi are the Basidiomycete decay fungi usually given credit for the decay. The central feature of the Shigo concept is that wood decay fungi are secondary organisms that follow pioneer organisms that modify the wood substrate. The pioneer organisms are supposedly responsible for detoxifying host-response phenols and modifying the pH of the wood.

The reader may think that the whole subject of infection and invasion must be thoroughly confusing to specialists, since we have three distinct concepts of the process. Actually, each concept is probably correct for some types of decay and may therefore be useful in categorizing types of decay in trees.

Before we accept any one or all three of these hypotheses, it may be informative to look at the evidence for each. The classic Hartig concept is founded on the observations of continuity of the decayed central column of the tree with the fruit body emerging from the stem at the base of decayed branches or from wounds. The validity of pathogenic relationship is verified by inoculation of trees with mycelium of pure cultures of decay fungi and the development of decay. Re-isolation from the decay-induced tissue completes Koch's normally accepted proof of pathogenicity.

The Haddow-Etheridge modification of the basic concept is in the area of infection courts. Culturing from large numbers of locations within the stem, branches, and twigs of firs recovered *Echinodontium tinctorium* from the obvious decayed central portion of the tree but also from small twigs and branches that were not obviously decayed. Also, a determination of the nuclear condition of the various isolates indicated that many of those compartmentalized in the twigs were haploid cultures. Presumably, they were recent single basidiospore infections that had not developed sufficiently to make contact and anastomose with another infection, resulting from a second spore.

A limited number of studies on basidiospore germination in the infection process of decay fungi have shown that basidiospores will germinate and infect wounds made in trees. Studies with *Chondrostereum purpureum* (*Stereum purpureum*), *Heterobasidion annosum* (*Fomes annosus*), and *Phellinus tremulae* (*Fomes igniarius* var. *populinus*) show that the spores will germinate and infect freshly made wounds. No one has demonstrated that the basidiospores of heart rot fungi will germinate on wood previously invaded by pioneer organisms.

It would be reasonable to assume that some fungi utilize stem or branch wounds as infection courts and that others initiate infection through the death of small twigs.

The Shigo succession concept is based primarily on evidence that fungi are compartmentalized during the invasion process. The concept further develops from many unsuccessful attempts to produce decay following inoculation with decay fungi. The inoculation wounds of many trees are invaded by a specific group of nondecay fungi. Note the difference between the terms "infection" and "invasion," because part of the difference between the Shigo and Haddow-Etheridge concepts relates to the difference between these two terms. Compartmentalization appears to be so general with invasion columns that it is assumed that compartmentalization is also the dominant factor influencing infection.

Isolations from large numbers of decay columns in northern hardwoods and mapping of locations where specific types of organisms were recovered led to the compartmentalization concept. On the margins of the decay and discolored zones, bacteria and Fungi Imperfecti were recovered. In the decayed wood behind the discoloration, other decay fungi were recovered. In the sound tissue outside the discolored wood, microorganisms were seldom recovered.

Isolations from large numbers of discoloration columns, resulting from inoculations with decay fungi, seldom recovered the decay fungus but commonly recovered *Phialophora* and *Trichocladium,* the same Fungi Imperfecti recovered from margins of decay columns of naturally infected stems. The lack of success in establishment of the decay fungus is assumed to be because the decay fungi cannot readily become established in freshly wounded wood. The wound response and pioneer invaders require a certain amount of time to prepare the substrate for invasion by the decay fungus. Others have inoculated trees with heart rot fungi and recovered the same fungi later, therefore showing the lack of need for the wound response and pioneer organisms.

An additional bit of evidence for prior modification of the substrate for development of decay fungi was provided by laboratory studies using blocks of wood from sound, discolored, and decayed tissues. Blocks of discolored and decayed tissue lost more weight when exposed to decay by specific decay fungi than did the sound block. The conclusion that decay progresses more rapidly in the modified wood substrate suggests a progressive succession of microorganisms in decay.

A note of caution in interpretation of isolation results is necessary to place some of the foregoing evidence in perspective. If an analogy can be made between the microorganism associations in wood and the macroorganism associations in a circus train, the roles of the specific organisms may be clearer. In the circus train, a large number of animals are compartmentalized into small cages. If one were to look at total numbers of organisms in portions of the train, it would appear that the lions, tigers, and other wild animals dominate but these animals are kept in their places by just a few dominant human beings. The real controllers are the human beings, even though they may be outnumbered by the caged animals. How would some nonbiased entity envision the dominant organism and the roles of the specific organisms if he or she were to come upon the circus train after a major derailment in the middle of the night? The nonbiased entity would probably easily recognize that the lions and tigers are now in control of the organisms that make up the train.

The placing of a chip of wood on agar media is like breaking up the circus train. Fungi (Fungi Imperfecti) that are aggressive and able to utilize the abundant sugar provided by the media rapidly overgrow decay fungi that have more specialized enzymes for slow decomposition of relatively inert substrates such as cellulose and lignin. Is the *Phialophora* a lion or a lion tamer in the original wood? We have no way of knowing at the present time, so that we (as nonbiased entities) must exercise care in assigning importance to one organism rather than another.

It is my suggestion that, with our present state of understanding of the infection and invasion phenomenon, we accept the possibility that any one of the three concepts may be correct, at least with regard to certain decay phenomena. Additional sophistication of our techniques and further studies of many more decay processes are necessary before we can make judgments regarding which concept should take the dominant place in introductory forest pathology teaching.

Why spend so much time on the concepts of infection and invasion? Because we do not understand these processes, we are very limited in our recommendations for prevention and control of losses due to heart rot. Avoiding wounding or treating wounds, for example, may or may not really make any difference. At any rate, it is sometimes impossible to avoid infection. Every tree is wounded many times. Why do some have heart rot at an early age and others seem to be free of decay? Once we can answer these questions, we will be in a much better position to make control recommendations.

Sap Rot and Products Decay "Disease" Cycle

More is known about the decay process in sap rot and products decay systems because it is a simpler system, in which the wood substrate is not responding as in the case of heart rot. Infection takes place when basidiospores come in contact with substrate, provided that there is sufficient moisture (above fiber saturation), an absence of spore germination inhibitors, moderate temperatures, and oxygen. The rate of invasion and enzymatic digestion of wood is dependent upon these same variables. Sporulation may require previous dikaryotization with a second compatible strain of the fungus as with some of the heart rot fungi. Spore dispersal by air currents completes the cycle.

It is interesting to note that, if appropriate moisture, temperature, lack of inhibitors, and oxygen requirements are met, decay will take place. The dispersal of spores must be very efficient, as spores are almost universally present.

SYMPTOMS AND EFFECTS OF DECAY

Weight Loss

Extracellular cellulase and lignin-degrading enzymes of the fungus break up the woody cell wall into fragments which can be further degraded inside the fungus to CO_2 and H_2O. This metabolism of wall material results in a weight loss to the wood.

Table 14-1 presents the changes in cell-wall chemical composition resulting from decay by a white rot fungus, *Coriolus versicolor (Polyporus versicolor)*, and a brown rot fungus, *Poria placenta (Poria monticola)*. Note that in the case of the white rot, when the test specimen has lost about 50% of its original weight, the lignin as well as the cellulose is reduced by about 50%. With brown rot, the lignin is almost unchanged from the original.

TABLE 14-1 Composition of Sweetgum Sapwood in Progressive Stages of Decay, Based on the Moisture-free Weight of the Original Sound Wood (condensed from Cowling, 1961)

Type of decay-causal organism	Average weight loss %	Percent of original cell wall		
		Cellulose	Hemicellulose	Lignin
White rot,				
Coriolus versicolor	0[a]	52	25	23
	25	40	19	17
	55	23	11	11
Brown rot,				
Poria placenta	0[a]	52	25	23
	20	40	17	23
	45	22	10	23

[a]Sound wood.

Strength Loss

The cellulase enzymes of the brown and white rot fungi operate somewhat differently, as can be seen by measuring the degree of polymerization of the cellulose at various periods of exposure to white and brown rot fungi (Fig. 14-11). The brown rot fungi utilize endocellulases to rapidly split the cellulose chains, thereby producing many short fragments. The white rot fungi split a limited number of cellulose chains and degrade the split chains by exocellulases much more thoroughly.

The splitting of many of the long chains of cellulose by brown rot fungi produce very dramatic strength losses within a very short period of exposure. White rot fungi take much longer to degrade enough structural elements to produce comparable strength losses. Decay fungi cause strength losses which may cause a living tree to break or a branch to break and fall. In buildings and wood products, strength losses may result in breakage of telephone poles, failure of bridges, collapse of rafter or floor joists, and many other structural failures.

FIGURE 14-11 Changes in the degree of polymerization of holocellulose from sweetgum sapwood with progressive stages of decay. (From Cowling, 1961.)

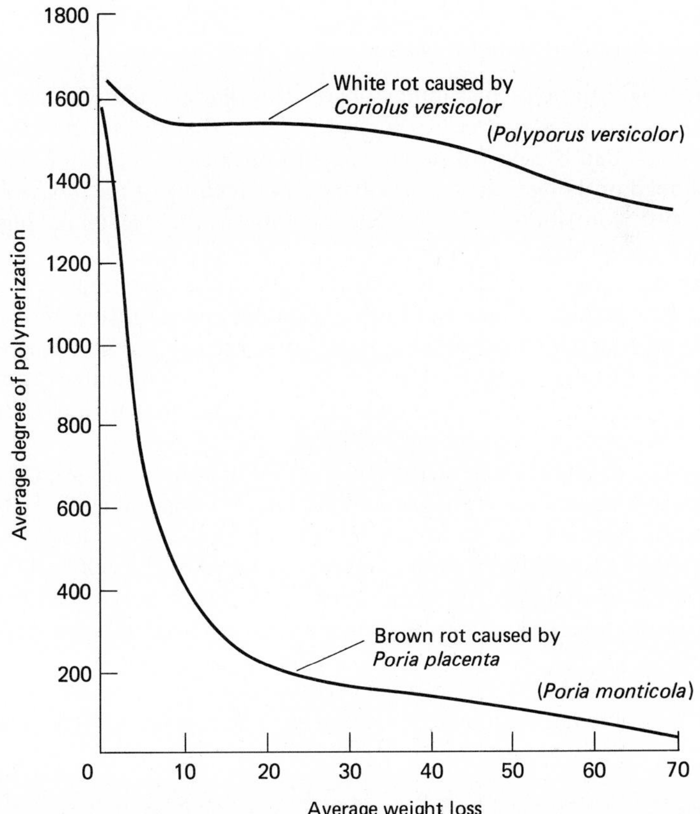

Reduced Standing Timber Volume

Heart rot fungi produce serious defect and volume reductions while a tree is still growing. If one is measuring volume increment added to trees yearly by growth, one should subtract the increasing increment of defect caused by heart rot fungi to get a realistic estimate of growth. There is a point in the life of trees when the amount of increment produced by growth is exactly equal to the amount of wood destroyed by heart rot fungi. This point is a theoretical age of maximum volume of usable wood and is referred to as the pathological rotation age. For obvious reasons, the forest manager does not allow timber to reach pathological rotation, and realistically would harvest the timber long before this point was reached. Intensive forest management involving short rotations obviously does not need to be concerned with pathological rotation, because harvesting occurs well before decay becomes a major problem. But trees that are not harvested for timber, but left for recreational or aesthetic reasons, will go past pathological rotation. At or near the pathological rotation age, the trees are becoming more and more weakened, as decay consumes more wood than is being laid down by growth. The manager of such forests must recognize the increasing hazard resulting from the weakening of the trees growing beyond pathological rotation.

Sap Rot Cleanup of Wood Residues

Sap rot fungi can be used as indicators of trees that are in the process of dying. Weakened trees or trees with large wounds that do not readily callus over are invaded by sap rot fungi. They do not progress much beyond the area of dead or dying tissues, but will cause weakening of the invaded tissue and possibly contribute to the overall weakening of the tree, climaxing in the breakage of the stem or a branch.

Sap rot fungi are important contributors to the decomposition of logging slash. Cleanup activities of this type should be recognized and encouraged. If logging debris is broken up and limbs are cut and scattered so that the debris is close to the ground, the environment is ideally suited for decomposition by sap rot fungi. Large stems, and limbs that stick up into the air are more slowly decomposed by sap rot fungi.

Sap rot fungi, in their eagerness to clean up wood residues, become a problem to foresters attempting to salvage fire- or insect-killed trees. A large amount of research has been conducted and is presently under way to evaluate the rate of deterioration of balsam fir killed by the defoliation of the spruce bud worm. Sap rot fungi such as *Stereum sanquinolentum* and *Hirschioporus abietinus (Polyporus abietinus)* are serious decomposers of the killed fir.

Sap rot fungi also exceed their usefulness by invading logs of timber stored for extended periods of time and by invading wood products if the moisture content is maintained above fiber saturation. The sap rot fungi of wood products will be discussed later in the chapter.

Increased Permeability of Wood

Wood decay fungi increase the permeability of the wood to water and other liquids. Increased permeability to water enhances the wettability of wood used in construction, thereby enhancing the potential development of future decay. Increased permeability increases the amount of paint needed if the wood is used in siding and house trim.

Increased Electrical Conductivity of Wood

Increased electrical conductivity is associated with wood decay. Decay appears to cause accumulation of ions in the discolored wood around the decaying region. It is based upon this increased electrical conductivity that an electrical measuring device, a Shigometer, is able to detect decay and discoloration in living trees (Fig. 14-12). The Shigometer measures resistance to a pulsed electric current. A twisted two-wire insulated probe is inserted into a

FIGURE 14-12 Shigometer used to detect decay in trees by measuring the resistance to a pulsed electric current.

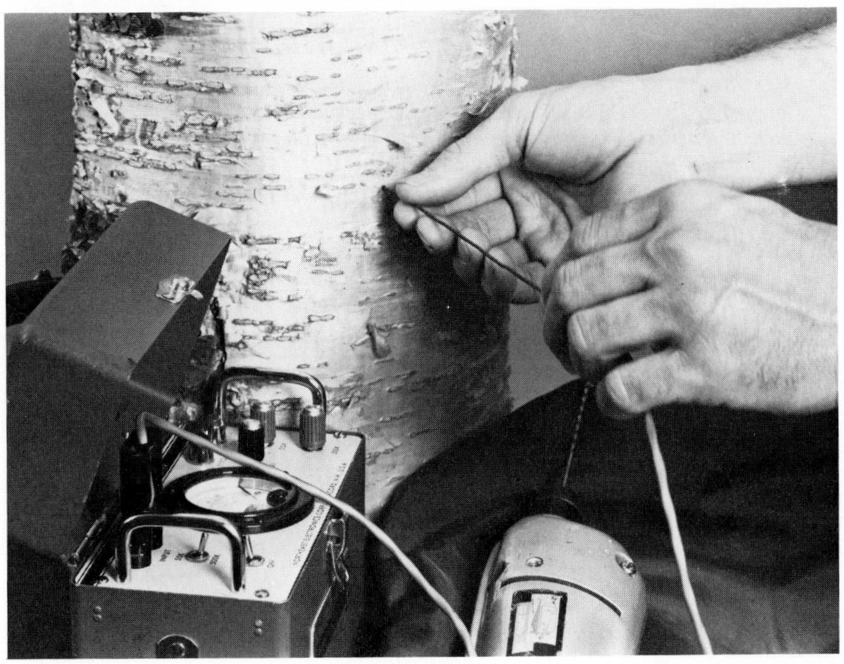

3/32-inch hole drilled into the tree. The tips of the wires are uninsulated and therefore one conducts a pulsed current into the tree while the other conducts the return flow of current back to a portable voltmeter for the measurement of resistance.

The probe is slowly inserted into the hole while the meter is being monitored. A rapid decrease in electrical resistance followed by an increase occurs whenever the probe contacts move from an area of sound tissue into a discolored zone and then into a decay column.

Changes in Wood Volume

Wood decay causes changes in the volume of wood, particularly with brown rot fungi (Fig. 14-13). Irregular volume changes upon drying result in the warping and bending of wood.

Changes in Pulping Quality

The pulping quality of wood is affected most by brown rot fungi. The increased lignin residue and rapid cellulose depolymerization of brown rot fungi make pulping of brown rotted wood very impractical. Wood degraded by white rot fungi is not seriously objectionable for pulping purposes unless the wood is bought on a volume basis. A given volume of wood decayed by white rot fungi has less weight and will yield less pulp than the same volume

FIGURE 14-13 Note changes in wood volume with decay in these experimental test blocks. The upper row of blocks decayed by a brown rotter were initially the same size as the lower row of blocks decayed by a white rotter. (Photograph compliments of Dr. Robert Zabel.)

of sound wood. If wood is bought on a weight basis, there is very little difference between the weight and quality of pulp produced from sound or from white-rot-decayed wood.

Discoloration of Wood

An objectionable feature of decayed wood is the discoloration associated with decay. Wood that is to be used for furniture or decorative purposes is generally of higher value if discolorations and decay are absent. The term "generally" must be included because in some instances defects due to decay are highly prized. One example is the use of pecky cypress for paneling. The brown rot fungus causes cavities in cypress that are of decorative value. Another example of the use of decayed wood in preference to sound is the use of wood decayed by *Phellinus pini* (*Fomes pini*). This white rot fungus produces white pockets in the wood which gives a pleasing appearance to paneling. To sell this type of decayed wood, it is important not to call it rot. *Phellinus pini*-decayed wood sells under trade names like "driftwood" or other more emotionally appealing titles.

Reduced Caloric Value

The last property of wood affected by decay that will be mentioned is the caloric value. Anyone who has cut or purchased firewood recognizes the desirability of sound wood over decayed. A cord of sound wood will produce more heat than a cord of decayed wood. Actually, if the wood were bought on a weight basis rather than by volume, there would not be much difference between the caloric value of sound wood and that of decayed wood. Compared to its other uses, wood used for heating purposes is not very significant at present, but with the increasing cost of fossil fuels, one can safely assume that the potential of wood as a renewable fuel resource will become much more important.

METHODS FOR RECOGNITION OF DECAY

Symptoms of decay were discussed in the preceding section. We expand on these in this section and categorize them as visual macroscopic indicators, microscopic indicators, cultural isolation, and physical test indicators used for recognition of decay in trees, logs, lumber, and wooden structures.

The ability to recognize symptoms of decay in trees, logs, lumber, and wooden structures can be an important means of avoiding economic losses. The buyer of timber must be able to recognize decay in trees, because the amount of decay will affect the value of the logs. The park manager or city forester must be able to recognize decay, to prevent damage to life and property as a result of breakage of defective trees. The log buyer at the sawmill or pulp mill must recognize decay, because of the effects on yield and quality of the products produced from the logs. The lumber dealer and building contractor must recognize decay, because decayed lumber will have less structural strength or require excessive amounts of paint to finish. The homeowner should be able to recognize decay when buying a house and in maintaining it, to avoid expensive replacement costs of defective parts of the structure.

Usually, decay can be detected by visual macroscopic symptoms but under special circumstances it may be necessary to verify the macroscopic characters with microscopic characters, cultural isolation and identification of the causal agent, and physical tests of the wood. Wood used in aircraft or in locations where freedom from defects is very important are examples of situations in which all of the characteristics used to detect decay might be used.

Macroscopic Characteristics of Decay

Decay of trees Decay in trees is associated with heart rot fungi, which decompose both the heartwood and sapwood portions of the living stem, and sap rot fungi, which decompose dead branches and dead portions of the stem.

The fruiting bodies of decay fungi, referred to as conks by field foresters, are positive indicators of advanced decay of both sap rot and heart rot fungi. If the conk can be identified as to species, it is possible to predict approximately how much of the tree is defective as a result of the decay (Fig. 14-14). For that reason, it is desirable for the forester to be able to recognize the fruit bodies of the common species of decay fungi present in a region.

The specific identification of decay fungi can sometimes be confusing to a beginner. You will note that two names have been used for most of the decay fungi mentioned in this book. The first name is the most currently accepted name for the fungus and the name in parentheses is the name utilized in much of the older literature. Unfortunately, it is necessary to be familiar with both.

Swollen punk knots and resinosis are indications of decay (Fig. 14-15). Decay fungi do not always fruit on the surface of every stem being decayed. During the early stages of invasion, fruiting is uncommon and certain decay fungi, such as *Phellinus pini* (*Fomes pini*) on eastern white pine, seldom fruit except on windthrown trees or logging slash. The presence of swollen knots around former branch stubs and the carbonaceous nature of the tissue behind the swelling are good symptoms. These knots do not heal over and often resin will exude from the wound.

FIGURE 14-14 Conk or fruit body of *Phellinus igniarius* on beech indicates a 10- to 14-ft (3 to 4.2 m) decay column. (Photograph compliments of Dr. S. B. Silverborg.)

FIGURE 14-15 Punk knot and resinosis on white pine, indicating internal discoloration and decay.

FIGURE 14-16 Decayed branch stub on beech, indicating internal discoloration and decay.

FIGURE 14-17 Basal scar indicator of internal discoloration and decay in sugar maple. (Photograph compliments of Dr. S. B. Silverborg.)

A decayed branch stub sticking out from a stem is a good indicator of some decay within the stem (Fig. 14-16). Branch stubs that are not points of infection for decay fungi are resin-impregnated and very hard.

Remnants of conks may be present. Many decay fungi produce annual fruit bodies that are consumed by insects and other fungi, so that only remnants of the conk remain. Careful examination of the stem and exposed roots is necessary to detect these remnants.

Basal fire scars are common points of infection by decay fungi and can always be suspected as indicators of decay. (Fig. 14-17). The amount of decay associated with the fire scar is directly proportional to the length of time since the fire.

Decay in logs Decay in logs may be from heart rot which developed in the living tree or may result from sap rot fungi becoming established in the logs during extended storage.

Heart rot in logs can be recognized by the same indicators as were used for the recognition of decay in trees.

Sap rot during extended storage can usually be recognized by the presence of fruit bodies of the decay fungus (Fig. 14-18). The fruit bodies are more numerous than those formed by heart rotters. They are formed at places other than branch stubs and old wounds, and the fruit body orientation shows that they were formed after harvest.

FIGURE 14-18 Sap rot decay of stored hardwood loss caused by *Coriolus versicolor.*

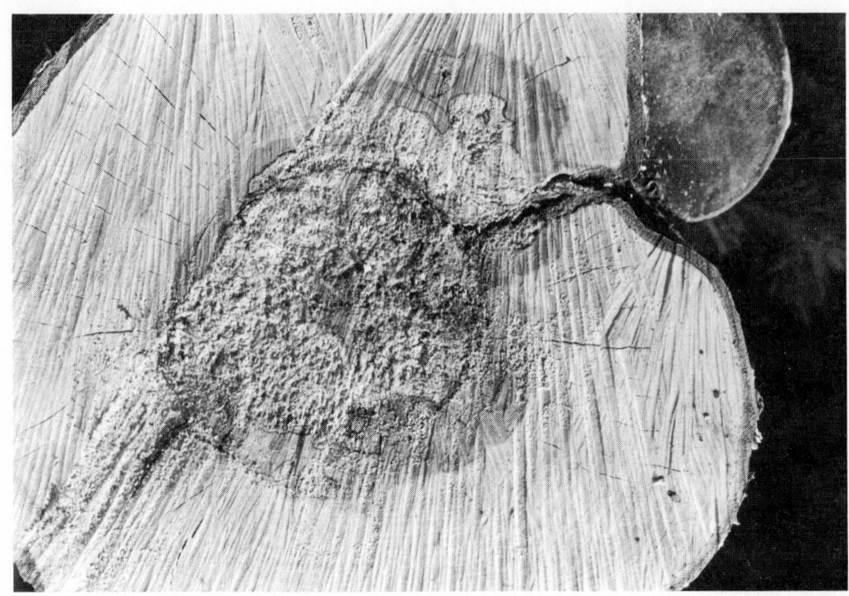

FIGURE 14-19 Indicators of decay in logs. Zone lines, discoloration, and roughened saw cut on the end of a log. The fruit body of *Phellinus igniarius* is also present on this birch log.

On the sawed ends, decay can be recognized by hollow cores, zone lines, discoloration patterns, and roughened saw cuts (Fig. 14-19).

Decay in lumber and primary wood products Decay of primary wood products may be the remnants of heart rot from the living tree (Fig. 14-20), later stages of sap rot infection of the stored logs, or beginning stages of infection by sap rot fungi of poorly seasoned lumber. Heart rot fungi do not survive long in lumber, so no further development of heart rot decay will occur in the lumber. Sap rot fungi can continue to decay the lumber as long as the moisture content is above fiber saturation. The moisture may result from moisture not dried out of the green lumber or may result from rain or groundwater being picked up because of improper piling.

Irregular color variations reduce the value of lumber. Lumber grading is based on the number and size of knots as well as color variations caused by decay and stain fungi (Fig. 14-21).

The zone lines that mark the border between decayed and sound wood are accumulations of thick walled pigmented hyphae and other pigmented materials of both the fungus and the wood and are excellent indicators of decay.

FIGURE 14-20 Defective board, showing white pocket rot caused in the living tree by the heart rot fungus *Phellinus pini.*

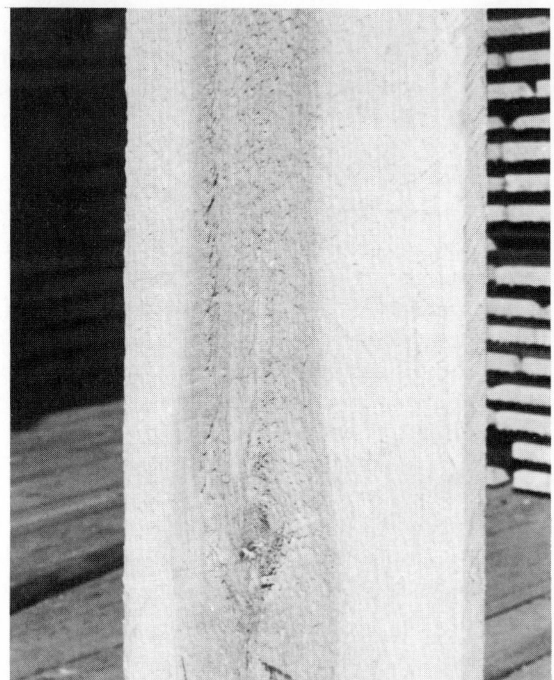

FIGURE 14-21 Red discoloration to this white pine board is associated with the early stages of *Phellinus pini* decay in the tree.

FIGURE 14-22 Mycelial fan which developed on this board. Moisture conditions may become favorable for decay fungi in tightly stacked lumber.

The punk knots caused by heart rot fungi and decayed branch stubs of living trees are observed as decayed knots in lumber.

Mycelial fans may occur (Fig. 14-22). Tightly stacked lumber does not dry out. The slight spaces between the boards are excellent places for decay fungi to produce sheets of mycelium called mycelial fans. The growth of the mycelium between the boards rapidly spreads the fungus throughout the stacked lumber.

Early stages of decay by both heart rot and sap rot fungi are not easily detected but may be very important concerns, depending upon the final use of the product. Roughened saw cuts and slight fiber pulling during planing (brashness) are indicators of incipient (early) decay.

Decay in buildings and wooden structures The most serious decay of wooden structures results from sap rot fungi invading wood that becomes wetted by rain, water-vapor condensation, or plumbing leaks.

Paint peeling is a good indicator of decay potential. Moisture condensation occurs in walls because of improper moisture barriers or because of leakage from the roof. Decay fungi will become established and cause structural damage whenever the moisture content of the wood rises above fiber saturation. Paint peeling is not the result of decay but an indicator of conditions ideal for decay.

A sagging roof line is a good indicator of improper construction and early phases of decay. If moisture condenses in the attic because of an improper moisture barrier between the attic and moisture-laden, warm rooms, or if moisture gets into the attic due to leakage around chimneys or because of defective roofing, decay will slowly weaken the roof joists and cause them to sag.

An early symptom of decay is discoloration of a window frame. Condensation of moisture on the inside of windows, or moisture seeping into the window casings because of improper construction of a drip cap on the outside, wets the wood above fiber saturation.

Carpenter ants are good indicators of decay or at least indicators of conditions favorable for decay (Fig. 14-23). Carpenter ants need wood with a moisture content just above fiber saturation to maintain the proper atmospheric moisture for the development of their eggs. (Fig. 14-24). They do not eat wood. Wood for the carpenter ant is just an ideal place to maintain a constant environment. Ants will excavate wood that is decayed in preference to sound wood. Chemical control of the ants may be satisfying to the homeowner, but it does not change the development of the decay.

Where the floor joists sit on the concrete wall, moisture is sometimes wicked into the wood from the ground through the concrete. Discoloration of the joists is an early symptom of the problem. Decay and structural failure will follow if steps to prevent the wicking of moisture are not taken.

Collapse of steps or a deck is a symptom of advanced decay. Untreated wood, in contact with the ground or exposed to the elements, will eventually decay. If one does not recognize the problem before collapse occurs, a major replacement cost will develop.

FIGURE 14-23 Carpenter ants are good indicators of conditions favorable for decay. Shown are a queen and numerous worker ants. (Photograph compliments of Dr. John Simeone.)

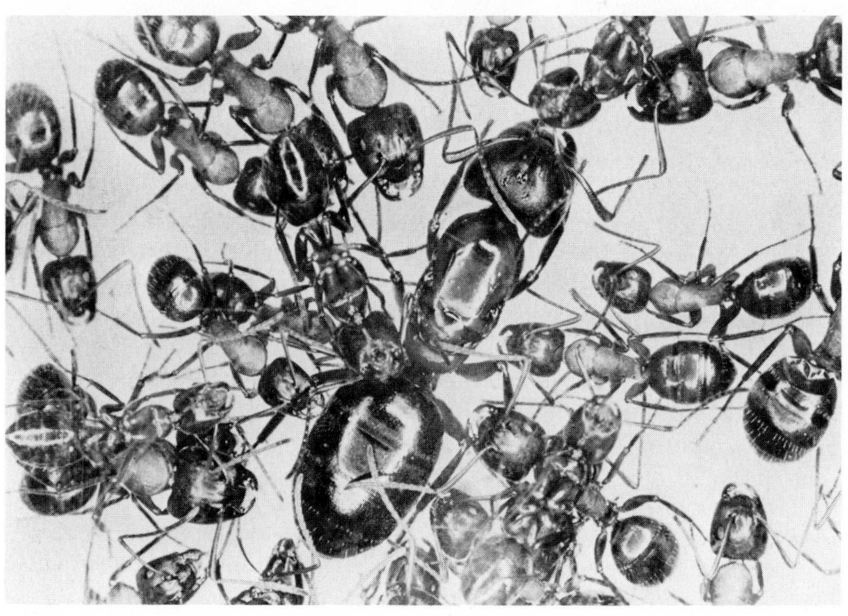

FIGURE 14-24 Samples of carpenter ant activity and decay of wood in a house with a serious moisture problem. (Photograph compliments of Dr. John Simeone.)

Microscopic Characteristics of Decay

Presence of characteristic hyphae of decay fungi Fungus hyphae can often be observed in thin sections of wood examined under the microscope. Some decay fungi have specific characteristics to their hyphae. Others cannot be distinguished from hyphae of other fungi that may invade the wood to utilize storage products of the ray and parenchyma cells.

Hyphae with clamp connections are characteristic of Basidiomycete fungi. If the hyphae have clamps, there is a good possibility that the fungus can cause decay. If it does not have clamps, it may or may not be a decay fungus.

Hyphae associated with wall lysis or wall degradation are excellent diagnostic characteristics of decay.

Hyphae in zone lines are usually thick-walled and pigmented. These hyphae are relatively persistent indicators of decay.

Cell-wall degradation Hyphae are seldom seen in thoroughly decayed wood, so the effects of former hyphae on cell walls are used for diagnostic purposes. The fungi that decay wood appear to recycle the limited nitrogen available in the wood. Hyphae in tissues that are thoroughly decayed are probably reabsorbed and the nutrients transported out to the periphery of the colony, where invasion of new tissues is in progress.

Bore holes are produced by hyphae to penetrate from one cell to another. The initial hole is smaller than the diameter of the hypha, which has to restrict its diameter to pass from one cell to another. Enzymatic activity of the hyphae later enlarge bore holes to many times the original diameter.

Cell-wall erosion or thinning occurs as a result of enzymatic activity of fungus hyphae within the lumen of the cells. Brown rot fungi produce ero-

(a)

(b)

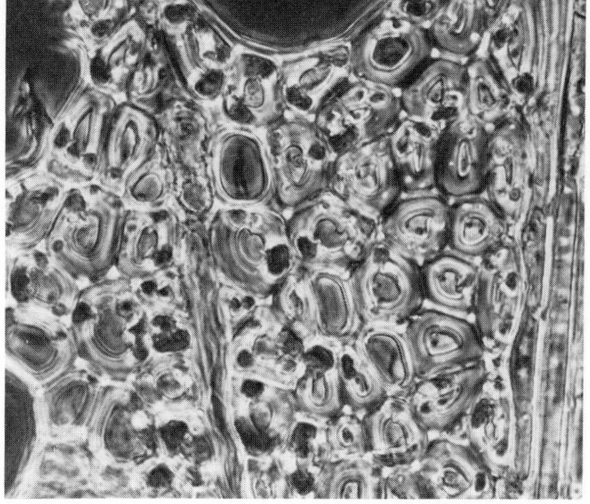

FIGURE 14-25 Erosion and soft rot of beech fibers caused by *Chaetomium globosum.* (a) Longitudinal section, showing diamond-shaped cavities in the secondary wall. (b) Cross section showing cavities in the secondary wall. (Photographs compliments of Alirio Perez-Mogollon.)

sions and cracks of the S_2 layer of the secondary wall. White rot fungi produce trench-like cavities of the secondary cell wall where the hyphae make contact with the wall. Soft rot fungi usually produce diamond-shaped cavities in the S_2 layer of conifers and trench-like cavities in the S_3 and S_2 layers of hardwoods (Fig. 14-25).

Isolation from Wood and Identification of Decay Fungi in Pure Culture

Isolation procedures Procedures for the isolation of fungi from wood are relatively simple. Small chips of wood are aseptically removed from wood and transferred to agar media. Aseptic techniques and media vary depending upon the training and preferences of the researcher. Large numbers of fungi will be recovered from almost any decayed wood. Variations in procedures attempt to selectively isolate decay fungi in preference to others.

Identification of decay fungi in culture Decay fungi are identified in culture based on a number of cultural characteristics. A publication by Nobels (1965) provides a reference for the cultural identification of decay fungi. A more recent reference by Stalpers (1978) can also be used. The growth rate is used as one of the characteristics for identification of the fungi. The color of the mycelial mat in culture is often characteristic but in some instances may be rather variable. Anatomical characteristics of the

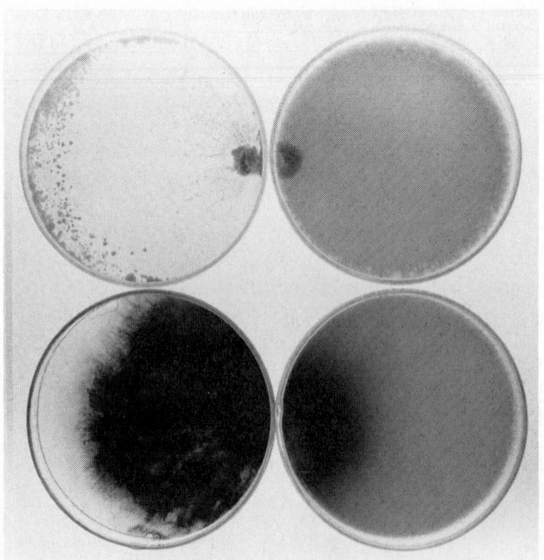

FIGURE 14-26 Phenol oxidase test. Note the extensive darkening of the medium caused by the white rot fungus in the lower right plate containing gallic acid. The left plates show the same fungi growing on malt extract agar.

fungus as seen with the microscope are important. The presence or absence of clamp connections, asexual spores (oidia), specialized hyphae, and other mycelial characteristics are used for identification. A phenol oxidase reaction on media containing an oxidizable phenolic derivative-like gallic or tannic acid is used to separate white rot fungi from brown rot fungi (Fig. 14-26). The odor of the culture is sometimes characteristic. Sometimes the fungus will fruit in culture, so that fruit body characteristics can be utilized to identify the fungus.

Physical Tests for Recognition of Decay

Limitations A serious limitation of physical tests for the verification of decay is that normal nondecayed wood is rather variable in its physical characteristics. Therefore, for a physical test to be valid, the wood must be significantly weakened below the expected range of variation for sound wood.

Specific tests Specific weight loss, resistance to penetration (toughness), resistance to fracture, and many other tests are used.

Changes in electrical properties can be used for both trees and wood products such as poles. Use of the Shigometer for this purpose has already been discussed.

X-ray and fluorescent microscopic analysis of fibers will show the effects of decay fungal enzymes on crystalline regions of the cellulose molecule and decomposition of lignin, respectively.

IDENTIFICATION OF SPECIES OF DECAY FUNGI BASED ON FRUIT BODIES

Early taxonomists of Homobasidiomycetes observed differences in the shapes of fruit body spore-producing surfaces. They classified the fungi into genera based on the shape of the spore-producing surface (hymenium) and the overall shape and texture of the fruit body. Mushrooms with a hymenium covering gill-like plates were grouped into many genera in the order Agaricales. Other Homobasidiomycetes with hymenial surfaces lining pores or covering smooth or spiny surfaces were placed in the order Polyporales. The Polyporales were subdivided into these common genera: *Polyporus,* with the hymenial surface lining pores in fleshy or leathery annual bracket fruit bodies; *Fomes,* with the hymenial surface lining pores of woody perennial bracket fruit bodies; *Daedalia,* with the hymenial surface lining short slit-like pores; *Poria,* with the hymenium lining pores but the fruit body resupinate; *Hydnum,* with the hymenial surface covering teeth-like projections from annual fruit bodies; *Echinodontium,* with hymenial surface covering teeth-like projections from perennial fruit bodies; and *Thelephora* and *Stereum,* with a flat hymenium occurring on a smooth leathery fruit body.

Most of the forest pathology literature relating to decay fungi has utilized this original Friesian classification system. Overholts' *Polyporaceae of the United States, Alaska and Canada* (1953) is an excellent reference using this system.

In recent years, mycologists have recognized that the former hymenial and annual or perennial criteria used for the classification of genera of the Polyporales are not mutually exclusive groups nor do they group the fungi phylogenetically. Some very similar fungi are grouped into different genera because the hymenial surfaces are different.

Microscopic examination of fruit bodies reveals that three types of hyphae make up some fruit bodies (trimitic); see Fig. 14-27. Others are made up of two types (dimitic), and a third group are made up of one type (monomitic). Other microscopic characters, such as color and shape of spores and the presence of sterile structures in the hymenium, are now considered important. European mycologists have led the movement toward a new classification of the Polyporales into many more genera than were originally recognized. The system is still in a state of flux, but many pathologists have shifted over to the new system. Keys by Pegler (1967), Domanski (1973), and Shaffer (1968) utilize larger numbers of genera for classifying the Polyporales.

It will be necessary to recognize both systems. The decay fungi described in this book will be listed using both names.

FIGURE 14-27 Hyphal types in fruit bodies of decay fungi.

1. Generative hyphae: Clamped and thin walled

2. Skeletal hyphae: Unbranched, thick walled, may originate from clamped generative hyphae

3. Binding hyphae: Highly branched and thick walled

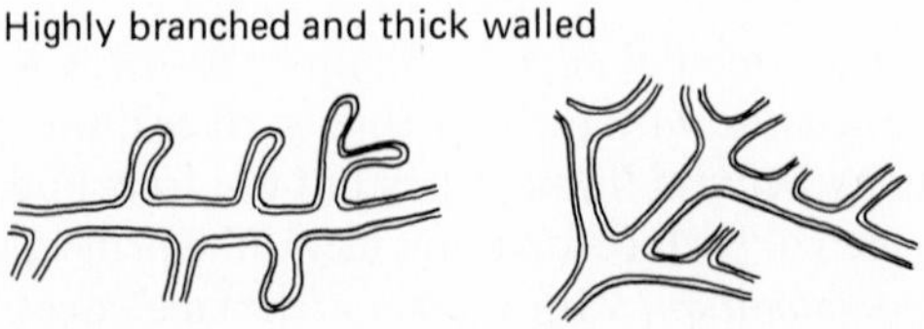

EXAMPLES OF HEART ROT AND SAP ROT DECAYS

Every tree species has at least one, but a very limited number of decay fungi that can cause heart rot in the living tree. Once the tree dies, or on dead branches of the living tree, a large number of decay fungi can survive. This type of observation, as well as the observation that the fungi that cause heart rot do not persist very long once the tree dies, lead to the conclusion that there is some specialization in the type of reaction that goes on between a heart rot fungus and its tree host.

The nature of the specialized interaction between heart rot fungi and host is not known, but one can speculate that tolerance of phenolic defense compounds produced by the tree is involved. Another aspect of the specialized interaction may involve the sources of the limiting nitrogen. In the living tree, heart rot fungi may parasitize living cells of the sapwood and cambium to derive sufficient nitrogen for hyphal growth and sporulation. Nitrogen in the living tree is replenished by the transport system of the tree. Nitrogen in the dead tree is static or replenished by diffusion of nitrogen from the soil or in rainwater.

A relatively complete discussion of many of the decays caused by fungi is provided in the textbook *Forest Pathology* by Boyce (1961). Therefore, rather than repeat Boyce's treatment, a few short statements will be made for the most common decays.

Hepting (1971) provides a very complete list of the decay fungi associated with forest and shade trees in the United States. By using Hepting's book to determine a list of possible names of fungi, the amateur can then use technical mycological books to identify fruit bodies. If more detail is necessary, one can refer to Boyce or to the references listed by Hepting.

Phellinus is one of the most important genera of heart rot fungi. The genus is characterized by fungi causing white rot. The fruit body is leathery to woody, usually perennial, with a brown context (interior) and pore layer. The hyphal system is dimitic and without clamps.

Phellinus igniarius (*Fomes igniarius*) is a very common heart rot pathogen of poplars, birches, oaks, maples, and beech. *Phellinus tremulae* (*Fomes igniarius* var. *populinus*) is a rather similar fungus with a wintergreen odor (Fig. 14-28). *Phellinus tremulae* is particularly serious as a decay fungus of trembling and bigtooth aspens. A black zone line around the stringy white rot decay column is usually characteristic of both these decays. The decayed wood loses most of its strength but is usually acceptable as pulpwood.

Phellinus pini (*Fomes pini*) is a major heart rot pathogen of most conifers (Fig. 14-29). The fruit bodies range from large bracket structures, to aggregates of small brackets, to resupinate (flat against the surface). The early stages of decay produce a distinctive red discoloration in the wood. At this

FIGURE 14-28
Phellinus tremulae on aspen.

FIGURE 14-29
Phellinus pini on Douglas fir.

stage the disease is known as red rot. Advanced decay forms white pockets of rot. Construction-grade lumber and lower grades of plywood can be made from the decayed wood. As already mentioned, decorative paneling can be made from *P. pini*-decayed wood.

Phaeolus schweinitzii (*Polyporus schweinitzii*) is one of the more common root and butt rot fungi of conifers (Fig. 14-30). The genus contains only one fungus, which is a brown rotter, with a spongy annual pored fruit body. The fruit body usually arises from the main roots of infected trees by means of a stipe (stalk). The context of the fruit body is rusty yellow to brown and is made up of a dimitic hyphal system. The upper surface is tomentose (velvety), yellow to purple brown to black in color. *Phaeolus schweinitzii*-infected conifer stands are very seriously damaged during high winds, and caution should therefore be exercised when developing recreation sites in infected stands. The volume lost to *P. schweinitzii* is large if the trees are windthrown. If the trees can be harvested before they fall, the volume lost is usually minimal and restricted to the lower portion, usually less than 8 feet (2.4m), of the bole of smaller trees and somewhat more in larger trees.

Three other root and butt rotters, *Heterobasidion annosum* (*Fomes annosus*), *Armillariella mellea* (*Armillaria mellea*), and *Phellinus weirii* (*Poria weirii*), will be discussed more fully in Chapter 16.

FIGURE 14-30 *Phaeolus schweinitzii* on spruce root.

FIGURE 14-31 *Inonotus obliguus* sterile conk on birch.

FIGURE 14-32 *Inonotus glomeratus* sterile conk and canker on sugar maple.

FIGURE 14-33 *Oxyporus populinus* on sugar maple.

Inonotus is a genus with fleshy, brown, annual, pored fruit bodies. Basidiospores are distinctly rust-colored. The decay is usually white rot. Two species of interest in the Northeast are *Inonotus obliquus* (*Poria obliqua*) on birches (Fig. 14-31) and *Inonotus glomeratus* (*Polyporus glomeratus*) on maples and beech (Fig. 14-32). Decay by these fungi is usually identified by the presence of a sterile conk on the tree. The cinder-like growth forms at a branch stub. Fertile forms of the fungi develop after the trees die. The decayed wood is acceptable for pulping purposes but is too weak to use for lumber.

Oxyporus populinus (*Fomes connatus*) is a white rotter of maples (Fig. 14-33). The fruit bodies are white, pored, and perennial. The hyphal system is monomitic. The fruit body is often found associated with Eutypella cankers and is often covered on the upper surface with moss, giving the fruit body a green appearance. Decay of *O. populinus* often results in a hollow tree of no merchantable value.

Laetiporus sulphureus (*Polyporus sulphureus*) is a brown heart rotter of many hardwoods, including oak, cherry, locust, and others (Fig. 14-34). The pored fruit body is annual, yellow, and consists of numerous large brackets. The context is fleshy and made up of a dimitic hyphal system. Decay by this fungus rapidly destroys the wood, which eventually disintegrates into a brown powder.

FIGURE 14-34 *Laetiporus sulphureus* on cherry.

FIGURE 14-35 *Echinodontium tinctorium* on western hemlock.

FIGURE 14-36 *Ganoderma applanatum* on elm stump.

In the Pacific Northwest, *Echinodontium tinctorium* is one of the more destructive heart rotters of firs and hemlock (Fig. 14-35). The toothed perennial fruit body is reddish brown in color. The brown stringy rot caused by this fungus extends up to 16 feet (4.9m) in either direction from the fruit body. In many instances, multiple infections render the infected tree useless as a forest product.

Ganoderma applanatum (*Fomes applanatus*) is an extremely common sap rot and heart rot fungus of hardwoods (Fig. 14-36). The large perennial fruit bodies have a white pore surface and a smooth dark upper surface. The context is red-brown and consists of a trimitic hyphal system. The white rot decay is usually restricted to the lower portions of the tree and, in living trees, is often responsible for weakening the tree enough to cause it to fall over in a heavy wind.

Many additional heart rots could be mentioned, but the above are some of the more common ones.

It is sometimes difficult for a beginner to distinguish heart rots from sap rots. As a general rule, heart rot fungi fruit at the base of branch stubs, whereas sap rot fungi produce fruit bodies anywhere on the stem. The sap rot fruit bodies are often numerous, whereas the heart rots are only numerous in stands and on individual trees when the infection levels are unusually high.

The number of sap rot decay fungi is much larger than the number of heart rot fungi. Therefore, a limited presentation on a few of them requires some arbitrary selection criteria. I will comment briefly on a few common ones from the Northeast. These are also present in other regions but may not represent the most common fungi in those regions.

FIGURE 14-37 *Coriolus versicolor* on dead maple.

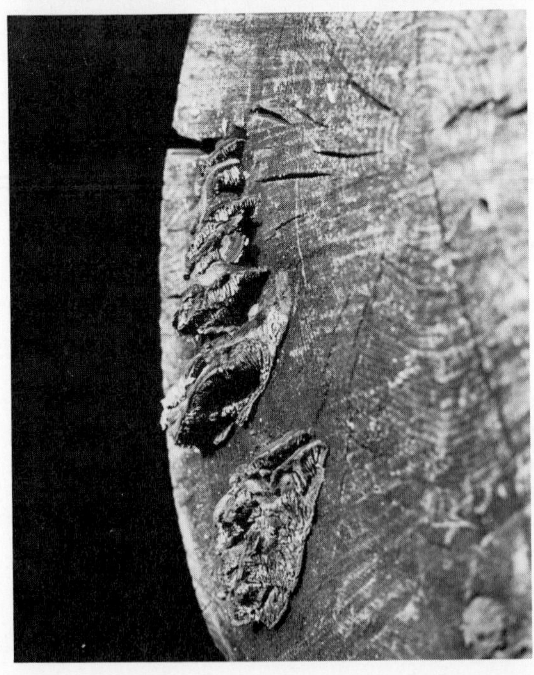

FIGURE 14-38 *Gleophyllum trabeum* on discarded pole.

Coriolus versicolor (*Polyporus versicolor*) is one of a large group of white rot decay fungi (Fig. 14-37). The genus is characterized by annual, pored, leathery, corky or woody fruit bodies. The hyphal system is trimitic. The thin (less than 5 mm) fruit bodies have a light-colored context. *Coriolus versicolor* is a small, thin, aggregating, bracket fungus with concentric zonation of gray, green, brown, and black bands on the upper surface and a white pore surface. It occurs as a sap rot fungus primarily on hardwoods, and is an important decay fungus of hardwood products.

Gleophyllum saepiarium (*Lenzites saepiaria*) and *Gleophyllum trabeum* (*L. trabea*) are two common, very similar looking, sap rot fungi of both conifers and hardwoods (Fig. 14-38). The brown rot decay by these fungi often occurs on fenceposts, poles, and other wood structures. The fruit bodies are corky, flat, somewhat zonate and tomentose above, with a plate-like hymenial surface. The context is yellow-orange and made up of a trimitic hyphal system in *G. sepiarium* and a dimitic hyphal system in *G. trabeum.* These fungi are very important products decay fungi.

Fomitopsis pinicola (*Fomes pinicola*) is a brown rot fungus of both conifers and hardwoods (Fig. 14-39). It is primarily a sap rot but can act as a weak heart rot pathogen. The perennial, pored, woody fruit body is made up of a trimitic hyphal system. The context is a straw-yellow in color. The lower surface is yellow-cream and the upper surface is black with a yellow to red outer band. The brown rot decay caused by this fungus breaks the wood into large blocks. White hyphal mats often occur between the blocks.

FIGURE 14-39 *Fomitopsis pinicola* on dead spruce.

FIGURE 14-40 *Fomes fomentarius* on dead beech.

FIGURE 14-41 *Ustulina vulgaris* on maple.

Fomes fomentarius is the only species retained in the genus *Fomes* in the new classification system. *Fomes fomentarius* is a white rotter, with a thick, woody, perennial fruit body (Fig. 14-40). Pores are very distinctive and extend 2 to 7 mm into the fruit body. The upper surface is gray to black, the lower surface is a gray to brown, and the context is brown. The hoof-like shape of the fruit body is somewhat characteristic. This white rotter is common on birch and beech, but also occurs on many other hardwoods. As with other sap rot fungi, the presence of fruit bodies of *F. fomentarius* generally indicates that the tree is dead and extensively decayed.

The last sap rot fungus presented in this limited group of examples is an Ascomycete decay fungus. *Ustulina vulgaris* is a common decay fungus of maple and beech (Fig. 14-41). The most characteristic symptom of the decay is the extensive zone-line development in the infected wood. In the field, advanced decay will disintegrate most of the wood, leaving the zone lines intact. The decayed stump section of the tree looks fire-charred. The fungus fruits by forming perithecia in black stroma on the surface of the decayed stump.

ROLE OF HEART ROT IN PLANT SUCCESSIONS

A topic worth considering is the role of heart rots in natural plant community successions. If nature is left unchecked, the heart rot fungi are the main means by which mature trees are finally brought down. The heart rot fungi of aspen cause mature trees to die and break off, allowing other species to come through. Everyone recognizes aspens as pioneer species, but the same can be said for many pines and redwoods. These species become established on fire-disturbed sites. The few giants that survive the fire are the source of seed as well as the source of fungus inoculum for infection of the next generation by heart rot fungi. The infection of young saplings guarantees that eventually the heart rot fungi of the lower stem will decay more wood than is produced by the cambium (recall the pathological rotation concept). The only known biological agent other than man that is capable of bringing down the giant redwoods is a heart rot fungus, *Poria sequoiae.* A fruitful area for a pathologically oriented ecologist is to investigate the relative importance of heart rot fungi in the development of present-day stand composition, as compared to the more traditionally important agents, such as logging and fire.

CONTROL OF HEART ROT DECAY

Our limited knowledge of infection courts and infection processes of heart rot fungi limits our ability to make good control recommendations.

The association of decay infection or invasion with wounds makes avoiding wounds a possible recommendation. Conscious efforts to prevent wounds in selective or partial cut-logging operations can avoid a great many future decay problems. But all wounds cannot be avoided.

Although tree wound dressings are sold which allegedly prevent infection in the wounds, there is no good evidence for or against the alleged therapeutic or preventive benefits of these materials.

Many trees do a good job of natural healing of wounds. It would appear that some of this ability is genetically controlled, because in a clonal species such as aspen some clones are more infected than others. If we could determine precisely how the tree prevents infection, we could perhaps help the tree by some type of topical dressing when wounding cannot be avoided.

Proper pruning practices and shaping up of irregular wounds enhances healing and presumably reduces infection by decay fungi. Living branches should be pruned flush with the stem. Tearing of the bark below the pruned branch should be avoided by first undercutting the branch. Callus tissues around a dead branch should not be cut into when pruning dead branches. Loose bark should be removed from around an irregular wound to form a smooth diamond-shaped wound.

Selection of which trees to cut and which trees to leave when thinning a young vegetatively propagated (coppice) stand can avoid decay later. Cut all high stump sprouts and selectively save the low ones. In companion sprouts, cut one only if it is small; if the sprouts are larger than 4 inches, either cut both or leave both.

An increase in decay can be expected if one selectively logs the dominant trees in the stand. The formerly suppressed trees may respond to increased light with rapid growth, but they have a high probability of having been infected with decay. As suppressed trees, they were less capable of providing an adequate defense against infection and invasion by decay fungi. The established decay in suppressed trees limits the production of high-quality wood.

Rapidly growing young trees can resist infection and invasion by decay fungi, so that overmature, slow-growing trees should be selected against in both forest and urban tree populations.

EXAMPLES OF WOOD DECAY IN THE HOME

Decay can cause expensive repair bills for the homeowner. It is paradoxical that decay is so common in the home because simple means of prevention have been well known for years. That decay need not destroy a home is attested by the fact that well-designed buildings of early settlers of this country are still standing hundreds of years later, because the wood that was properly used 200 or more years ago is still sound.

Why, then, is decay so common? One reason is that modern builders and homeowners fail to recognize how to prevent decay. An understanding of decay in the home and its prevention by builders and the general public would go a long way toward reducing expensive repair bills and conservation of our wood resource.

Symptoms of Decay in Buildings

Some of the symptoms of decay in buildings were covered with the general treatment of recognition of decay earlier in this chapter. These will now be reviewed and expanded.

A common symptom of decay is the musty smell often associated with basements and in other cases with rooms in the upper part of the house. The musty smell may come from fungi which are working over an old pair of boots sitting in the corner, or from fungi degrading the floor joists (Fig. 14-42). One cannot tell from the smell, but if the moisture is sufficient for fungi to degrade the boots, it is also sufficient for fungi to decay the wood.

Black carpenter ants have the same environmental requirements as decay fungi. Instead of killing off the ants with chemical insecticides, appreciate the ants for showing you that you have a decay problem. If you eliminate the environmental conditions that caused the ants to take up residence within your home, they will leave to find a better place to live. The carpenter ant is a predator of other insects and should be applauded for his good work rather than condemned for pointing out your decay hazards.

FIGURE 14-42 Decay of floor joists in the crawl space below a house.

Peeling and discoloration of painted siding can be covered up with aluminum or fiberglass siding. This treatment, like the killing of the ants, is a cosmetic treatment of the symptoms and not control of the problem. The paint is peeling because there is water in the walls. Determine the cause and take care of the moisture before hiding it behind aluminum.

Sagging of floors and roof occur before final breakage. Recognize the cause and take steps to correct the problem before the whole building collapses.

The last indicator of decay in the home, which appears well after the musty smell, ants, paint peeling, and sagging of floors, are the fruit bodies of the decay fungi. By the time you have fruit bodies of fungi, you have an expensive repair problem. I have seen mushrooms growing up through a plush wall-to-wall carpeted bathroom and have had to tell the homeowner that it would not make any difference if he kept the mushrooms picked—that he should, instead, have the leaky toilet fixed and replace the floor. Paint peeling and a collapsing patio on a 5-year old, $70,000 house cannot be repaired simply when obvious fruit bodies of *Gleophyllum trabeum* are present. The deck must be replaced.

Decay Hazard Zones and Control Treatments (Fig. 14-43)

Attic The attic is a decay hazard zone because of the condensation of water vapor from heated rooms on the cold roof rafters and roof boards. Obviously, the condensation problem is associated with homes in northern climates, but leaks around chimneys and because of improper flashing of the valleys of the roof are common to all climates. Condensation can be controlled by a vapor barrier in the ceiling of heated rooms to prevent water vapor from escaping. Aluminum-foil-backed insulation or plastic sheeting between the ceiling joists and the ceiling boards are the best methods. A good coat of oil-base paint is almost as effective.

All leaks in the roof should be immediately repaired. The old adage that you cannot repair the roof when it is raining and do not need to when it is not does not take into account that fungi continue to operate rain or shine.

Another prevention associated with attic problems is to provide adequate ventilation to dissipate any moisture that might build up because of the first two problems, as well as from condensation of water vapor from trapped moisture-laden air during cooling. Adequate ventilation requires placement of louvers on the side walls near the roof ridge and under the eaves of the overhang.

Walls Moisture in the siding and framing may result from: (1) condensation of water vapor, as in the attic problem; (2) water leaking into the walls from ice backed up on the roof overhang; (3) rain wetting of the

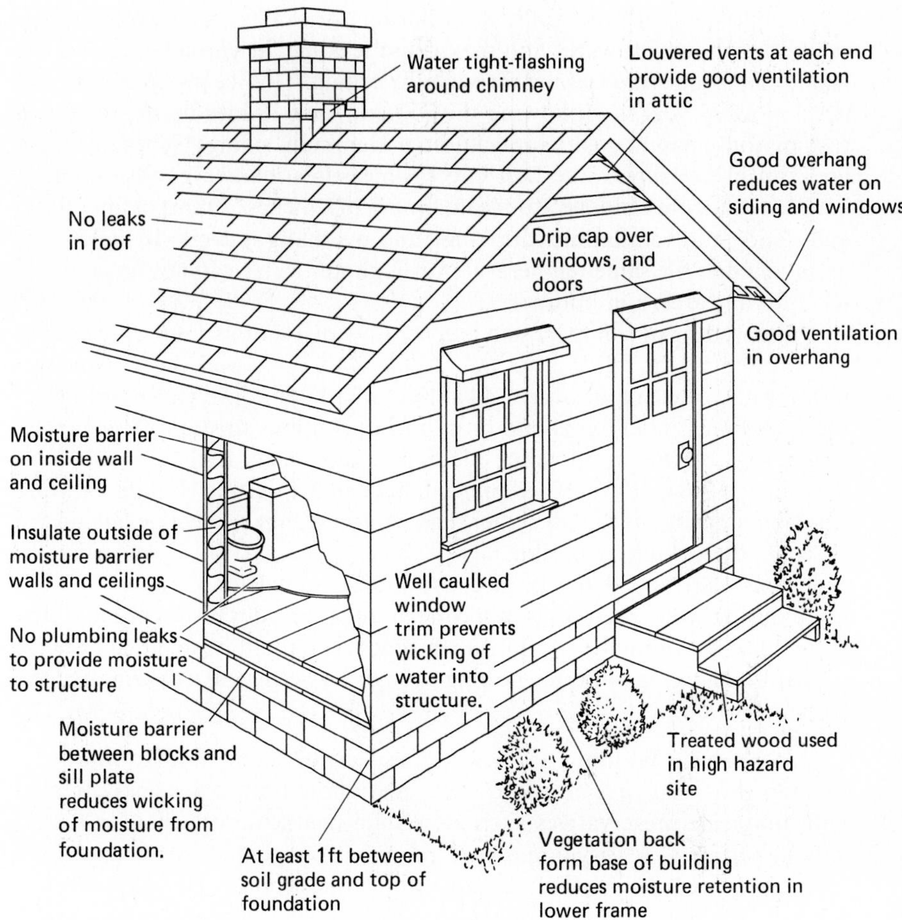

FIGURE 14-43 Prevention of decay problems in homes.

siding; and (4) moisture trapped in the walls during the construction of a new home.

Control of condensation in the walls is by use of foil-backed insulation, plastic sheeting, and oil-base paint. The first two are preferred. Paint is somewhat less efficient, but better than nothing.

Water leaking into the walls from ice dams on the overhang of the roof occurs during winters when heavy snowfall is followed by a period of very cold weather. In poorly insulated homes, heat from the rooms rising into the attic causes the snow to melt and water to run off. The roof overhang does not get warmed by the escaping heat, so water freezes in the gutters and overhang. As more and more ice is built up, the water may work back under the shingles and into the walls. Rain during a warming trend would accentuate the amount of water and may cause leakage far enough back on the roof to have moisture leaking through the ceiling. This ice problem does not develop every year, so following the winters when it does occur, there is a rash of house painting and application of aluminum siding. Neither of these are control measures. The best control measures involve good insulation between the rooms and the attic (a minimum of 8 inches (20 cm) of fiberglass) and good attic ventilation with ridge and overhang louvers. If the roof is maintained at the same temperature throughout, water will not form in one location and freeze in another.

The wetting of siding by rain is generally not a serious source of water in the walls unless the siding is not tight or has cracks that wick the moisture into the walls. Wetting of untreated siding usually does not cause problems. Many old unpainted barns and farm houses continue to stand. Mechanical weathering of the wood takes place but very little decay occurs. Roof overhangs of 18 in. (46 cm) will prevent much of the wetting by rain. Painting will help moisture to run off rather than penetrate, but is effective only in the absence of cracking or peeling.

Moisture trapped in walls because of unseasoned lumber or wet lumber used during construction is a problem of short duration, because the material will eventually dry out. Such practices should be avoided because the period of drying time may be long enough for decay and weakening of the structure to occur.

Windows Window frames are especially vulnerable to moisture buildup. For that reason, wood used in window frames is usually treated with fungicide preservatives such as sodium pentachlorophenate, a water-soluble form of a very important preservative, pentachlorophenol. The wood is treated during the manufacture of the windows. It is an effective preservative if the window frames are not exposed to excessive amounts of moisture. Any preservative will, with time, leach out of the treated piece, and then the material is vulnerable to decay. On the outside, proper construction of a moisture drip cap over windows is important. The window trim and frame should be sealed into the siding by calking of all cracks where water might leak in. On the inside of the house, it is important to adjust or control the level of humidification. Present-day humidifiers are capable of producing enough humidity to almost peel the wallpaper from the walls. It is important to reduce the humidity as the outside temperature decreases. If moisture consistently condenses on the windows, it is too high.

Plates, joists, and siding Plates, joists, and siding are sometimes decayed because water is wicked into the material from the soil through the concrete foundation. A minimum distance of exposed foundation above the soil grade of 1 ft, (30 cm), and treatment of the contact between the wood and the foundation with a waterproof substance such as asphalt, will prevent the problem.

Moisture can also accumulate in the joist plates and lower siding because vegetation close to buildings prevents adequate ventilation and drying. Caution should be exercised in planting dense decorative vegetation around the foundation. Usually, this type of problem occurs because the plantings are allowed to grow too large, so that a little maintenance and pruning of foundation vegetation is all that is necessary to avoid decay of the lower walls.

Subflooring and joists The subflooring and floor joists are often seriously decayed because moisture seeps into the material because of faulty plumbing. Early repair of faulty plumbing will avoid costly replacement repairs to the floor and structural members.

Moisture can accumulate in subflooring and floor joists in structures with a crawl space rather than a basement. Evaporation of water from the soil, followed by condensation on the floor, can be avoided by placing a moisture barrier such as asphalt paper or plastic on top of the soil. Another good precaution is to provide ventilation louvers to the crawl space to allow the moisture to escape.

Decorative wood Wooden steps, patios, and decorative wood used in the exterior of homes represent a special decay hazard because one cannot prevent wetting of the material. Construction methods that allow rapid drying of the wetted wood are necessary. The flooring of patios or steps should be loose enough to allow ventilation to dry it out. Avoid small cracks or joints between various members where moisture will wick into the wood and not easily dry out. Any wood that is in direct contact with the ground will decay rapidly if it is not treated with a preservative. Therefore, use treated wood any place where moisture will be a problem.

CONTROL OF WOOD DECAY IN PRODUCTS

Wood Preservatives

A number of excellent preservatives are available for different types of wood use. Railroad ties and the lower ends of telephone poles are exposed to extreme hazards and are pressure-treated with creosote. Creosote was once the unusable residue from coal coaking and was quite variable in its toxicity to decay fungi. Today, good control over creosote production results in a very fungitoxic compound with very good resistance to leaching from treated wood.

Creosote has an objectionable smell, cannot be painted over, and will produce burns on your skin. Therefore, other preservatives are used where the wood is used in buildings. Pentachlorophenol is the most commonly used preservative where objections to creosote prevent its use. We have already mentioned the use of a water-soluble form of pentachlorophenol in windows. Other preservatives are copper naphthenate, which gives a green color to the wood; chromated zinc chloride; and chromated copper arsenate.

Wood is preferably treated with a preservative using a pressure system for pushing as much preservative into the wood as is possible. In the absence of a pressure treatment, one can dip or brush the preservative onto the pieces of wood. Microorganisms in the soil, as well as water leaching, will eventually remove the preservative from any treated wood. Therefore, the only disadvantage of the dip and brush treatments is that not as much preservative is impregnated into the wood, so that pieces treated by these methods will decay before pressure-treated wood.

TABLE 14-2 Natural Decay Resistance of Domestic Woods (from Sheffer and Cowling, 1966)

Exceptionally resistant[a]	Resistant	Moderately resistant	Nonresistant
Baldcypress (old growth)	Catalpa	Baldcypress	Alder
	Cedar	Douglas fir	Ashes
Black locust	Black cherry	Honeylocust	Basswood
Red mulberry	Chestnut	Western larch	Beech
Osage-orange	Arizona cypress	Swamp chestnut oak	Birches
Pacific yew	Junipers	Eastern white pine	Buckeye
	White and bur oaks	Longleaf pine	Butternut
	Redwood	Slash pine	Elms
	Sassafras	Tamarack	Hackberry
	Black walnut		Hemlocks
			Hickories
			Magnolia
			Maples
			Red and black oaks
			Most pines
			Poplars
			Spruces
			Sweetgum
			Sycamore
			Willows
			Yellow-poplar

[a]Sapwood of even the most resistant woods is equivalent to nonresistant woods, so categories are based on heartwood only.

It is not possible to effectively pressure-treat all wood. For example, the western spruce and fir lumber presently sold as construction-grade lumber in the Northeast is almost impossible to properly treat. The preservative does not penetrate the wood. A good preservative treatment should have total penetration of the preservative. Southern yellow pine sapwood is readily penetrated by preservatives and is therefore often specified where maximum durability is desired.

Natural Resistance of Wood to Decay

The heartwood of a number of species of trees is impregnated with toxic compounds that provide natural decay resistance to sap rot fungi.

Table 14-2 lists the natural decay resistance of a number of our domestic woods. The list of nonresistant woods contains those without heartwood, such as poplars, birches, and maples. The sapwood of all species is equally susceptible to decay by sap rot fungi. Therefore, it is important to specify heartwood when buying naturally durable wood.

REFERENCES

AMBURGEY, T. L. 1974. Wood-inhabiting fungi prevention and control. Pest Control *42*:22–25.

ANDERSON, L. O. 1972. Condensation problems: their prevention and solution. USDA For. Serv. Res. Pap. FPL 132. 37 pp.

ANONYMOUS. 1974. Wood handbook: wood as an engineering material. Agricultural handbook 72, USDA Forest Service.

BIESTERTELDT, R. C., T. L. AMBURGEY, and L. H. WILLIAMS. 1973. Finding and keeping a healthy house. USDA For. Serv. Gen. Tech. Rep. SO-1. 19 pp.

BOYCE, J. S. 1961. Forest pathology, 3rd ed. McGraw-Hill Book Company, New York. 572 pp.

BROOKS, F. T., and W. C. MORE. 1923. On the invasion of woody tissues by wound parasites. Proc. Camb. Philos. Soc. (Biol. Sci.) *1*:56–58.

COWLING, E. B. 1961. Comparative biochemistry of the decay of sweetgum sapwood by white rot and brown rot fungi. USDA For. Serv. Agr. Tech. Bull. 1258. 79 pp.

COWLING, E. B. and W. BROWN. 1969. Structural features of cellulosic materials in relation to enzymatic hydrolysis. *In* Cellulases and their applications. Adv. Chem. Ser. 95. American Chemical Society, Washington, D.C., pp. 152–187.

DOMANSKI, S. 1972. Fungi, Polyporaceae I (resupinate), Mucronoporaceae

I (resupinate). Translation of the book "Grzyby," published in 1965 by Panstwowe Wydawnictwo Naukowe. Available from the U.S. Department of Commerce, National Technical Information Service, Springfield, Va.

DOMANSKI, S., H. ORTOS, and A. SKIRGIELLO. 1973. Fungi, Polyporaceae II (pileatae), Mucronoporaceae II (pileatae). Translation of the book "Grzyby," Vol. 3, published in 1967 by Panstwowe Wydawnictwo Naukowe. Available from the U.S. Department of Commerce, National Technical Information Service, Springfield, Va.

ETHERIDGE, D. E., and H. M. CRAIG. 1976. Factors influencing infection and initiation of decay by the Indian paint fungus (*Echinodontium tinctorium*) in western hemlock. Can. J. For. Res. *6*:299–318.

FERGUS, C. L. 1960. Illustrated genera of wood decay fungi. Burgess Publishing Company, Minneapolis.

GJOVIK, L. R. and R. H. BAECHLER. 1977. Selection, production, procurement and use of preservative-treated wood, supplementing federal specification TT-W-571. USDA For. Serv. Gen. Tech. Rep. FPL-15. 36 pp.

HADDOW, W. R. 1938. The disease caused by *Trametes pini* (Thore) Fries in white pine (*Pinus strobus* L.). R. Can. Inst. Trans. *47*:21–80.

HART, J. H., and D. M. SHRIMPTON. 1979. Role of stilbenes in resistance of wood to decay. Phytopathology *69*:1138–1143.

HARTIG, R. 1874. Important diseases of forest trees. Phytopathological Classics 12. English transl. by W. Merrill, D. H. Lambert, and W. Liese, 1975. American Phytopathological Society, St. Paul. 120 pp.

HEPTING, G. H. 1971. Diseases of forest and shade trees of the United States. Agricultural handbook 386, USDA Forest Service. 658 pp.

HIGHLEY, T. L. 1975. Properties of cellulases of two brown-rot fungi and two white-rot fungi. Wood and Fiber *6*:275–281.

HIGHLEY, T. L. 1976. Hemicellulases of white and brown rot fungi in relation to host preferences. Mater. Org. *11*:33–46.

HIGHLEY, T. L. 1977. Degradation of cellulose by culture filtrates of *Poria placenta.* Mater. Org. *12*:161–174.

HIGHLEY, T. L., and T. K. KIRK. 1979. Mechanisms of wood decay and unique features of heartrots. Phytopathology *69*:1151–1157.

KING, K. W., and M. I. VESSAL. 1969. Enzymes of the cellulase complex. *In* Cellulases and their applications. Adv. Chem. Ser. 95. American Chemical Society, Washington, D.C., pp. 7–25.

KIRK, T. K. 1975. Lignin-degrading enzyme system. Biotechnol. Bioeng. Symp. *5*:139–150.

KIRK, T. K. 1975. Chemistry of lignin degradation by wood destroying fungi. *In* Biological transformation of wood by microorganisms, ed. W. Liese. Springer-Verlag, New York, pp. 153–164.

KIRK, T. K., and J. M. HARKIN. 1973. Lignin biodegradation and the bioconversion of wood. Am. Inst. Chem. Eng. Symp. Ser. *69*:124–126.

KIRK, T. K., and T. L. HIGHLEY. 1973. Quantitative changes in structural components of conifer woods during decay by white and brown rot fungi. Phytopathology *63*:1338–1342.

KOENIGS, J. W. 1974. Hydrogen peroxide and iron: a proposed system for decomposition of wood by brown-rot Basidiomycetes. Wood and Fiber *6*:66–77.

LANE, P. H., and T. C. SCHEFFER. 1960. Water sprays protect hardwood logs from stain and decay. For. Prod. J. *10*:277–282.

LARSEN, M. J., M. F. JURGENSEN, A. E. HARVEY, and J. C. WARD. 1978. Dinitrogen fixation associated with sporophores of *Fomitopsis pinicola, Fomes fomentarius* and *Echinodontium tinctorium.* Mycologia *70*:1217–1222.

LIESE, W. 1970. Ultrastructural aspects of woody tissue disintegration. Annu. Rev. Phytopathol. *8*:231–258.

LOWE, J. L. 1942. The Polyporaceae of New York State (except *Poria*). N.Y. State Coll. For. Syracuse Univ. Tech. Publ. 60. 128 pp.

LOWE, J. L. 1957. Polyporaceae of North America. The genus *Fomes.* N.Y. State Univ. Coll. For. Syracuse Univ. 97 pp.

LOWE, J. L. 1966. Polyporaceae of North America. The genus *Poria.* State Univ. Coll. For. Syracuse Univ. Tech. Publ. 90. 183 pp.

LOWE, J. L. 1975. Polyporaceae of North America. The genus *Tyromyces.* Mycotaxon *2*:1–82.

MANION, P. D., and D. W. FRENCH. 1968. Inoculation of living aspen trees with basidiospores of *Fomes igniarius* var. *populinus.* Phytopathology *58*:1302–1304.

MANION, P. D., and R. A. ZABEL. 1979. Stem decay perspectives—an introduction to tree defense and decay patterns. Phytopathology *69*:1136–1138.

MERRILL, W. 1970. Spore germination and host penetration by heartrotting Hymenomycetes. Annu. Rev. Phytopathol. *8*:281–300.

MERRILL, W., and E. B. COWLING. 1966. Role of nitrogen in wood deterioration: amount and distribution of nitrogen in fungi. Phytopathology *56*:1083–1090.

MERRILL, W., and A. L. SHIGO. 1979. An expanded concept of tree decay. Phytopathology *69*:1158-1160.

NOBLES, M. K. 1965. Identification of cultures of wood inhabiting Hymenomycetes. Can. J. Bot. *43*:1097–1139.

NYLAND, R. D., and W. J. GABRIEL. 1971. Logging damage to partially cut hardwood stands in New York State. State Univ. Coll. For. Syracuse Univ. AFRI Res. Rep. 5. 38 pp.

OVERHOLTS, L. O. 1953. The Polyporaceae of the United States, Alaska and Canada. University of Michigan Press, Ann Arbor. 466 pp.

PARTRIDGE, A. D., and D. L. MILLER. 1974. Major wood decays in the inland northwest. Univ. Idaho, Idaho, Res. Found. Nat. Res. Ser. 3. 125 pp.

PEGLER, D. N. 1967. Polyporaceae. Part II, with a key to world genera. Bull. Br. Mycol. Soc. *1*:17–36.

PEGLER, D. W. 1973. Aphyllophorales IV. Poriod Families. *In* The fungi: an advanced treatise, ed. G. C. Ainsworth, F. K. Sparrow, and A. S. Sussman. Academic Press, Inc., New York, pp. 397–420.

ROTH, E. R., and B. SLEETH. 1939. Butt rot in unburned sprout oak stands. USDA Tech. Bull. 684. 42 pp.

SCHEFFER, T. C., and E. B. COWLING. 1966. Natural resistance of wood to microbial deterioration. Annu. Rev. Phytopathol. *4*:147–170.

SCHEFFER, T. C., and A. F. VERRALL. 1973. Principles for protecting wood buildings from decay. USDA For. Serv. Res. Pap. FPL 190. 56 pp.

SHAFFER, R. L. 1968. Keys to genera of higher fungi, 2nd ed. Univ. Mich. Biol. Sta., Ann Arbor.

SHAIN, L. 1979. Dynamic responses of differentiated sapwood to injury and infection. Phytopathology *69*:1143–1147.

SHIGO, A. L. 1967. Succession of organisms in discoloration and decay of wood. Int. Rev. For. Res. *2*:237–299.

SHIGO, A. L., and W. E. HILLIS. 1973. Heartwood, discolored wood, and microorganisms in living trees. Annu. Rev. Phytopathol. *11*:197–222.

SHIGO, A. L., and E. vH. LARSON. 1969. A photo guide to the patterns of discoloration and decay in living northern hardwood trees. USDA For. Serv. Res. Pap. NE-127. 100 p.

SHIGO, A. L., and H. G. MARX. 1977. Compartmentalization of decay in trees. USDA For. Serv. Agr. Inf. Bull. 405. 73 pp.

SHIGO, A. L., and E. M. SHARON. 1970. Mapping columns of discolored and decayed tissues in sugar maple, *Acer saccharum.* Phytopathology *60*:232–237.

SHIGO, A. L., and A. SHIGO. 1974. Detection of discoloration and decay in living trees and utility poles. USDA For. Serv. Res. Pap. NE 294. 11 pp.

SHIGO, A. L., and C. L. WILSON. 1971. Are tree wound dressings beneficial? Arborist's News *36*:85–88.

SHORTLE, W. C. 1979. Mechanisms of compartmentalization of decay in living trees. Phytopathology *69*:1147–1151.

SHORTLE, W. C., and E. B. COWLING. 1978. Development of discoloration, decay, and microorganisms following wounding of sweetgum and yellow poplar trees. Phytopathology *68*:609–616.

SHORTLE, W. C., and E. B. COWLING. 1978. Interaction of live sapwood and fungi commonly found in discolored and decayed wood. Phytopathology *68*:617–623.

SILVERBORG, S. B. 1951. Wood decay in houses. N.Y. State Univ. Coll. For. Syracuse Univ. 21 pp.

SILVERBORG, S. B. 1959. Rate of decay in northern hardwoods following artificial inoculation with some common heartrot fungi. For. Sci. *5*:223–228.

STALPERS, J. A. 1978. Identification of wood-inhabiting Aphyllophorales in pure culture. Studies in Mycology 16. Centraalbureau voor Schimmelculture, Baarn, The Netherlands. 248 pp.

TOOLE, E. R. 1965. Deterioration of hardwood logging slash in the south. USDA For. Serv. Tech. Bull. 1328. 27 pp.

WAGENER, W. W., and R. W. DAVIDSON. 1954. Heart rots in living trees. Bot. Rev. *20*:61–134.

ZABEL, R., and R. ST. GEORGE. 1960. Wood protection from fungi and insects during storage and use. Proc. Fifth World For. Congr. *3*:1530–1540.

15

FUNGI AS AGENTS OF TREE DISEASES: WOOD STAIN

- Types of wood stains and mode of action
- Wood stain "disease" cycle
- Symptoms of wood stain
- Examples of wood stain and control

Wood stains are generally problems of wood products, but blue stain fungi have a number of features in common with fungi that cause wilts. The Ascomycete genus *Ceratocystis* contains both wilt and stain fungi. Insect associations for dissemination are common to both groups. Invasion of parenchyma cells of dead or dying trees is common to both groups.

Association of the blue stain fungus *Ceratocystis pini* with *Dendroctonus frontalis* (southern pine beetle) and *D. brevicomis* (western pine beetle) is so general there have been questions since 1928 as to which is really the cause of the death of pines. Stain fungi will cause death in the absence of beetles in inoculation experiments, but one finds neither blue stain fungus nor beetles alone causing the death of trees in the field; both are involved.

Unlike wood decay, wood stains usually do not produce degrading because of strength loss but because of the appearance of the wood. Therefore, we are mainly concerned about stains in wood used for decorative purposes rather than for structural purposes alone.

TYPES OF WOOD STAINS AND MODE OF ACTION

Wood stains can be caused by incipient stages of decay fungi, by stain fungi, or strictly by chemical reactions in the absence of microorganisms. Chemically induced stains will be discussed in this chapter because the degrading of lumber is similar in concern to that of fungal-induced stains.

Incipient Decay

Early stages of invasion by wood decay fungi, incipient decay, usually produce stain (Fig. 15-1). Some stains are caused by the decay fungus, as in the case of *Phellinus pini (Fomes pini)*, red rot. Others are caused by nondecay fungi and bacteria associated with the decay fungi. Decay was discussed in Chapter 14 and will not be elaborated upon except to contrast and compare decay and blue stain.

Blue Stains

Stains caused by blue stain fungi (Figs. 15-2 to 15-4) will be contrasted with decay fungi to point out specific details.

Recall that decay fungi were generally from the class Basidiomycetes. Stain fungi are generally from the classes Ascomycetes and Fungi Imperfecti. The stain fungi invade parenchyma cells and utilize the cell contents. The decay fungi, in contrast, invade and utilize the cell-wall materials of structural cells. Stain is usually restricted to the sapwood because, in trees with both sapwood and heartwood, the heartwood does not contain many living

FIGURE 15-1 Incipient decay in pine as evidenced by the irregular dark zone lines.

FIGURE 15-2 Blue stain in a pine log.

FIGURE 15-3 Pine lumber cut from blue-stained logs.

FIGURE 15-4 Blue stain resulting from invasion of pine lumber after it was cut.

storage cells. The stain fungi have similar moisture, oxygen, and temperature requirements to the decay fungi, so the wood decay concepts of prevention relating to controlling environmental conditions apply also to biologically induced stain.

Bark beetles are often associated with blue stain fungi. They serve as vectors and provide wounds for infection by the fungi.

Chemical Stains

Chemical stains vary, depending upon the species of tree from which the wood is cut. Chemical stains may occur during the drying process because of the release of enzymes from the living cells during the cutting (Figs. 15-5 and 15-6). Maple and birch lumber will sometimes develop oxidation stains. They may result from chemical oxidation and reactions of phenols and sugars in the wood of white pine. Other chemical stains of oak occur because of chemical reactions between iron and tannins in wood to produce a dark iron tannate stain.

FIGURE 15-5 Chocolate brown or coffee stain of pine resulting from reaction of sugars and phenols during kiln drying.

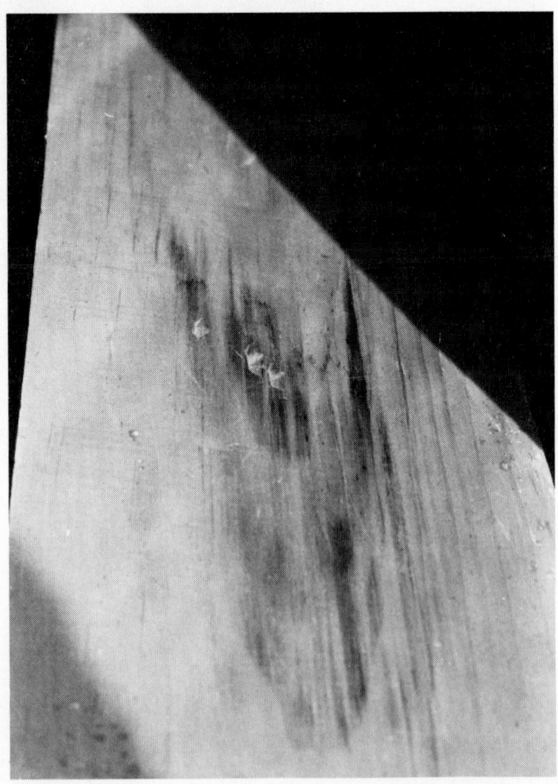

FIGURE 15-6 Sticker stain of sugar maple is seen as a slight color change going across the board at the top of the pointer. This deep penetrating, yet subtle color change is associated with changes in chemical reaction in the area under the stickers used to separate the boards during air drying of the rough lumber.

Molds

Many fungi can grow on the surface and discolor wood. These are particularly important in humid climates, where fungi grow on almost any surface (Fig. 15-7).

Weathering

Ultraviolet light, leaching of pigments by water, and mechanical abrasion of exposed wood produces a gray weathering stain to wood surfaces. Sometimes this is desirable and other times it is undesirable. A bright red-brown redwood fence or patio will gray and need to be stained to renew the original appearance. On the other hand, old weathered barn boards are a premium decorative wood for internal paneling. It all depends on your point of view.

FIGURE 15-7 Mold fungi growing on the surface of this sugar maple board during drying may affect the clearness of the finished board.

Ambrosia Stains

Wood-boring beetles, in contrast to bark beetles, may feed on or have requirements for specific fungi. Some ambrosia beetles have developed specialized organs called mycangia for storing and carrying specific fungi. They inoculate the wood with their fungi and tend the crop like farmers, weeding and fertilizing as necessary. Localized stain associated with small ambrosia beetle bore holes is characteristic of this type of stain.

WOOD STAIN "DISEASE" CYCLE

Infection of wood by blue stain fungi can occur because of bark beetle activity in the trees or stored logs, or by contamination of boards by saws or handling equipment, by airborne or rain-splashed spores in the stacked lumber, or by contact with infected stickers used in the loose stacking of boards.

Invasion and utilization of stored materials of sapwood rays and parenchyma cells results in bluish or grayish radial streaking. The fungus hyphae generally penetrate from one cell to another through pits but occasionally produce bore holes.

As with wood decay, moisture is critical for stain fungi development. The minimum moisture necessary is fiber saturation. Ponding or sprinkling of logs reduces the oxygen and therefore inhibits the blue stain fungi.

Sporulation by conidia and/or ascospores may occur generally over the surface of moist lumber or may occur specifically in beetle galleries of infected logs and trees.

The asexual and sexual fruiting structures are ideally developed for insect dissemination of spores. Conidiophores are aggregated into stalks called coremia, with sticky masses of spores produced at the tip. Perithecia with long necks do not shoot off ascospores but accumulate mature spores in a sticky matrix at the top of the stalk. Both of these are ideally suited for projection into insect galleries to contaminate the insect vectors.

SYMPTOMS OF WOOD STAIN

Recognition of stain is sometimes complicated by natural color variations in the wood, so it is necessary to understand wood anatomy and normal color variation.

Radiating distinct blue to gray streaks in sapwood, when observed in cross section, may apppear as interrupted lines of general blue to gray discoloration on the flat faces of boards. The color is due to pigmented hyphae and reaction products of the fungus invading the wood.

The chemical stains range from very dark pigment in the case of iron tannate stain of oak to very subtle coloring in the case of sticker stain of sugar maple. Sticker stain is recognized as a slight light-reflectance difference in the wood surface in the immediate area where it made contact with stickers. After surfacing and finishing, the band of different-colored wood is evident only when light reflects from the board at an oblique angle, so such stained boards sometimes are used in the manufacture of medium-quality furniture.

EXAMPLES OF WOOD STAIN AND CONTROL

Blue stain is prevented by minimizing the time between when the tree is cut and the board is dried below fiber saturation. Water ponding or sprinklers can be used to reduce stain during storage of logs. Rapid kiln-drying prevents stain during drying. If wood is air-dried, application of a fungicide dip to the boards as they come off the green chain is suggested.

Intercontinental shipment of green lumber in ships presents a serious stain problem. There is no need to dry the wood, because weight is no problem for the ships. Therefore, treatment of the lumber with tetrachlorophenol is used. Good treatment and packaging to protect the lumber from leaching of preservative will prevent serious deterioration even with up to 2 years' storage. Poor treatment, or allowing the preservative to leach out by exposure to rain, results in serious losses by stain, mold, and decay within 6 to 9 months.

Chemical stains are individually specific problems. Treatment of green lumber with sodium azide prevents oxidation staining of hardwood lumber.

A chocolate brown stain of white pine lumber occurs because of chemical reactions of sugars and phenols of the wood. If one uses moderate kiln-drying schedules, the stain can be prevented. Temperatures below 150°F and humidity below 65% are recommended.

Another common chemical stain occurs when iron comes in contact with moist oak lumber. A reaction between the iron and tannins of the wood produces a dark ink-like stain. Control is to avoid contact between wet oak lumber and iron materials.

REFERENCES

BATRA, L. R. 1963. Ecology of ambrosia fungi and their dissemination by beetles. Trans. Kans. Acad. Sci. *66*:213–236.

CAMPBELL, R. N. 1959. Fungus sap-stains of hardwoods. South. Lumberman *199*:115–120.

CZERJESI, A. J., and J. W. ROFF. 1975. Toxicity tests of some chemicals against certain wood-staining fungi. Int. Biodetect. Bull. *11*:90–96.

MILLER, J. M., and F. P. KEEN. 1960. Biology and control of western pine beetle. USDA For. Serv. Misc. Publ. 800. 381 pp.

ROFF, J. W., A. J. CZERJESI, and G. W. SWANN. 1974. Prevention of sap stain and mold in packaged lumber. Dept. Environ. Can. For. Serv. Publ. 1325. 43 pp.

ZABEL, R. A. 1953. Lumber stains and their control in northern white pine. J. For. Prod. Res. Soc. *3*:36–38.

ZABEL, R. A., and C. H. FOSTER. 1949. Effectiveness of stain control compounds on white pine seasoned in New York. N.Y. State Coll. For. Tech. Publ. 71.

16

FUNGI AS AGENTS OF TREE DISEASES: ROOT ROT

- Types of root rot
- Mode of action of root rot fungi
- Root rot disease cycle
- Symptoms of root rot
- Methods for diagnosis of root rot
- Examples of root rot and control
- Forest practices and root diseases

When an apparently healthy, mature tree or group of mature trees die for no apparent reason, we are often at a loss for an explanation. The difficulty in identifying the cause sometimes occurs because we are often looking at only half the tree. This preoccupation with the aboveground half of the tree is logistically imposed on us because of the difficulty in observing, collecting, and testing the root systems of trees. Although we have a great deal of information on a few root rot problems, we have just barely scratched the surface.

TYPES OF ROOT ROT

Root rots of seedlings in the nursery have some properties in common with root rots to be discussed in this chapter, but the tree nursery is a unique enough cultural system to warrant a specific chapter. Therefore, root rots of seedlings are discussed in Chapter 21.

We recognize two types of root rots. Structural root rots are caused by Basidiomycete decay fungi that parasitize the cambium of roots as well as cause decay in xylem cells. Decay is similar to other wood decays, but parasitism of cambium leading to death is somewhat unique for decay fungi.

The second type of root rot occurs on small feeder roots. It is caused by Phycomycete and Imperfect fungi that parasitize the feeder roots. The association of mycorrhizal fungi with feeder roots obviously places feeder root rot fungi in direct competition with mycorrhizal fungi, thereby producing a complicated three-organism interaction system.

MODE OF ACTION OF ROOT ROT FUNGI

Fungi that cause structural root rots have cellulase and ligninase enzymes typical of other decay fungi. The infected xylem root tissue is probably compartmentalized just like heart rot decay columns. Therefore, many trees are infected but never show typical aboveground symptoms. Sometimes the compartmentalization is slowly formed so that a major portion of the root system is affected. If enough of the tree root system is functionally destroyed, the tree will slowly deteriorate and die. Usually, affected trees break off during wind storms.

The feeder roots of plants are not permanent roots. They often function for less than a single season. For a tree to remain healthy, it must regenerate and maintain a certain amount of feeder root uptake surface. Degeneration of some feeder roots is therefore a natural cyclic process. A large number of soil microorganisms are involved in the decomposition of feeder roots.

Feeder root rots become a problem when soil conditions cause a buildup of specific feeder root decomposers that are mildly parasitic. Fungal pathogens of feeder roots are a normal part of the soil microflora. Diseases of feeder roots are usually associated with mineral imbalance, poor soil aeration, or other factors that affect the balance between feeder root development and degeneration. Senescence may not be balanced by regeneration and the trees slowly decline and die.

ROOT ROT DISEASE CYCLE

The disease cycles of structural root rots follow the same basic disease cycle as those of heart rot decay fungi. Infection of wounds by wind-disseminated basidiospores is followed by invasion and decay of the wood. The host response attempts to compartmentalize the infected tissue. At some point, sporulation completes the cycle.

Secondary spread of root rot fungi from tree to tree may occur where the roots of trees contact each other. *Armillariella mellea (Armillaria mellea)* has another mechanism for secondary spread. Rhizomorphs or aggregates of fungal hyphae that look like shoestrings grow out into the soil. If the rhizomorph makes contact with a susceptible root, it may initiate infection.

The rhizomorphs are the main means for the spread for *A. mellea.* In a western ponderosa pine stand, the fungus has spread for hundreds of years by rhizomorphs alone. This conclusion was based on laboratory tests for incompatibility factors between a number of isolates of the fungus from the same area. If two isolates freely merge when grown together in a petri dish, this indicates a lack of incompatibility factors and a probable common vegetative origin for the isolates.

The disease cycle of a typical Phycomycete feeder root pathogen involves infection of wounds by motile zoospores. The zoospore, being motile, can move toward specific infection points. Gradients in concentrations of chemicals diffusing from senescing roots provide stimuli for zoospores to follow.

Germination, infection, and invasion of the weakened root proceeds in sequence to quickly utilize the substrate.

Soil moisture and temperature conditions are often ideal for fungi and other microorganisms. Therefore, great numbers of microorganisms compete for any available substrate. Evolutionary adaptions for successful competitive survival in soil are many. Rapid colonization of a substrate to exclude other competitors is one. Production of antibiotics that inhibit competitors is another. The ability to recognize chemical stimuli for triggering of spore formation, spore germination, or movement toward a point source is a third. The ability to produce a dormant stage to successfully survive periods during which a nutrient supply is lacking is a fourth adaptation of successful soil competitors.

If we again pick up the disease cycle of our feeder root pathogen, we see that two evolutionary adaptations have already been described: utilization of chemical stimuli and rapid invasion. The organism reproduces asexually by means of zoospores in sporangia or may go dormant by producing a sexual oospore.

The oospore forms from the fusion of two modified hyphal structures, a small antheridium and a large oogonium. Transfer of nuclei from the antheridium to the oogonium brings compatible genomes together. Fusion of nuclei and meiosis completes the sexual process. Germination of the oospore to form a sporangium and zoospores reinitiates the active growth phase.

SYMPTOMS OF ROOT ROT

Initially, root rots look like nutrient deficiencies. With reduced water and mineral uptake, the foliage becomes smaller and yellowed. Growth is reduced, so that foliage may appear tufted at ends of branches.

It is difficult to observe symptoms of small feeder root pathogens on the roots, but decay of structural roots becomes obvious when trees fall over or if root systems are excavated.

METHODS FOR DIAGNOSIS OF ROOT ROT

The aboveground symptoms, given above, provide some indication of root rots, but they also suggest other problems, related to nutrients, lack of mycorrhizae, nematodes, air pollution, and so on. The appearance of decayed roots is useful for diagnosis. These details are provided below, together with specific treatment for various root rots.

The presence of characteristic fruit bodies or fungus structures of the causal organism are good diagnostic characters. Isolation and identification of the organism in culture is additional diagnostic evidence sometimes used when other diagnostic characters are lacking.

The methods we have described apply primarily to structural root decay. The fruiting structures of feeder root pathogens are microscopic and are therefore not useful as diagnostic criteria in the field. Isolation of microorganisms from roots on culture media seldom recovers the plant pathogen, because of the wide array of competitive saprophytes, which rapidly colonize the media. Therefore, a very useful technique for the diagnosis of feeder root pathogens is to use a bait to selectively trap the pathogen and then culture from the bait. A number of baits have been used. For *Phytophthora cinnamomi* root rot, baits like apple or germinated lupine seed are used. A sample of soil is placed in a hole bored in the apple or a sample of soil mixed in water is used to float the germinating lupine seeds. After a few days, the fruit or the lupine root is surface-sterilized and then the fungi are isolated on culture media from browned tissues.

EXAMPLES OF ROOT ROT AND CONTROL

Armillariella mellea

Armillariella mellea (Armillaria mellea) causes a root rot of both hardwoods and conifers. *A. mellea* root rot is usually but not exclusively characterized by common association with weakened hosts. It is a contributing factor in many declines, so it will be mentioned again in Chapter 18. It can also become a problem when replanting orchards or plantations where there are extensive stumps remaining from the former plantation (Fig. 16-1). This fungus is a good sap rot fungus on dead stumps and dying trees but is not satisfied with its saprophytic existence. Rhizomorphs (aggregates of fungus hyphae), which look like shoe strings, move out into the soil in search of additional food. If they contact a healthy root, penetration and death of the cambium may occur, thereby providing another dead root and eventually an entire root system, for its saprophytic existence.

FIGURE 16-1 Regeneration in this Oregon forest is being killed by *Armillariella mellea.*

FIGURE 16-2 Light-brown-colored *Armillariella mellea* mushrooms are common in the fall in northeastern conifer as well as in hardwood forests. (Photograph compliments of Dr. Savel Silverborg.)

Armillariella mellea is a gilled mushroom fungus that develops most prolifically in the fall of the year (Fig. 16-2). The white to tan mushroom is recognized by its white spores and a ring of veil tissue remaining on the stipe once the gills are exposed.

Trees infected by *A. mellea* develop thin, chlorotic crowns. The cambium at the base of the tree is invaded by a mycelial fan that can be readily recognized by chopping some of the bark away (Fig. 16-3). Advanced infection results in the death of trees.

FIGURE 16-3 White mycelial fan exposed by chopping some of the bark away is typical of *Armillariella mellea* infection. Extensive resin production, not readily visible in the photograph, is also a symptom of infection in conifers.

FIGURE 16-4 Black shoestring-like rhizomorphs of *Armillariella mellea.*

Armillariella mellea is often a problem that develops on hardwood sites that are regenerated artificially with conifers. Chemical or mechanical killing of poor-quality hardwoods followed by planting with conifers is theoretically one means of increasing productivity. The hardwood stumps are ideal sites for the development of *A. mellea,* which then moves out by means of rhizomorphs to infect the developing conifers (Fig 16-4).

There are no specific control recommendations for *A. mellea* except to keep trees in a vigorous condition. A second suggestion is to avoid the accumulation of large numbers of stumps or dying trees in the vicinity of future crop trees.

Heterobasidion annosum

Heterobasidion annosum (Fomes annosus) is potentially an important root rot pathogen of conifers wherever they are grown. In the natural stand, the fungus is a typical root and butt rot pathogen of mature trees, much like *Phaeolus schweinitzii.* The root systems are decayed such that the tree is liable to windthrow.

In plantations of closely spaced trees, roots are often grafted to each other to produce one large root system with many stems. Thinning of the plantation results in a large number of cut stumps, which are really many wounds to the large, multistemmed tree. Wounds are the normal point of infection for *H. annosum,* so the freshly exposed stumps provide ideal infection sites. Development of the fungus in the stump and associated root system causes death and decomposition of the infected tissue. Expansion of the area of invaded root system occurs through root grafts and root contacts. The root system supplying a number of stems is consumed, so that these stems die or are windthrown. The fungus continues to expand through the root system of the multiple-stemmed tree for a number of years. For some reason, the infection centers eventually stop expanding.

The amount of timber lost is related to the number of cut stumps where infections take place and the time after wounding. Losses in plantations are also related to soil conditions and the amount of inoculum present. Plantations on light sandy soil are more severely affected by *H. annosum.* The severity of the annosus problem is increased if inoculum is readily available from infection by previous cutting or adjacent stands.

The characteristic symptom of an annosus infection is an opening in the stand, with trees ranging from dead and windthrown in the center to chlorotic, thin-crowned trees on the periphery (Fig. 16-5). As the fungus slowly consumes more and more of the root system, the stems show chlorosis and the poor growth that is characteristic of nutrient deficiencies. Examination of the decay of affected trees should reveal a stringy white rot, characteristic of the decay of the fungus *H. annosum* (Fig. 16-6). Fruit bodies of the fungus are brown on the top with a white pore surface underneath.

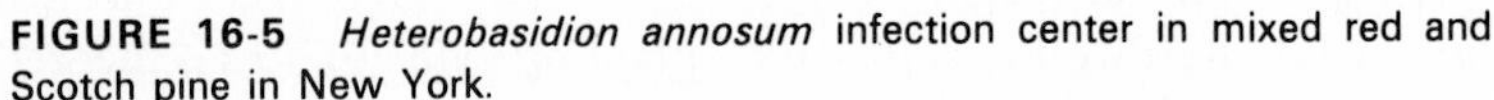

FIGURE 16-5 *Heterobasidion annosum* infection center in mixed red and Scotch pine in New York.

FIGURE 16-6 Fruiting body of *Heterobasidion annosum* on the root system of a fallen tree. Note also the stringy root rot characteristic of annosus decay.

They are formed on the stems under the duff or often on the underside of the stump and roots of fallen trees (Fig. 16-7).

It is important to be able to recognize the presence of *H. annosum,* because if you fail to recognize the disease and take corrective measures, the future of the present stand, as well as future stands you may expect to grow on the site, is in serious jeopardy. Regenerated stands on formerly infected sites will become infected from the old infection centers, so one must know where infection occurs.

If *H. annosum* is present in the vicinity of a plantation, it is advisable to treat stumps with urea, borax, or sodium nitrate at the time the stems are cut. Which treatment to use is a matter of local preference. The main concern is treatment immediately after cutting. No treatment provides an absolute guarantee that infection will not take place, so one should thin as few times as is reasonably acceptable for proper development of the stand.

Another possible treatment of the future is to inoculate with competitive microorganisms. *Peniophora gigantea* is a decay fungus that is effective as a stump treatment. The material is not generally available now, but it may be an alternative to chemical treatments in the future.

FIGURE 16-7 *Heterobasidion annosum* fruit bodies observed by scratching back the duff from the base of a symptomatic tree. (Photograph compliments of Dr. Savel Silverborg.)

There are no good recommendations as to what to do for a seriously infected stand. Therefore, it is very important for forest managers to recognize the presence and seriousness of this problem if they are to avoid destroying the productivity of conifer lands.

Phellinus weirii

Phellinus weirii (Poria weirii) is another root rot pathogen. It lacks the wide geographic distribution and host range of the former root rots but makes up for these by disrupting the potential future production of one of the most important timber species of North America, Douglas fir, in the Pacific Northwest (Figs. 16-8 to 16-11).

Like the former two pathogens *P. weirii* is a normal root rot fungus of mature trees. It becomes a serious threat to second-growth Douglas fir about 50 years after the stand is established. Ten- to 20-in.-diameter (25 to 51cm) rapidly growing trees are killed and fall like matchsticks, producing a very serious problem in maintaining full stocking for maximum production from the site.

The infection centers often appear to center on a large stump left from the previously cut stand. Aerial photographs of some infection centers appear ring-like. The inner portion of the ring or the center of the infection is

FIGURE 16-8 *Phellinus weirii*-killed Douglas fir in Oregon.

FIGURE 16-9 *Phellinus weirii* initially established in the old Douglas fir stump in Oregon is responsible for mortality in these 10- to 15-inch (26-38 cm) Douglas fir trees 50 years later.

FIGURE 16-10 Resin exuding from the base of this Douglas fir is a symptom of infection by *Phellinus weirii.*

FIGURE 16-11 Fallen tree on Vancouver Island, showing extensive stringy brown rot of the root system typical of *Phellinus weirii.*

invaded by hardwoods and brush species. The distinct, outer rim of the ring is the killing front.

The serious destruction of future stands by this disease has only recently been considered by foresters and pathologists of the Pacific Northwest. Much about this disease is yet to be uncovered. No good prevention or control measure is known. If one can recognize the infection centers before cutting, these should not be allowed to regenerate to Douglas fir. Pines or hemlock are suggested for replanting in known infection centers.

A different approach to control or understanding of the disease is presently being considered. It involves the role of red alder in affecting soil nitrogen. Mention of this was made in Chapter 7. The form of nitrogen significantly affects microorganisms, populations, and interactions as well as tree vigor. This is obviously a difficult phenomenon to understand enough about to be able to manipulate, but if it can be worked out, it may represent a significant step forward in the management of trees with minimum losses due to root pathogens.

Phytophthora cinnamomi

Feeder root disease by *P. cinnamomi* is destroying extensive natural forest vegetation in Australia and New Zealand. A forest type that was originally composed of better than 100 species is being reduced to two to three brush species by this pathogen.

It is not known if this pathogen was introduced into the area. The behavior in natural stands appears like an introduction, but the association of killing fronts by this disease with road construction or in poorly drained sites suggests environmental disturbance and expansion of a native pathogen. Road construction may expand the range of the pathogen into natural areas, but it also produces site modifications and stressed trees due to changed drainage patterns and more exposure. A killing front moves out from the road like a wave into the natural stand, well back from the original disturbance. The death of trees may be enough of a site disturbance to maintain the disturbance and expansion of killing front into the natural vegetation.

The role of site factors in *P. cinnamomi* root rot in the southeastern United States is also evident. In the Southeast, a disease called littleleaf is presently assumed by many to be caused by *P. cinnamomi.* I depart from many forest pathologists and suggest that this disease is better understood as a decline. The significant role of soil nutrition and soil oxygen on plant health should not be underestimated by pathologists who are traditionally looking for a single specific pathogen to explain diseases. Additional discussion of the role of *P. cinnamomi* as a pathogen or contributing factor in a decline is provided in Chapter 18.

FOREST PRACTICES AND ROOT DISEASES

H. annosum, A. mellea, and *P. weirii,* are important diseases only because of modern forest practice. Modern forest practice emphasizes maximum fiber production per acre, and this is translated into tightly spaced pure stands or conversion to a pure stand, usually conifers. Another aspect of modern forest practice is tc grow the same species over and over on the same site. Stop and think of how many conifer species grow as climax types in pure stands. Most are pioneer species developing pure stands only after major site changes, usually caused by fire. These root rot fungi and others will continue to be serious problems until more natural forest management systems are developed.

REFERENCES

ARTMAN, J. D. and W. J. STAMBAUGH. 1970. A practical approach to the application of *Peniophora gigantea* for control of *Fomes annosus.* Plant Dis. Rep *54:*799–802.

CHILDS, T. W. 1970. Laminated root rot of Douglas fir in western Oregon and Washington. USDA For. Serv. Res. Pap. PNW-102. 27 pp.

FROELICH, R. C., E. G. KUHLMAN, C. S. HODGES, M. J. WEISS, and J. D. NICHOLS. 1977. *Fomes annosus* root rot in the south, guideline for prevention. USDA For Serv. Southeast. Area State and Private Forestry. 17 pp.

GIBBS, J.N. 1967. The role of host vigour in susceptibility of pines to *Fomes annosus.* Annu. Bot. *31:*803–815.

GROSS, H. L. 1970. Root diseases of forest trees in Ontario. Can. For. Serv. Ont. Reg. For. Res. Lab., Sault Ste. Marie, Inf. Rep. O-X-137. 16 pp.

HODGES, C. S. 1969. Modes of infection and spread of *Fomes annosus.* Annu. Rev. Phytopathol. *7:*247–266.

HODGES, C.W. 1974. Cost of treating stumps to prevent infection by *Fomes annosus.* J. For. *72:*402–404.

HODGES, C. W. 1974. Symptomatology and spread of *Fomes annosus* in southern pine plantations. USDA For. Serv. Res. Pap. SE-114. 10 pp.

HOUSTON, D. R., and H. G. ENO. 1969. Use of soil fumigants to control spread of *Fomes annosus.* USDA For. Serv. Res. Pap. NE-123. 23 pp.

HUNTLY, J. H., J. D. CAFLEY, and E. JORGENSEN. 1961. Armillaria root rot in Ontario. For. Chron. *37:*228–236.

LEAPHART, C. D. 1963. Armillaria root rot. USDA For. Serv. For. Pest Leafl. 78. 8 pp.

NEWHOOK, F. J. and F. D. PODGER. 1972. The role of *Phytophthora cinnamomi* in Australia and New Zealand forests. Annu. Rev. Phytopathol. *10:*299–326.

PODGER, F. D. 1975. The role of *Phytophthora cinnamomi* in dieback disease of Australia eucalyp forests. *In* Biology and control of soil-burn plant pathogens, Ed. G. W. Bruehl. The American Phytopathological Society, St. Paul, Minn. pp. 27–36.

REDFERN, D. B. 1975. The influence of food base on rhizomorph growth and pathogenicity of *Armillaria mellea* isolates. *In* Biology and control of soil-born plant pathogens, G. W. Bruehl. The American Phytopathological Society, St. Paul, Minn., pp. 69–73.

RISHBETH, J. 1975. Stump inoculation: A biological control of *Fomes annosus. In* Biology and control of soil-born plant pathogens, Ed. G. W. Bruehl. The American Phytopathological Society. St. Paul, Minn., pp. 158–162.

ROSS, E. W. 1973 *Fomes annosus* in the southeastern United States: Relation of environmental and biotic factors to stump colonization and losses in the residual stand. USDA For. Serv. Tech. Bull. 1459. 26 pp.

SHAW III, C. G. and L. F. ROTH. 1976. Persistence and distribution of a clone of *Armillaria mellea* in a ponderosa pine forest. Phytopathology *66:*1210–1213.

SINCLAIR, W. A. 1964. Root- and butt-rot of conifers caused by *Fomes annosus,* with special reference to inoculum dispersal and control of the disease in New York. Cornell Univ. Agri. Exp. Stat. Mem. 391. 54 pp.

TRAPPE, J. M. 1972. Regulation of soil organisms by red alder: potential biological system for control of *Poria weirii. In* Managing young forests in the Douglas fir region, ed. A. B. Berg. Ore. State Univ. School For., Corvallis, Ore., pp. 35–51.

17

PARASITIC FLOWERING PLANTS AS AGENTS OF TREE DISEASES

- Types of parasitic flowering plants
- Mode of action of parasitic flowering plants
- Disease cycle of parasitic flowering plants
- Symptoms of diseases caused by parasitic flowering plants
- Methods for recognition of diseases caused by parasitic flowering plants
- Examples of parasitic flowering plants and control

Coevolution of closely associated flowering plants has produced a number of obligate associations. Evolution of the parasitism of one plant on another has generally developed into a balanced relationship wherein the host population is not seriously affected by the parasite. But the imposition of human influence on natural processes of "undisturbed" areas, cultivation of trees in plantations, and management of natural forests for fiber upsets the balances to produce a number of problems.

TYPES OF PARASITIC FLOWERING PLANTS

All parasitic flowering plants are Dicotyledons. No parasitic Monocotyledons or Gymnosperms are known. It is assumed that evolution toward parasitism has occurred at least eight times in unrelated groups of Dicotyledons.

The absence of one common ancestor for all parasitic plants has produced a number of different types of specific associations. Nevertheless, a certain amount of common structure and function is found in most parasites. The following groups of plants have some representative parasites. The individual parasitic representatives within each group may have more in common with nonparasitic members of the same group than with other parasites of other groups.

1. The order Santalales has a number of parasitic flowering plants in the families Loranthaceae (mistletoe), Viscaceae (dwarf mistletoe), Santalaceae (sandalwood), Olacaceae, and Myzodendraceae.
2. The families Scrophulariaceae and Orobanchaceae (figwarts and broomrapes) have 26 parasitic genera. In the Northeast, *Melampyrum* (cow-wheat), *Castilleja* (Indian paintbrush), *Euphrasia* (eyebright), *Gerardia* (downy false foxglove), and *Pedicularis* (lousewort) are in this group. *Striga* (witchweed) and *Senna* (seymeria) are two southern parasitic plants in this group.
3. The families Rafflesiaceae and Hydnaraceae are mostly tropical parasitic flowering plants. *Rafflesia* produces the largest known flower (1 meter in diameter) from a subterranean root parasitic system.
4. The family Balanophoraceae is made up of small groups of tropical fungus-like plants. Tuberous rhizomes characterize these root parasites.
5. The genus *Cuscuta* (dodders), in the otherwise autotrophic family Convolvulaceae, resembles nonparasitic plants more than parasitic plants, but have evolved modifications typical of plant parasites. There are about 158 species of this widely distributed group of annual plants.

6. The genus *Cassytha* (scrub dodder) in the otherwise autrotrophic family Lauraceae is similar to dodder. In contrast to dodder, it is perennial and regionally confined to coastal areas.
7. The family Lennaceae is a small family of xerophytic parasitic plants found only in America.
8. The family Krameriaceae is a small group of xerophytic parasitic plants found mostly in Mexico.

Some of the plant parasites are totally dependent upon the host for preformed photosynthates (holoparasitic), while others are only partially dependent upon the host for photosynthates, because they still maintain some photosynthetic leaf surface (hemiparasites). Often the leaf is much reduced, so that photosynthetic capacity is minimal. Parasitic plants may be perennial or annual. Some, like sandalwood, are trees. Some parasitic plants are associated with roots. Others occur as stem parasites. A major portion of the life cycle of a parasite may be spent below the ground (subterranean) or within the host (endophytic). Systemic infection of meristems of hosts sometimes results in major expansion and development of the endophytic system.

MODE OF ACTION OF PARASITIC FLOWERING PLANTS

The mode of action of all parasitic plants revolves around the haustorium or parasitic nutrient-uptake organ. These organs of parasitic plants have no obvious morphological link to such organs as the roots and stems of autotrophic plants.

The haustorium functions like a root but has no root cap. Root hairs are absent or very reduced. Phloem is absent or highly reduced. The organ is composed mostly of vessel members and a few tracheids.

Haustoria may be compact organs that penetrate localized regions of host root or shoot tissues, or they may be much branched and ramify generally throughout host tissues. In some dwarf mistletoes, the haustorium or endophytic system grows just behind the apical meristem. In this way, the haustoria develop systemically in the most recent tissues.

It is not fully understood what causes haustoria to penetrate plants or what the haustoria remove from the plant. Haustorial cells align with host phloem or xylem elements, with emphasis on xylem vessels, and produce a continuity bridge between the two plants through pits or through resorbed host vessel walls.

Through the haustoria, elements are transferred from one plant to another. One might assume that carbohydrates and other complex organic materials are absorbed from phloem elements, but organic materials such as sugars, nitrogenous compounds, enzymes, and minerals, as well as water, can also be translocated from xylem vessels.

The continuity of the parasite with the host plant is sufficient to allow transfer of viruses and mycoplasmas between plants. Dodders are used by researchers to establish host transfer. A mycoplasma disease, sandal spike, is acquired by the sandal plant from a number of host trees.

There is generally a low degree of host specificity for flowering plant parasites, although many are restricted to a limited host range by ecological circumstances. Mechanisms of resistance other than escape are very poorly understood.

DISEASE CYCLE OF PARASITIC FLOWERING PLANTS

Seeds of parasites may be small, numerous, and wind-disseminated or large, few, and disseminated by birds. Dwarf mistletoe seeds have a unique forcible discharge mechanism for propelling the seeds up to 30 ft (9m), further if helped by the wind.

Germinating seeds form a radical that may be chemotactically attracted toward the host plant. Upon contact with the host, touch (thigmotropic) or chemical (chemotrophic) stimuli induce a cushion or holdfast to form a fine, intrusive organ to enzymatically and mechanically penetrate the host.

A primary haustorium or endophytic system develops from this penetration peg of the radical tip. Secondary haustoria may develop in some parasites from branching of the radical.

The endophytic existence for the parasite may be lengthy with a brief period of shoot growth for flowering, or the parasite may produce annual or perennial shoots with photosynthetic capacity.

SYMPTOMS OF DISEASES CAUSED BY PARASITIC FLOWERING PLANTS

Effects on parasitized hosts range from excessive hypertrophy and witches' brooms to almost no effect. Reduced reproductive capacity, chlorosis, and reduced growth are common symptoms.

METHODS FOR RECOGNITION OF DISEASES CAUSED BY PARASITIC FLOWERING PLANTS

Recognition of diseases caused by parasitic plants always involves observation and identification of the parasite. Specific symptoms such as witches' brooming, branch swelling, or remnants of shoots of the parasite may be used for diagnostic purposes if one is rather familiar with a specific disease. Caution should be exercised in the diagnosis of parasitic plant diseases in the absence of an identifiable parasite because fungi, viruses, mycoplasmas, and other parasites, as well as environmental and genetic factors, will also induce these types of symptoms.

EXAMPLES OF PARASITIC FLOWERING PLANTS AND CONTROL

Although Indian paintbrush and many other flowering plants of the Northeast are parasites of other plants, little is known of their effects on the parasitized hosts.

Witchweed

The root parasite witchweed (*Striga asiatica*) is an introduced plant from the Old World Tropics which represents a serious threat to corn, wheat, and other cultivated grasses and sedges of the South. The seeds of this parasite are difficult to separate from the grain, so that it is sometimes distributed with contaminated seed lots. Control is through use of herbicides and crop rotation. Because crabgrass is a host for witchweed, control may have to extend to areas surrounding fields.

Senna Seymeria

Seymeria cassioides is a newly recognized parasitic threat to slash pine and other southern pines. It weakens and sometimes kills pine seedlings if it becomes established in an area. It develops rapidly on reseeded or replanted sites where a few residual pines of a previous crop represent foci for infection centers. Control is by herbicide treatment of the succulent annual.

Dodders

Dodders (*Cuscuta* spp.) are of cosmopolitan occurrence and often result in large economic losses in field crops, particularly in the tropics (Figs. 17-1 and 17-2). Dodder seeds resemble seeds of many commercial crops, particularly legumes and flax. Before rigid methods of seed control were introduced, dodder seeds were often common in commercial seed lots; consequently, dodder species have been introduced into areas where they were not native. Dodders are seldom of importance in forest pathology. They have caused some problems in nurseries, particularly on black locust, green ash, and poplar.

Morphologically, dodders are highly reduced plants having reduced leaves, a coiled habit, little vascular development, and a reduced chlorophyll content.

FIGURE 17-1 Coiling shoots of dodder on understory forest vegetation.

FIGURE 17-2 Closeup of coiling shoot and haustoria penetrating red oak seedlings in the understory forest vegetation.

The radical from the germinating seed penetrates the ground sufficiently to support the seed and coiling shoot as it searches for a suitable host. Growth up to 35 cm over a 7-week period is possible before finding a host. Once contact is made with a host, the shoot coils around the host and produces penetration haustoria. The contact with the ground deteriorates and up to half a mile (0.8 km) of coiling shoot may develop.

Dodders have little or no ability for photosynthesis and thus depend on the host as a carbohydrate source. Their effects on the host often result in hypertrophies causing witches' brooms and galls.

Preventing the introduction of dodder seeds into nurseries is an effective method of control. Sprays are available, but these are often injurious to the host plants. Soil fumigation, where economical, may be desirable.

Leafy Mistletoe

The true or leafy mistletoes have a long history and a considerable folklore. There are several species found in North America. The genus *Phoradendron* is the most significant to forest pathology (Figs. 17-3 to 17-5). In the arid and semiarid Southwest, about a dozen species of this genus occur on a wide variety of hosts. They are most frequently associated with hardwoods, but several species occur on junipers, cypress, incense cedar, and true firs.

FIGURE 17-3 *Phoradendron villosum* on emery oak.

FIGURE 17-4 Closeup of leafy mistletoe.

FIGURE 17-5 Leafy mistletoe disease cycle.

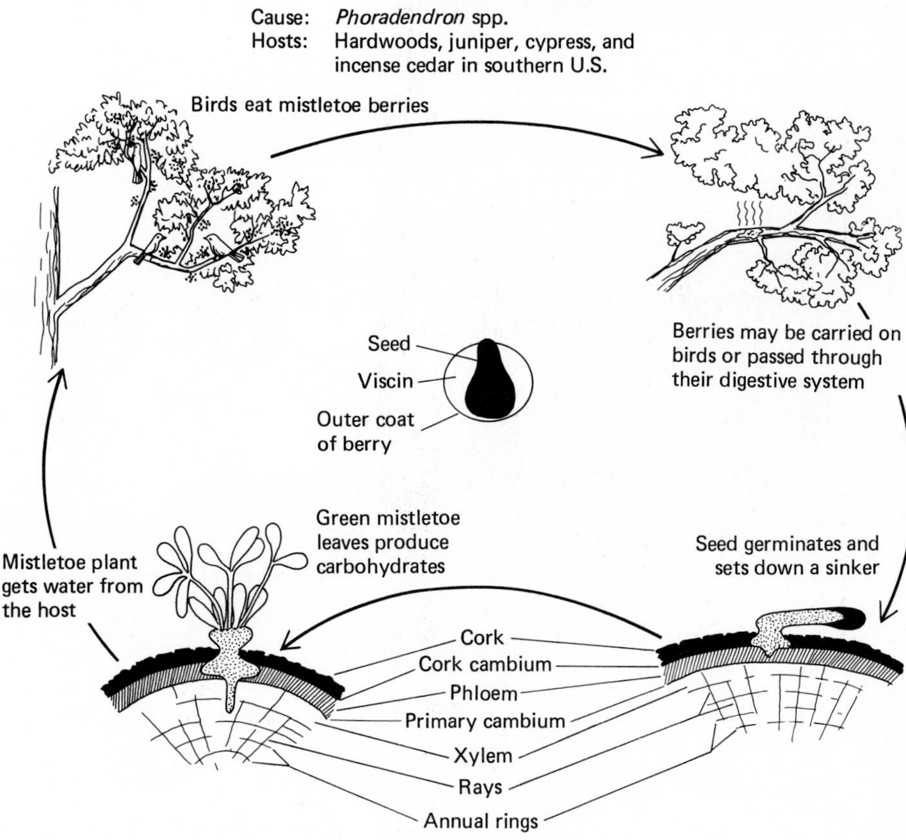

Cause: *Phoradendron* spp.
Hosts: Hardwoods, juniper, cypress, and incense cedar in southern U.S.

Birds eat mistletoe berries

Berries may be carried on birds or passed through their digestive system

Seed

Viscin

Outer coat of berry

Seed germinates and sets down a sinker

Green mistletoe leaves produce carbohydrates

Mistletoe plant gets water from the host

Cork

Cork cambium

Phloem

Primary cambium

Xylem

Rays

Annual rings

The leafy mistletoes are limited to warmer climates and do not occur above the 40° latitude line (New Jersey–Oregon). Northward extension is probably limited by temperature.

These plants are morphologically well developed with moderate-size leaves and stems. Birds, feeding on mistletoe berries, disseminate the seeds, which pass through the birds undigested. Mistletoes support much of their carbohydrate requirements by photosynthesis. They use the host xylem as a water source and may therefore become a problem when water availability to the host is limited. Infected portions of the host often exhibit galls and brooms, and the uninfected host tissues may become deformed with reduced growth. The death of the host is rarely due to the mistletoe alone. Economic losses, in terms of volume and usable product, may occur.

Control is by pruning, fire, and sprays where economical, such as for high-value crops such as fruit crops. In Oklahoma and a few other areas, the fruiting mistletoe may be gathered and sold as a Christmas ornamental.

Dwarf Mistletoe

Dwarf mistletoes (*Arceuthobium* spp.) are the most serious parasitic higher plants in North America. About 20 species, usually host-specific, are important. These organisms cause large amounts of economic losses and may be the most serious problem in western conifer forests (Fig. 17-6 to 17-9). They are a major problem on ponderosa pine, lodgepole pine, Douglas fir, true firs, western larch, and other western conifers. In the East, black spruce may be severely affected. Losses may result from mortality of older trees, growth reduction, poor quality of product, and an overall reduction in host vigor.

FIGURE 17-6 Brown shoots of *Arceuthobium gillii* on Chihuahua pine in Arizona.

FIGURE 17-7 Douglas fir branch systemically infected with *Arceuthobium douglasii* in Arizona.

FIGURE 17-8 Yellow orange shoots of *Arceuthobium vaginatum* on ponderosa pine in Arizona.

FIGURE 17-9 Closeup of *Arceuthobium vaginatum* on the main stem of a ponderosa pine in Arizona.

The portion of the tree invaded by the parasite becomes swollen, and witches' brooms sometimes occur. Physiological stimulation of infected portions result in reduced growth in other portions of the host. The disease occurs as patches of infected trees within a stand. Local spread of the disease occurs as a result of the forcible ejection of a viscous seed from the host (Fig. 17-10). Seeds shot from the host may have a range of about 100 to 200 ft (30-61m), depending on the height at which they were dispersed and on the wind. The seeds adhere to objects they strike, because of their sticky viscous coat (Fig. 17-11). When the viscous coating dries, it acts as a glue enabling the seed to hold fast to the host until germination and penetration occur. Seed dispersal usually occurs in the fall or late summer, but germination is usually delayed until spring. Once penetration of the host occurs, the parasite develops an extensive haustorial absorptive system, resulting in the production of "sinkers," which penetrate into the xylem of the host. They are subsequently embedded by additional growth of xylem tracheids. Vessel elements of the sinkers connect with tracheids of the host xylem, and the parasite derives most of its carbohydrate and probably all of its water and minerals from the host. The mistletoe life cycle, from infection to seed production, takes about 4 to 5 years (Fig. 17-12). Long-range dispersal, by external transport on birds, serves to introduce the parasite into new stands.

Control is by eradication of infected hosts by clear-cutting. Pruning and selective removal of large trees with crown infections may be practiced in high-value lightly to moderately infected stands.

Recognition that fire control is associated with the expanding importance of mistletoe has contributed somewhat to a second look at the role of fires in timber management. In some instances, the best management practice would be a major burn.

Understanding the dynamics of seed dispersal and the disease cycle has developed a method for learning to live with minimal losses due to mistletoe.

FIGURE 17-10
Closeup of *Arceuthobium* berries.

FIGURE 17-11
Seed of *Arceuthobium* sticking to pine needle.

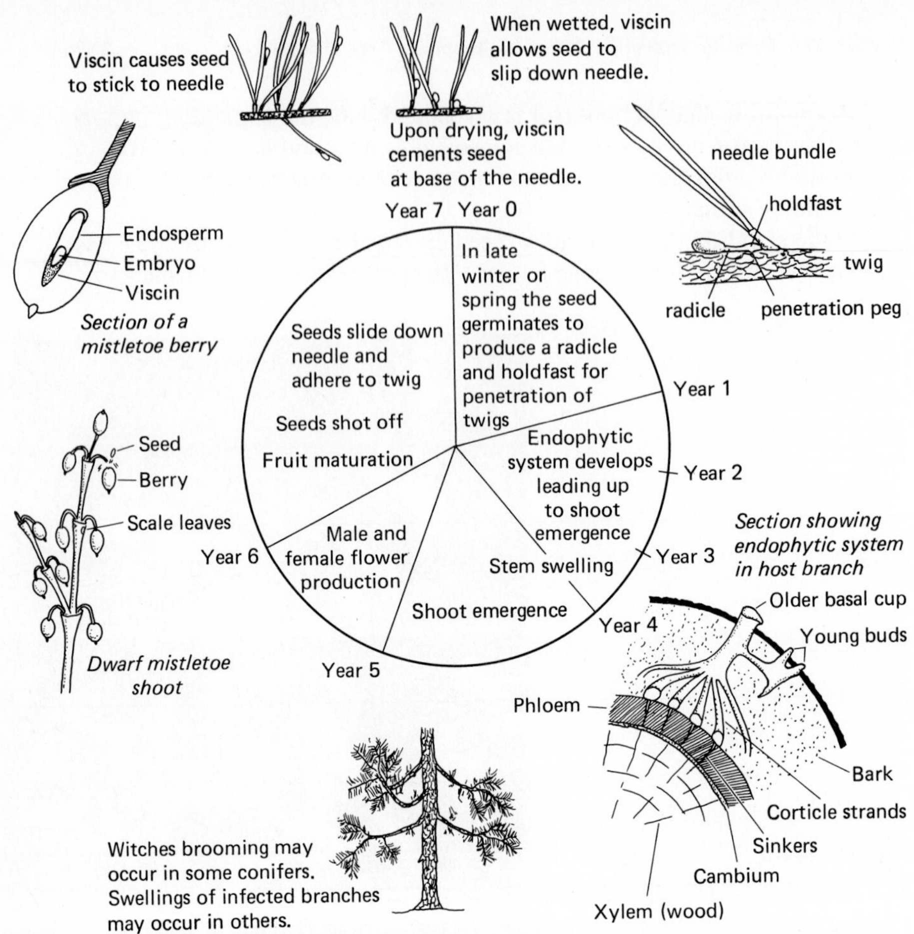

FIGURE 17-12 Dwarf mistletoe disease cycle.

The growth impact is minimal unless the plants are highly infected. Highly infected plants generally develop from understory plants below an infected overstory. Therefore, to reduce the impact on managed timber, it is important to selectively remove those trees in which the upper crown is significantly affected by mistletoe. The program involves first removal of infected overstory trees, followed by thinning of infected understory to remove most of the seriously affected trees. Pruning is used only to reduce infection if removal reduces stocking below adequate levels. Follow-up evaluations and additional sanitation cuts should keep the disease down. This type of intensive management may permit continued economic production from mistletoe-infected stands.

Extensive evaluation of dwarf mistletoe development in lodgepole and ponderosa pine forests of the Southwest has led to an integration of timber yield and loss figures into a computer-based management option system. The SWYLD (southwestern yield data) program provides a resource manager with projected yields for various management options. If a particular stand has a high incidence of upper-crown infection, the yields over the short and long run can be predicted for various levels of cutting. Lower crown infection of varying intensity results in different impacts on yields, depending upon thinning and pruning practices. Yields from two-storied stands can also be predicted, based on the intensity of the overstory infection and various cutting and thinning options.

The SWYLD integration of growth and disease-loss relationships, in relation to management decisions, represents one of the most advanced online contributions that forest pathologists have made to forest management. By recognizing the impact of various management practices on disease, it develops a philosophy of minimizing losses by learning to live with a disease agent. This philosophy should be a universal doctrine of both pathologists and forest resource managers.

REFERENCES

GILL, L. S., and F. G. HAWKSWORTH. 1961. The mistletoes, a literature review. USDA For. Serv. Tech. Bull. 1242. 87 pp.

GRAHAM, D. P. 1967. A training aid on dwarf mistletoe and its control. Issued by USDA For. Serv. Pac. Northwest Reg. Insect Disease Control Br., 49 pp.

HAWKSWORTH, F. G. 1973. Dwarf mistletoe and its role in lodgepole pine ecosystems. *In* Management of lodgepole pine ecosystems, symposium, ed. D. M. Baumgartner. Wash. State Univ., Pullman, Wash., pp. 342–358.

HAWKSWORTH, F. G. 1977. The 6-class dwarf mistletoe rating system. USDA For. Serv. Gen. Tech. Rep. RM-48. 7 pp.

HAWKSWORTH, F. G., T E. HINDS, D. W. JOHNSON, and T. D. LANDIS. 1977. Silvicultural control of dwarf mistletoe in young lodgepole pine stands. USDA For. Serv. Tech. Rep. R2-10. 11 pp.

HAWKSWORTH, F. G., and D. WIENS. 1972. Biology and classification of dwarf mistletoes (*Arceuthobium*). Agricultural Handbook 401. USDA Forest Service. 234 pp.

KUIJT, J. 1969. The biology of parasitic flowering plants. University of California Press, Berkeley, Calif. 246 pp.

MALCOLM, W. M. 1966. Root parasitism of *Castilleja coccinea.* Ecology *47*:179–186.

MANN, W. F., H. E. GRELEN, and B. C. WILLIAMSON. 1969. *Seymeria cassioides,* a parasitic weed on slash pine. For. Sci. *15*:318–319.

SCHARPF, R. F., and J. R. PARMETER, JR., eds. 1978. Proceedings of the symposium on dwarf mistletoe control through forest management. USDA For. Serv. Gen. Tech. Rep. PSW-31. 190 pp.

18

DECLINE DISEASES OF COMPLEX BIOTIC AND ABIOTIC ORIGIN

- The decline syndrome
- Common denominators of declines
- Symptoms of declines
- Examples of declines
- Ecological role of declines

Preceding chapters have dealt with single-causal-factor abiotic and biotic diseases. This chapter develops a third category of disease, the declines, which are caused by the interaction of a number of interchangeable, specifically ordered abiotic and biotic factors to produce a gradual general deterioration, often ending in the death of trees.

The concept of declines as a distinct category of disease is not totally accepted by forest pathologists. Some would assume that the category decline is a collection of diseases with incompletely understood etiology. Declines are poorly understood, and therefore some of the diseases we call declines may eventually be shown, by additional investigation, to have specific single causal agents.

Hepting (1963) suggests the involvement of significant climatic changes in declines. He presents evidence of a climatic warming trend during the middle part of this century. It is very difficult to verify the role of climatic changes in disease because it is difficult to even document climatic change.

Plant development or deterioration is intimately tied to weather conditions. Weather is just a short-term reflection of climate. Most declines are associated with some type of local or regional climatic conditions. Therefore, it is very reasonable to accept the role of continental climatic changes in declines.

The possibility of ecological succession involvement in declines is presented at the end of this chapter. It is well to recognize from the very beginning that we have a limited conceptual base on which to build our presentation of declines.

THE DECLINE SYNDROME

Details of specific declines will be covered later, but it is useful for development of the concept at this point to refer to a specific decline. Attempts to understand maple decline from standard pathogen-disease or cause-and-effect relationships have demonstrated the singular roles of soil moisture, soil aeration, nematodes, *Armillariella mellea* root rot, sugar maple borers, high temperatures, Verticillium wilt, canker fungi, insect defoliators, root excavation, salt, injuries, and air pollution (Fig. 18-1).

A complex decline disease may involve three or more sets of factors (Fig. 18-2). The first set of factors are generally static or nonchanging factors such as the climate, the soil type or site, the genetic potential of the tree, and the age of the tree. These are the *predisposing factors* that weaken a plant growing in the the wrong location. Such factors put a permanent stress on the plant and predispose it to the actions of other factors. The second group of factors are called *incitants*. These are short in duration and may be physical or biological in nature. Examples of incitants are insect defoliators, late spring frost, drought, and salt spray.These generally produce a drastic injury. The plant attempts to recover but has difficulty because of the predisposing stress of the environment. The third group of factors, called *contributing*, finally begin to appear. Bark beetles, canker fungi, root and sap rot fungi, viruses, and mycoplasmas produce noticeable symptoms and signs on the weakened host. These organisms are persistent and are often blamed for the condition of the host. They are better understood as indicators of weakened hosts. Eventually, the plant dies or is rendered useless as an ornamental or forest tree.

FIGURE 18-1 Summary of the factors associated with maple decline.

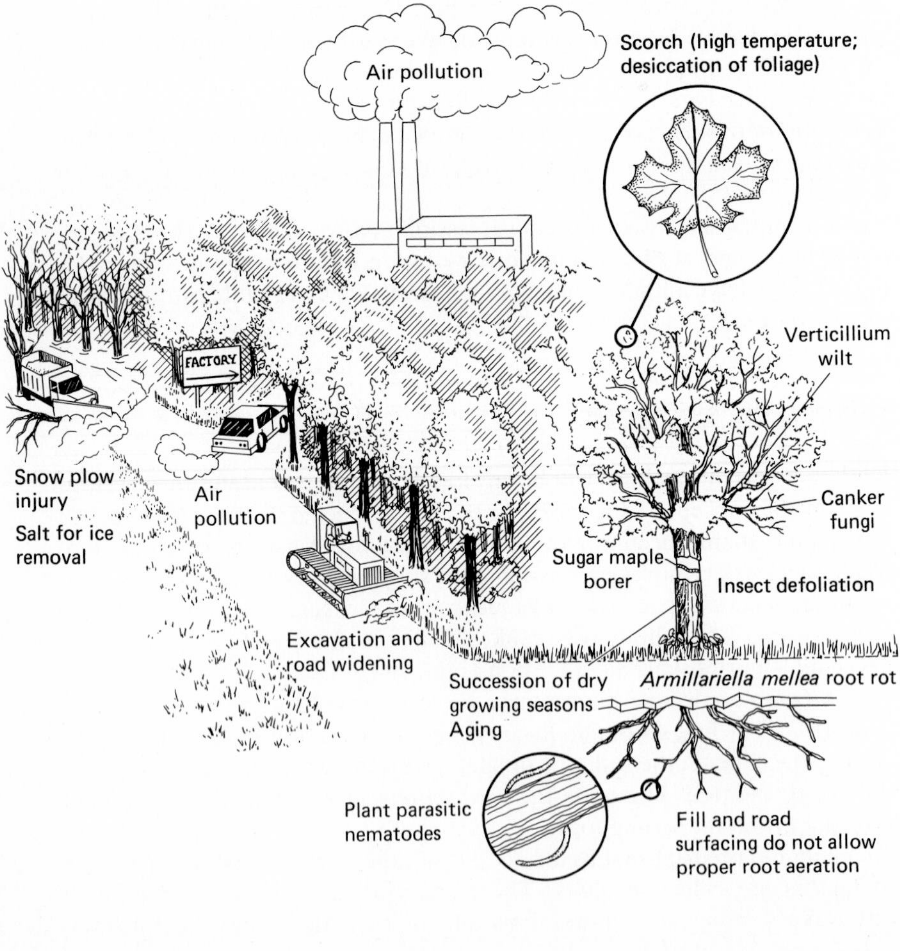

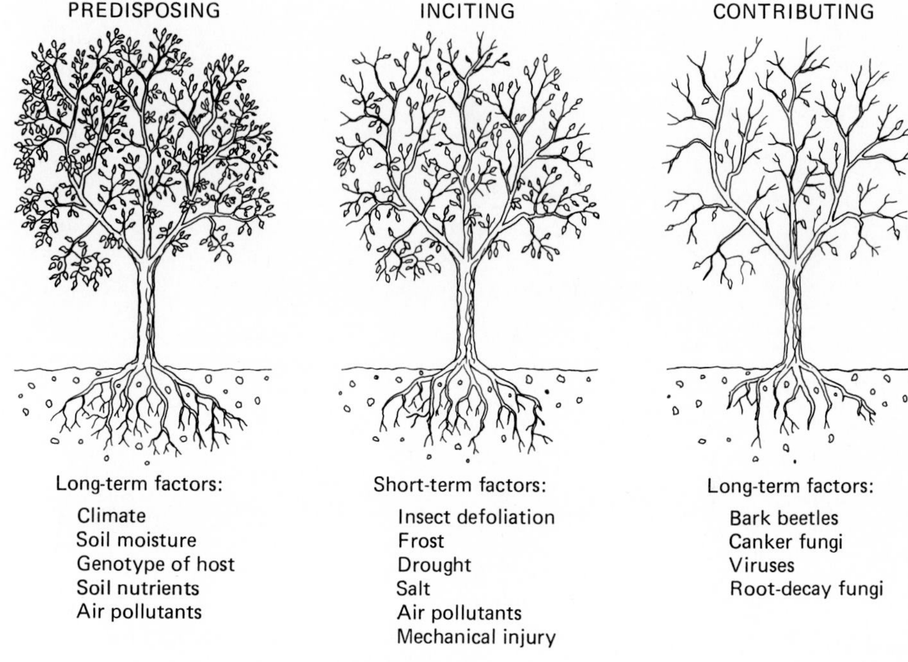

FIGURE 18-2 Categories of factors influencing declines.

I like to think of the relationships between the various factors as a dynamic spiral obstacle course (Fig. 18-3). The factors are like barriers in the course. During the life of any tree, it will continuously and repeatedly encounter an array of stress factors. The predisposing factors nudge the tree inward to the second level of the spiral. The inciting factors accentuate the inward spiral of predisposed trees. The contributing factors of the inner spiral compete for a niche on the weakened and dying tree.

COMMON DENOMINATORS OF DECLINES

1. Many interchangeable factors are involved in the decline syndrome. This implies that more than one factor can produce a given effect and, therefore, if punch number one does not produce an effect, punch number two or three may be involved.

2. At least three factors are involved in a decline, one each from the categories of predisposing, inciting, and contributing factors.

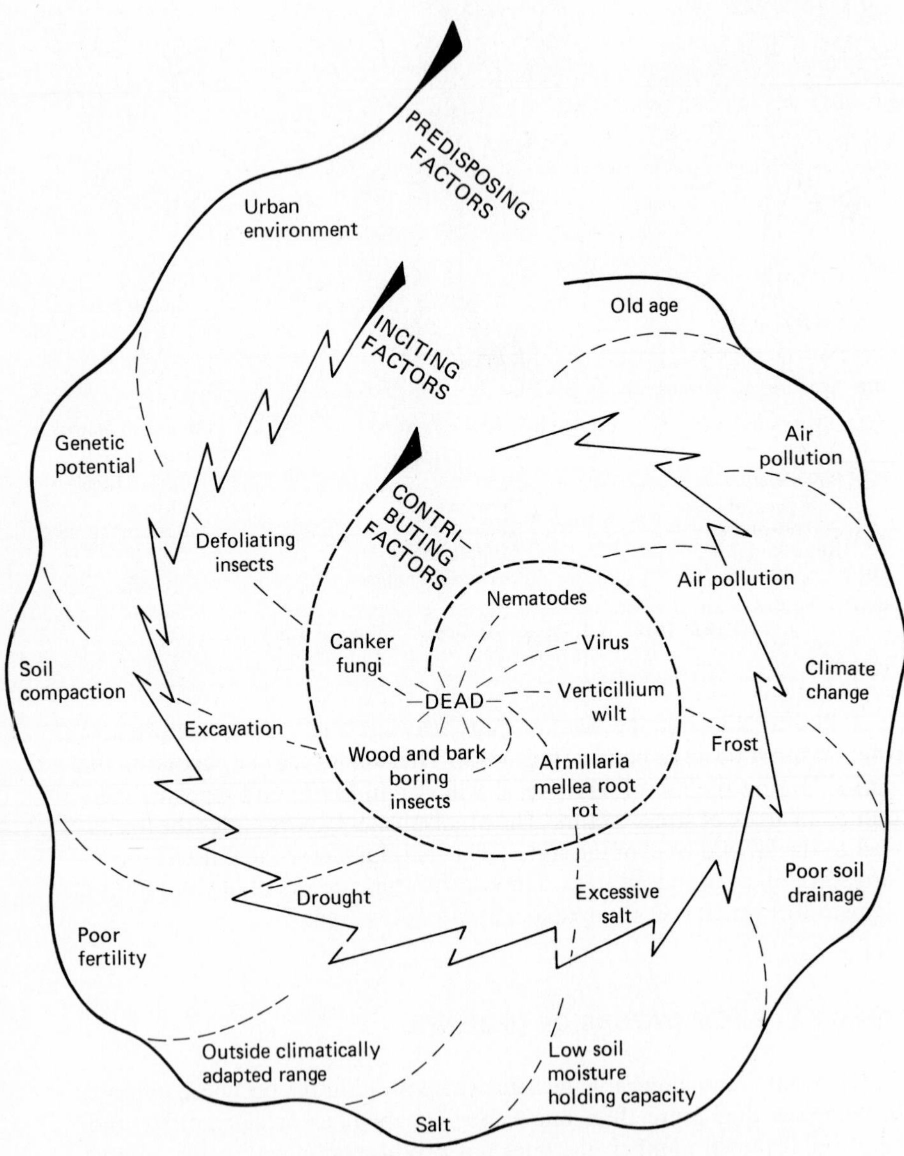

FIGURE 18-3 Decline disease spiral.

3. Weak nonaggressive fungus pathogens and insects are often involved as contributing factors and may be given more credit for the condition than they really deserve.

4. Climatic or site factors are almost always major predisposing or inciting factors in the decline syndrome. Examination of site or climatic history of a decline syndrome will often show a weak, but nevertheless real correlation with site or climate. The reason an absolute correlation does not occur is because of the importance of sequential timing of the inciting and contributing factors.

5. Feeder roots and mycorrhizae (see Chapter 9) degenerate prior to onset of symptoms in the aboveground portion of the trees. Quantification of root conditions is a difficult task, particularly with large trees. A balance normally develops between uptake capacity and demands upon the root system by the crown of the tree. Degeneration and regeneration of feeder roots and mycorrhizae is a continual process necessary to keep pace with parasitic microorganisms in the soil and seasonal growth demands of the crown. Imbalance in the regenerative capacity of roots or excessive demands of the top will accentuate degeneration of root.

6. Declining trees usually have a serious depletion of storage-reserve carbohydrates. Storage reserves are used to start spring growth or regenerate fungus- or insect-defoliated leaves. Excessive continuous demands on the reserves deplete the storage reserves and limit the tree's ability to respond.

7. *Armillariella mellea* was discussed more fully in Chapter 16. This common contributing factor in declines needs specific mention. There are very few decay fungi that can both decay xylem, like sap rot fungi, and cause the death of cambium in living trees, like parasites. This unique capacity allows *A. mellea* to become established as a decay fungus on weakened or dead trees and expand its sphere of influence by parasitizing and causing the death of both weakened and relatively healthy trees in the immediate vicinity.

8. Viruses are another common denominator of at least two hardwood declines. Viruses were more fully discussed in Chapter 6. I am only speculating on the role of viruses when I name them as contributing factors. Therefore, an explanation is in order of why viruses are given only contributing-factor status, rather than incitant-factor status. The study of viruses of native vegetation has only been explored extensively as it relates to agricultural crops. Weeds are common reservoirs of viruses that plague agricultural crops. There is no reason to expect that trees are less commonly virus-infected than weeds are. If we make the assumption that virus presence in trees is the rule rather than the exception, then to find viruses in ash dieback and birch dieback is not unexpected. But why are symptoms of viruses more commonly observed under these conditions? Viruses often

produce no symptoms. If most plants of a given species have viruses, it is difficult to predict what a nonvirused plant will look like. Environmental and health conditions of plants will influence the symptoms and effects produced on plants by viruses.

To minimize the role of virus on trees is a misinterpretation of viruses. I would equate viruses to the nonproductive members of a household, such as dogs, cats, and other pets. In times of plenty (ideal environment) the consuming aspects of pets (viruses) is not noticeable in a smooth-running, balanced household. But in times of stress (drought), the consumption of food by pets may reduce the amount available for the productive members of the household (photosynthetic leaves, transporting phloem, storage parenchyma, absorbing roots). The effects of viruses may therefore be accentuated by conditions that reduce the vigor of the already infected host.

9. Another item that is common to declines is the age of the host. Declines generally occur in mature trees at or after the age when juvenile recuperative vigor is lost. The transition from a rapidly growing competitive juvenile to a slow-growing, reproductive, mature individual is gradual. Trees affected by declines generally have produced the majority of their height growth and are mature, reproductive individuals with less potential for recovery from inciting factors than that of juvenile trees.

SYMPTOMS OF DECLINES

1. Reduced growth is the first noticeable symptom. Shoot growth is shorter and diameter growth is reduced.
2. Reduced growth produces shorter internodes and therefore often a tufted appearance of foliage near the ends of twigs.
3. Roots and mycorrhizae degenerate prior to the appearance of above-ground symptoms, but these symptoms are usually not recognized until top symptoms are seen.
4. Chemical analysis of the roots shows a reduction in stored food reserves.
5. Premature fall coloration in late summer or early fall is common in hardwoods.
6. Yellowing and reduced-size foliage, similar to that caused by mineral deficiencies, is observed.
7. Twigs and branches often die during the winter, and facultative parasitic fungi often invade the dying twigs and accentuate the expansion of dead areas.
8. Portions of the crown die, giving an irregular or asymmetric form to the crown.

9. In hardwoods, water sprouting from adventitious buds on the main stem and large branches produces clusters of foliage.
10. Root rot decay fungi, such as *Armillariella mellea,* parasitize the root system and accentuate the decline syndrome. Rhizomorph and mycelial fan signs of *A. mellea* are common.

EXAMPLES OF DECLINES

A few general comments are necessary to explain my interpretation of the factors affecting declines. One characteristic that is common to all declines is lack of agreement among various researchers on the cause and importance of the specific factors implicated in declines. This controversial nature of declines is probably even more characteristic than the common denominators listed above. Each investigation attempts to find the "cause." With at least three factors involved and an interchangeability of the individual factors with others, often on a regional basis, it is not unlikely that controversies over interpretation of "cause" arise. No one has satisfactorily demonstrated pathogenicity or cause relationships using Koch's rules for proof of pathogenicity for any of the declines. I will present my interpretation of the main factors associated with specific declines, but it must be remembered that these do not exhaust the various factors involved with each of the problems, nor do they necessarily represent the opinions of the original researchers in each case.

Birch Dieback

Classic birch dieback emerged as a forest problem between 1930 and 1950 in New Brunswick and Nova Scotia, Canada, and the northeastern United States (Figs. 18-4 and 18-5). Birch dieback is still a potential forest problem and is also very common on ornamental birches used throughout the north central and northeastern states. The factors of classic birch dieback are as follows:

Predisposing factors Average summer temperatures in eastern Canada increased 1°C over a 10- to 20-year period.

Inciting factors Stand opening by logging, as well as increased soil temperatures, cause an increase in air temperature. Birch rootlet mortality increased from 6% to 60% as a result of a 2°C increase in soil temperatures above the normal summer temperature. Leaf miner, leaf skeletonizing, and other foliage insects, as well as late spring or early fall frosts, incite birch dieback in localized situations.

FIGURE 18-4 Birch dieback in New York associated with stand opening by logging.

FIGURE 18-5 Birch dieback in eastern Canada.

Contributing Factors Bronze birch borer is a destructive wood-boring insect of weakened birches. *Armillariella mellea* contributes to intensification of birch dieback by invading the root systems of weakened trees. At least one virus, apple mosaic virus, has recently been associated with declining birch trees.

Ash Dieback

Ash dieback was first observed in the late 1930s but developed into a serious concern for foresters in the 1950s in the northeastern United States (Fig. 18-6). The disease is still commonly observed in New York, although most serious concern over the disease has waned.

Predisposing Factors Trees growing in exposed hedgerows were very commonly diseased. Although not explicitly considered important at the time, it now appears from the data available that heavier soils were common to ash dieback sites.

FIGURE 18-6 Ash dieback in small New York woodlot.

Inciting factors The 1950s was a particularly dry period for the northeastern United States. This drought was considered the most important factor in the initiation of ash dieback.

Contributing factors Two fungi, *Cytophoma pruinosa* and *Fusicoccum* spp., were commonly isolated from cankers on declining ash. The fungi, when inoculated in ash trees, caused cankers only in weakened trees. The involvement of tobacco mosaic virus, ash ringspot virus transmitted by *Xiphinema* nematodes, and the association of mycoplasma-like organisms in the phloem of some declining ash trees, suggest possible contributions by these agents.

Maple Decline

The beginnings of maple decline problem are difficult to trace, but serious interest was generated in the problem during the 1950s and early 1960s in the northeastern United States and eastern Canada (Figs. 18-7 and 18-8). Maple decline is the most serious problem of ornamental and roadside maples in the Northeast at the present time. Wood lots and sugar bushes also have maple decline.

FIGURE 18-7 Maple decline of a roadside sugar maple in New York.

FIGURE 18-8 Maple decline of a streetside Norway maple in Syracuse, New York.

Predisposing factors The roadside and urban conditions of soil compaction, impeded drainage, poor soil aeration, salt, heat, and air pollution are all predisposing site factors.

Inciting factors The drought of the 1950s was part of the maple decline problem. Logging of hardwood stands induces climatic changes for the remaining trees. The death of American elms in urban environments due to Dutch elm disease produced climatic effects on the maples similar to those caused by logging in natural stands. Defoliating insects are locally important incitants. Root damage from excavation for road improvements (gas, water, and sewer lines) and other construction are common inciting factors.

Contributing factors *Armillariella mellea* root rot accentuates the degeneration of weakened maples. The vascular wilt fungus *Verticillium* may also be considered a contributing factor, even though most forest pathologists assume that it is a pathogen capable of causing disease by itself. Association of *Verticillium* with declining trees speeds up the decline of the weakened trees.

Pole Blight of Western White Pine

Western white pine of the inland empire was severely affected by pole blight during the period from 1916 through the middle of the century.

Predisposing factors Shallow soils and soils with low moisture storage capacity were common to pole blight areas. Sites with less than 5 in. (13 cm) of storage capacity in the upper 3 ft (90 cm) of soil were subjected to pole blight.

Inciting factors By measuring annual growth rings of trees in the inland empire, it was determined that the period 1916–1940 experienced the most severe drought the area had sustained in 180 years.

Contributing factors *Armillariella mellea* root rot and *Leptographium* sp. were associated commonly with roots. *Europhium trinacriforme,* a canker fungus, was associated with elongate cankers on the stems of declining trees. None of these were capable of inducing pole blight but were contributing factors in the disease.

Littleleaf Disease

Abandoned cotton-growing areas in the Piedmont region of southeastern United States regenerated naturally to shortleaf pine. Decline of pole-sized trees (approximately 20 years of age) was first reported in the mid 1930s and by the 1940s was recognized as a serious threat to shortleaf pine (Fig. 18-9). Littleleaf continues to be a problem today.

Predisposing factors Sheet erosion during cotton cropping eliminated most of the top soil of the Piedmont. Fertility was depleted. Drainage is impeded commonly by an impervious hard-packed lower horizon. Periodic moisture stress followed by moisture excess are characteristics of the sites. Poor soil aeration results from excessive moisture.

Inciting factors Although emphasis has never been placed on the inciting factors for the disease, it is apparent from the literature that the disease does not progress at the same rate from year to year. The period 1946–1951 was marked by a rapid increase in the disease and, therefore, weather conditions over that period may be considered as potential inciting agents. Similar declines of exotic conifers and native vegetation of Australia appear to be associated with stand opening by road construction.

Contributing factors In my opinion, which is definitely not the majority opinion at the present time, *Phytophthora cinnamomi,* a feeder root

FIGURE 18-9 Littleleaf disease of shortleaf pine in Georgia. (Photograph compliments of Dr. Savel Silverborg.)

parasite, is at best a contributing factor. (See Chapter 16 for more discussion of *P. cinnamomi.*) The recovery of *P. cinnamomi* from sites where littleleaf was a problem provided a reasonable pathogen to account for the disease. But difficulties in the demonstration of pathogenicity using Koch's rules of proof have not dispelled the popular opinion that we now have the cause for littleleaf. The spread pattern for littleleaf is not typical of expanding foci expected for disease caused by root pathogens. *P. cinnamomi* is as ubiquitous to littleleaf as *Armillariella mellea* is to other declines. *A. mellea* is also found in littleleaf sites. Diseased as well as nondiseased stands are infested by *P. cinnamomi.* Therefore, the distinction of "causal agent" for *P. cinnamomi* is inappropriate.

Other root rot fungi in the *Pythium* complex, *Pythium irregulare* and *Pythium debaryanum,* are found in sites with or without littleleaf. Although these are shown to be pathogenic on shortleaf and loblolly pines, there is no reason to assume that they are any different from *P. cinnamomi.*

This list of declines is by no means complete. Cytospora canker and beech bark disease are discussed in Chapter 12. We could easily expand to include oak decline, sweetgum blight, and many others less well documented.

ECOLOGICAL ROLE OF DECLINES

A final consideration is the ecological role of declines. Trees are continually expanding their range to the limit of their site tolerance. Those on the periphery of their site tolerance are most affected by declines. It is interesting that declines occur well within the natural geographic range of the species and not on the periphery of the geographic range. Within the geographic range, variations between specific sites are the conditions that seem to affect declines.

Are we really observing natural ecological succession in action? If declines are nature's way of moving successions along or preventing expansion of site tolerance for tree species, it is rather futile to expect that control is possible. We may need to learn to recognize this as a type of natural balance phenomenon and adjust our expectations accordingly.

REFERENCES

CAMPBELL, W. A. 1961. Littleleaf disease of shortleaf pine: Present status and future needs. *In* Recent Advances in Botany. University of Toronto Press, Toronto, pp. 1529–1532.

CAMPBELL, W. H., O. L. COPELAND, JR., and G. H. HEPTING. 1953. Managing shortleaf pine in littleleaf disease areas. USDA For. Serv. S.E. For. Exp. Sta. Pap. 25. 12 pp.

HANSBROUGH, J. R., V. S. JENSEN, H. J. MACALONEY, and R. W. NASH. 1950. Excessive birch mortality in the Northeast. USDA For. Serv. Tree Pest Leafl. 52. 4 pp.

HEPTING, G. H. 1963. Climate and forest disease. Annu. Rev. Phytopathol. *1*:31–50.

LEAPHART, C.D., 1958. Pole blight—how it may influence western white pine management in light of current knowledge. J. For. *56*:746–751.

LEAPHART, C. D. and A. R. STAGE. 1971. Climate: a factor in the origin of the pole blight disease of *Pinus monticola* Dougl. Ecology *52*:229-239

OTROSINA, W, J. and D. H. MARX. 1975. Populations of *Phytophthora cinnamomi* and *Pythium* spp. under shortleaf and loblolly pines in littleleaf disease sites. Phytopathology *65*:1224–1229.

PARKER, A. K. 1957. The nature of the association of *Europhium trinacriforme* with pole blight lesions. Can. J. Bot. *35*:845–856.

PARKER, J. 1970. Effects of defoliation and drought on root food reserves in sugar maple seedlings. USDA For. Serv. Res. Pap. NE-169.

REDMOND, D. R. 1955. Studies in forest pathology XV. Rootlets, mycorrhiza, and soil temperature in relation to birch dieback. Can. J. Bot. *33*:595–627.

REDMOND, D. R. 1957. The future of birch from the viewpoint of diseases and insects. For. Chron. *33*:25–30.

ROSS, E. W. 1966. Ash dieback etiological and developmental studies. State University College of Forestry at Syracuse University, Tech. Pub. 88. 80 pp.

ROTH, E. R. 1954. Spread and intensification of littleleaf disease of pine. J. For. *52*:592–596.

SINCLAIR, W. A. 1964. Comparisons of recent declines of white ash, oak, and sugar maple in Northeastern woodlands. Cornell Plantations *20*:62–67.

SINCLAIR, W. A. 1966. Decline of hardwoods: possible causes. Proc. Inter. Shade Tree Conf. *42*:17–32.

SPAULDING, P. and H. J. MACALONEY. 1931. A study of organic factors concerned in the decadence of birch on cut-over lands in Northern New England. J. For. *29*:1134–1149.

STALEY, J. M. 1965. Decline and mortality of red and scarlet oaks. For. Sci. *11*:2–17.

STONE, E. L., R. R. MORROW, and D. S. WELCH. 1954. A malady of red pine on poorly drained sites. J. For. *52*:104–114.

WESTING, A. H. 1966. Sugar maple decline: an evaluation. Econ. Bot. *20*:196–212.

ZAK, B. 1957. Littleleaf of pine. USDA For. Serv. For. Pest Leafl. 20. 4 pp.

ZAK, B. 1961. Aeration and other soil factors affecting southern pines as related to littleleaf disease. USDA For. Serv. Tech. Bull. 1248. 30 pp.

PART THREE

OVERVIEW ASPECTS OF TREE DISEASES

Chapters 2 through 18 have emphasized tree diseases in relation to specific agents and parts of trees. Chapters 19 through 23 will develop more general topics, including plant disease epidemics, disease control through genetic resistance, diseases of seedlings in the nursery, pathological consideration of urban tree management, and pathological consideration of intensively managed forest plantations.

These chapters draw on examples of diseases already covered and expand into other diseases. They are intended to provide the reader with an applied take-home message for disease management in relation to forest and urban tree management.

19

PLANT DISEASE EPIDEMICS

- Calculation of disease increase rates
- Uses of epidemiological information
- Sample problems of disease epidemics
- Solution and discussion of problems

Up to this point, we have concerned ourselves primarily with types of organisms and nonliving agents that cause disease. A number of specific diseases have been emphasized to demonstrate the etiology (disease cycle) of disease-causing agents and control procedures. All of this has a very narrow application unless one develops a perspective regarding how a disease-causing agent produces disease in a population of plants. Control of plant disease in a single plant is like trying to fight brush fires by stepping on each burning ember. Control is best applied to and evaluated on a population of plants.

Sound understanding of disease epidemics and the influence of control procedures on the development of the epidemics would make presentation of our case to the general public a lot easier. Very few control measures completely stop a disease. If a disease can be slowed to the extent that replanting or regeneration compensates for the amount lost, the control measure is successful.

We often oversell the effectiveness of a control measure and then later disappoint the public because the disease is not eliminated. For example, in Syracuse, New York, the sanitation control measure for Dutch elm disease was accepted as the cure for DED. When it did not eliminate the problem, city officials became disillusioned with the value of controlling the disease and in a tight fiscal year cut the funds. Would it not have been better to demonstrate the effectiveness of the control procedure in increasing the life expectancy of the population of elms?

CALCULATION OF DISEASE INCREASE RATES

A disease epidemic is an increase in the amount of disease with time. Some diseases increase rapidly, others more slowly. Disease increase can sometimes be equated to money invested at compound interest. The basic formula for compound interest of money or disease is

$$r_1 = \frac{1}{t_2 - t_1} \log_e \frac{x_2}{x_1}$$

The rate of interest or increase (r), the length of time ($t_2 - t_1$), the amount of initial principal or initial disease (x_1), and the final amount of principal or disease (x_2) are the four terms of the formula. Obviously, this formula can be used to determine r, t_2, or x_2, depending upon which values you substitute for in the formula.

The basic formula applies to compound-interest diseases only when fewer than 5% of the population of plants are diseased. Beyond 5% diseased, actual disease increase is less than compound interest would predict. Why?

The compound-interest formula is based on an infinite supply of money or additional infectible plants. As more and more plants become diseased, the infectible population becomes smaller and more widely spaced. There differences in the spatial distribution of infectible plants change the ratio of successful infection to total inoculum production. As the population of healthy susceptible plants becomes smaller, the percent of inoculum infecting new hosts is reduced almost to zero.

To compensate for a changing population of infectible plants we apply a correction factor $x/(1 - x)$ instead of using x. The amount of disease (x) is expressed as a decimal proportion. If we insert $x/(1 - x)$ for the amount of disease terms and arrange the logit fraction into a subtraction, our working formula for disease becomes:

$$r_1 = \frac{1}{t_2 - t_1}\left(\log_e \frac{x_2}{1 - x_2} - \log_e \frac{x_1}{1 - x_1}\right)$$

$\text{Log}_e\, x/(1 - x)$ values from 0.001 to 0.999 can be calculated and are provided as a table of logit values (Table 19-2).

With a logit table available, it is a simple matter to look up the logit values and plug these into the formula

$$r_1 = \frac{1}{t_2 - t_1} \quad (\text{logit of } x_2 - \text{logit of } x_1).$$

Another way that a disease can increase is like money invested at simple interest. With simple interest one does not collect interest on previous interest dividends. The formula

$$QR = \frac{1}{t_2 - t_1}\left(\log_e \frac{1}{1 - x_2} - \log_e \frac{1}{1 - x_1}\right)$$

represents the simple-interest disease. Q represents the amount of inoculum. Note the correction factor $1/(1 - x)$ for simple-interest diseases. A physiological disease may follow a simple interest increase because the loss of an individual plant does not contribute inoculum to infect others. Verticillium wilt follows a simple-interest increase. It is the same as always removing the interest dividends in a savings account. Any dividends (in disease terms, disease losses) do not add to the principal and therefore do not influence the size of the next dividend.

If one plots the accumulative percent disease against time, either a sigmoid growth curve or a straight-line plot is obtained (Fig. 19-1). A compound-interest disease will generate a typical sigmoid growth-curve plot. The compound-interest disease-increase curve has a lag phase, a logarithmic phase, and a leveling-off phase.

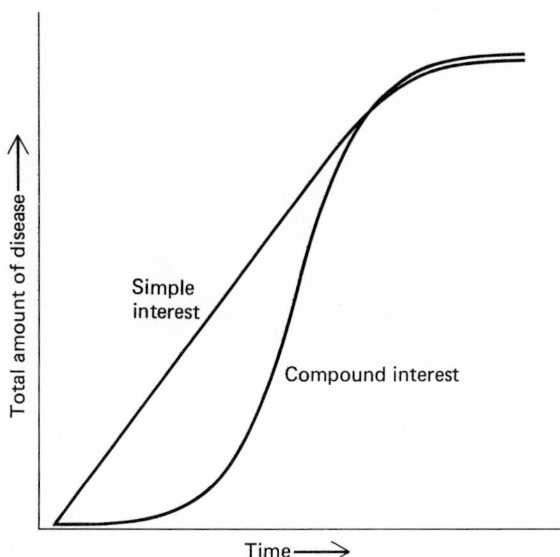

FIGURE 19-1 Disease increase plots for simple- and compound-interest diseases.

The simple-interest disease, in contrast, plots as a straight line. There is no log phase or compound-interest phase to a simple-interest disease plot. But, as with compound interest, there is a leveling-off phase due to a limited amount of infectible host population.

It is difficult to work with the curved lines of the compound interest plot, so we generally plot the logit of the cumulative amount of disease (Fig. 19-2). This technique produces a linear plot which can be compared to other linear plots. One can readily see differences with straight lines that are difficult to detect in the original curved-line plot.

In the real world, one does not necessarily know if a disease increases by compound or simple interest. The first test is to plot the cumulative disease increase. As already described, the two plots are rather distinctive.

A compound-interest disease is usually caused by a pathogen that reproduces on the infected host to produce additional inoculum for secondary spread. The secondary spread from infected plants results in foci of infection (pockets of infection).

The simple-interest disease does not produce inoculum on infected hosts for secondary disease spread. In the absence of secondary spread, the infection is usually random. Randomness or nonrandomness of infected individuals is one of the best indicators of simple-interest or compound-interest diseases, respectively.

Why would it be important to recognize simple- from compound-interest diseases? If one looks at the plots of compound- and simple-interest diseases, it is apparent that predicting future losses for a compound-interest

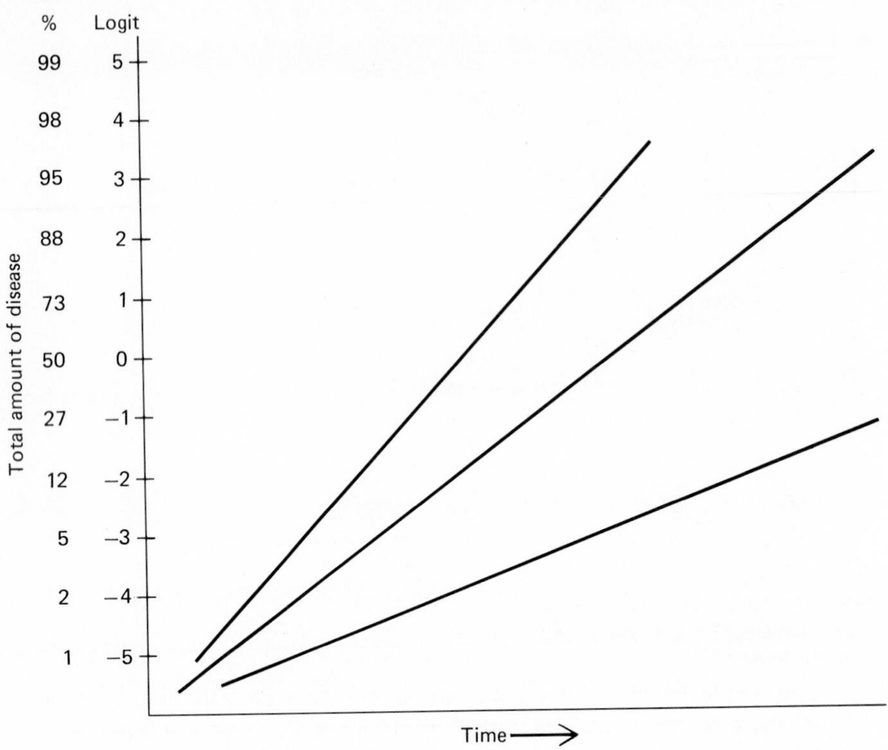

FIGURE 19-2 Compound-interest diseases plotted on a logit scale for percent disease.

disease using a simple-interest formula will greatly underestimate the actual loss. Similarly, if one uses a compound-interest disease formula to predict future losses of a simple-interest disease, one greatly exaggerates the actual losses.

USES OF EPIDEMIOLOGICAL INFORMATION

Table 19-1 presents rate of increases (r values) calculated for a number of diseases. Why does Dutch elm disease or oak wilt have different r values in different locations?

Note that Hypoxylon canker is assigned a simple-interest rate of increase (QR). Although Hypoxylon produces secondary inoculum on infected trees, the period of time between infection and sporulation is long, three or more years. In an even-aged stand, most of the aspen trees have passed the most susceptible period before secondary inoculum is produced. Therefore, Hypoxylon canker losses can be better predicted with a simple-interest disease increase rate.

TABLE 19-1 Rate of Disease Increase per Unit per Year for Several Tree Diseases.[a]

Disease	r_l	Location
White pine blister rust	0.50	British Columbia
	0.67	Minnesota
	0.50	Maine
Dutch elm disease	1.37	Illinois
	0.50	Connecticut
	0.25	Quebec
Chestnut blight	1.42	Pennsylvania
	1.10	Virginia
	0.83	Connecticut
Oak wilt	0.10	Pennsylvania
	0.31	West Virginia
	0.22	Illinois
	0.20	Tennessee
	0.36	Arkansas
Elm phloem necrosis	0.46	Indiana
Tympanis canker (red pine)	0.08–0.15	Connecticut
Polyporus tomentosus (spruce)	0.03–0.05	Saskatchewan
Cytospora canker (spruce)	0.14	Quebec
Hypoxylon canker (trembling aspen)	0.20 *QR*	New York

[a] r_l values based on Merrill, 1967; *QR* value based on Manion and Blume, 1975.

Differences may occur because of differing factors such as climate, density of hosts, and control practice. Specific differences in disease increase rates may give clues as to how effective certain control practices or how important climatic and site factors are on disease increases.

SAMPLE PROBLEMS OF DISEASE EPIDEMICS

It is difficult to really understand this chapter without actually doing some calculations of disease increase rates and then using these to make management decisions. For this reason, three sample problems are included. The first two are hypothetical problems, but the third is based on the Dutch elm disease epidemic for Syracuse, New York.

A word of caution before starting the calculations. Be certain to express percent disease (x) as a cumulative percent disease, not just the annual percent loss. Table 19-2 gives the logit values that you will need to solve these problems. The solution to the problems, plus a discussion, are presented at the end of the chapter.

TABLE 19-2 Logit Values [$\log_e (x/1 - x)$] for Decimal Proportions from 0.001 to 0.999. For x less than 0.50 (on left), logit is negative. For x greater than 0.50 (on right), logit is positive.

	Thousandths, for x (in left column)										
x	0	1	2	3	4	5	6	7	8	9	
0.00		6.91	6.21	5.81	5.52	5.29	5.11	4.95	4.82	4.70	4.60	0.99
0.01	4.60	4.50	4.41	4.33	4.25	4.18	4.12	4.06	4.00	3.94	3.89	0.98
0.02	3.89	3.84	3.79	3.74	3.71	3.66	3.62	3.58	3.55	3.51	3.48	0.97
0.03	3.48	3.44	3.41	3.38	3.35	3.32	3.29	3.26	3.23	3.20	3.18	0.96
0.04	3.18	3.15	3.13	3.10	3.08	3.06	3.03	3.01	2.99	2.97	2.94	0.95
0.05	2.94	2.92	2.90	2.88	2.86	2.84	2.82	2.81	2.79	2.77	2.75	0.94
0.06	2.75	2.73	2.72	2.70	2.68	2.67	2.65	2.63	2.62	2.60	2.59	0.93
0.07	2.59	2.57	2.56	2.54	2.53	2.51	2.50	2.48	2.47	2.45	2.44	0.92
0.08	2.44	2.43	2.42	2.40	2.39	2.38	2.36	2.35	2.34	2.33	2.31	0.91
0.09	2.31	2.30	2.29	2.28	2.27	2.25	2.24	2.23	2.22	2.21	2.20	0.90
0.10	2.20	2.19	2.18	2.16	2.15	2.14	2.13	2.12	2.11	2.10	2.09	0.89
0.11	2.09	2.08	2.07	2.06	2.05	2.04	2.03	2.02	2.01	2.00	1.99	0.88
0.12	1.99	1.98	1.97	1.96	1.96	1.95	1.94	1.93	1.92	1.91	1.90	0.87
0.13	1.90	1.89	1.88	1.87	1.87	1.86	1.85	1.84	1.83	1.82	1.82	0.86
0.14	1.82	1.81	1.80	1.79	1.78	1.77	1.77	1.76	1.75	1.74	1.73	0.85
0.15	1.73	1.73	1.72	1.71	1.70	1.70	1.69	1.68	1.67	1.67	1.66	0.84
0.16	1.66	1.65	1.64	1.64	1.63	1.62	1.61	1.61	1.60	1.59	1.59	0.83
0.17	1.59	1.58	1.57	1.56	1.56	1.55	1.54	1.54	1.53	1.52	1.52	0.82
0.18	1.52	1.51	1.50	1.50	1.49	1.48	1.48	1.47	1.46	1.46	1.45	0.81
0.19	1.45	1.44	1.44	1.43	1.42	1.42	1.41	1.41	1.40	1.39	1.39	0.80
0.20	1.39	1.38	1.37	1.37	1.36	1.36	1.35	1.34	1.34	1.33	1.32	0.79
0.21	1.32	1.32	1.31	1.31	1.30	1.30	1.29	1.28	1.28	1.27	1.27	0.78
0.22	1.27	1.26	1.25	1.25	1.24	1.24	1.23	1.22	1.22	1.21	1.21	0.77
0.23	1.21	1.20	1.20	1.19	1.19	1.18	1.17	1.17	1.16	1.16	1.15	0.76
0.24	1.15	1.15	1.14	1.14	1.13	1.13	1.12	1.11	1.11	1.10	1.10	0.75
		9	8	7	6	5	4	3	2	1	0,	x

Thousandths, for x (in left column)

x	0	1	2	3	4	5	6	7	8	9		
0.25	1.10	1.09	1.09	1.08	1.08	1.07	1.07	1.06	1.06	1.05	1.05	074
0.26	1.05	1.04	1.03	1.03	1.03	1.02	1.02	1.01	1.00	1.00	0.99	0.73
0.27	0.99	0.99	0.98	0.98	0.97	0.97	0.96	0.96	0.95	0.95	0.94	0.72
0.28	0.94	0.94	0.93	0.93	0.92	0.92	0.91	0.91	0.91	0.90	0.90	0.71
0.29	0.90	0.89	0.89	0.88	0.88	0.87	0.87	0.86	0.86	0.85	0.85	0.70
0.30	0.85	0.84	0.84	0.83	0.83	0.82	0.82	0.81	0.81	0.80	0.80	0.69
0.31	0.80	0.80	0.79	0.79	0.78	0.78	0.77	0.77	0.76	0.76	0.75	0.68
0.32	0.75	0.75	0.74	0.74	0.74	0.73	0.73	0.72	0.72	0.71	0.71	0.67
0.33	0.71	0.70	0.70	0.69	0.69	0.69	0.68	0.68	0.67	0.67	0.66	0.66
0.34	0.66	0.66	0.65	0.65	0.65	0.64	0.64	0.63	0.63	0.62	0.62	0.65
0.35	0.62	0.61	0.61	0.61	0.60	0.60	0.59	0.59	0.58	0.58	0.58	0.64
0.36	0.58	0.57	0.57	0.56	0.56	0.55	0.55	0.55	0.54	0.54	0.53	0.63
0.37	0.53	0.53	0.52	0.52	0.52	0.51	0.51	0.50	0.50	0.49	0.49	0.62
0.38	0.49	0.49	0.48	0.48	0.47	0.47	0.46	0.46	0.46	0.45	0.45	0.61
0.39	0.45	0.44	0.44	0.43	0.43	0.43	0.42	0.42	0.41	0.41	0.41	0.60
0.40	0.41	0.40	0.40	0.39	0.39	0.39	0.38	0.38	0.37	0.37	0.36	0.59
0.41	0.36	0.36	0.36	0.35	0.35	0.34	0.34	0.34	0.33	0.33	0.32	0.58
0.42	0.32	0.32	0.31	0.31	0.31	0.30	0.30	0.29	0.29	0.29	0.28	0.57
0.43	0.28	0.28	0.27	0.27	0.27	0.26	0.26	0.25	0.25	0.25	0.24	0.56
0.44	0.24	0.24	0.23	0.23	0.22	0.22	0.22	0.21	0.21	0.20	0.20	0.55
0.45	0.20	0.20	0.19	0.19	0.18	0.18	0.18	0.17	0.17	0.16	0.16	0.54
0.46	0.16	0.16	0.15	0.15	0.14	0.14	0.14	0.13	0.13	0.12	0.12	0.53
0.47	0.12	0.12	0.11	0.11	0.10	0.10	0.10	0.09	0.09	0.08	0.08	0.52
0.48	0.08	0.08	0.07	0.07	0.06	0.06	0.06	0.05	0.05	0.04	0.04	0.51
0.49	0.04	0.04	0.03	0.03	0.02	0.02	0.02	0.01	0.01	0.00	0.00	0.50
		9	8	7	6	5	4	3	2	1	0	x

Thousandths, for x (in right column)

Sample Problem 1: Cottonwood Quick Butt Rot

Mr. J. I. Fink, city forester for Burning Stump, Arkansas, decided to demonstrate to the city fathers the seriousness of the current epidemic of cottonwood quick butt rot. He recorded the number of cars crushed by falling trees during the summers of 1962, 1963, and 1964. Out of a population of 1000 cars, 10 were hit in 1962, 11 were hit in 1963, and 12 were hit in 1964. What is the disease increase rate of the car destruction epidemic, and if there is a very large population of cottonwoods, how long will it be before 20% of the cars are destroyed?

Sample Problem 2: Red Christmas Trees

Rather than develop an advertising campaign to push red, sparsely foliated Christmas trees, Ms. L. V. Schrooge decided to invest some money in fungicide sprays. Being concerned about unnecessary expenditures, she ordered an evaluation of various fungicides and application timings. The normal course of the foliage epidemic has an r value of 0.4 per unit per day, during the month of June. No spread occurs after July 1 or before June 1.

Ecologically Blessed is a fungicide whose action is short-lived (1 day) and is effective only against established needle infection.

Big Zap is a contact fungicide that kills anything and everything that moves or does not move except trees and is effective for a period of 30 days. It cannot penetrate the needles to affect the parasite once inside the plant.

Plot the normal course of the epidemic if a sampling determined that 1 in 1000 trees produced viable inoculum on June 1. What percent of the trees are infected by July 1?

What will happen if Ecologically Blessed is used as a spray treatment on the trees on June 15? Assume that the fungicide is 90% effective; that is, 1 in 10 trees do not receive sufficient treatment to kill internal pathogens.

What will happen if Big Zap is applied on June 10 and June 20? New growth of foliage causes some foliage to be unprotected, thereby reducing the effectiveness of the fungicide. The spray is, therefore, 100% effective only for 5 days after application. The cost of the fungicide prohibits additional applications.

Graph each treatment. Which is best? What is the effect of Big Zap on the rate of increase from June 1 to July 1? Would it make an difference if the plantation contained 10,000 trees or 1,000,000 trees?

Sample Problem 3: Dutch Elm Disease in Syracuse, New York

The yearly losses of American elms to Dutch elm disease are given in Table 19-3. Working space is provided for you to calculate the rate of increase (r_1) values for the disease epidemic.

TABLE 19-3 Yearly Losses of American Elms to Dutch Elm Disease in Syracuse, New York.[a]

Year	Number of trees killed	Accumulative losses to DED	Accumulative percent loss	Logit value	Rate of increase
(Minimum sanitation)					
1951	1	______	______	____	____
1952	7	______	______	____	____
1953	19	______	______	____	____
1954	181	______	______	____	____
1955	657	______	______	____	____
1956	864	______	______	____	____
1957	1066	______	______	____	____
(Maximum sanitation)					
1958	425	______	______	____	____
1959	817	______	______	____	____
1960	581	______	______	____	____
1961	510	______	______	____	____
1962	529	______	______	____	____
1963	751	______	______	____	____
1964	748	______	______	____	____
(No sanitation)					
1965	2597	______	______	____	____
1966	4033	______	______	____	____
1967	5687	______	______	____	____

[a]The original population contained 53,618 elms.

Calculate an average rate of increase (r_l) values for the three periods and determine the projected date for loss of 90% of the trees, based on the value for each period.

Can you justify sanitation as a control for DED based on these figures? The sanitation program involved removal of dead and dying elms. The rate of increase during the sanitation period reflects root graft spread and infection by European elm bark beetles produced on elm material outside the control area. Do you suppose that a more effective control program is justified?

SOLUTION AND DISCUSSION OF PROBLEMS

Cottonwood Quick Butt Rot If we assume that the number of cars is directly proportional to the number of trees that fall over due to the butt rot, the number of crushed cars is a good quantitative measure of the tree disease.

Stop and think about the disease cycle of a root rot. You should recall that the time interval between infection and inoculum production is long. Root rot diseases, because of the long time interval and the usual association with some environmental event, generally increase at a simple interest rate. Therefore, you should use the simple-interest formula.

If you use a compound-interest formula, you will see that the disease increase rate keeps changing. If the disease increase rate does not generally remain constant, you are probably using the wrong formula.

In this case, the rate of increase is 0.01 and it would take 21 years, or until 1983, to destroy 20% of the cars.

Cottonwood quick butt rot is a simple example, but it demonstrates how a root rot disease would progress through a population of non-regenerating trees. A recreation site or a street setting, where large mature trees are not being replaced, would be drastically changed by the loss of 20% of the trees. The obvious answer is to balance losses with replacements.

Red Christmas Trees The graphic plot of this disease epidemic, and the effects of control activities, are presented in Fig. 19-3. Note the utility of the logit scale of the vertical axis. This allows one to plot straight lines to more accurately predict future disease levels.

It should be evident that a disease epidemic with a 0.4 per unit per day rate of increase can totally infect the crop in one season. Neither fungicide is effective in reducing losses to a reasonable level using the present spray schedules. Many applications of fungicide would be needed to control the disease.

A more effective approach to control would be to look for resistant varieties or different growing conditions to reduce the rate of disease increase. If the disease increase rate could be slowed somewhat, the application of fungicides could be economically justified.

Dutch Elm Disease in Syracuse, New York An understanding of the Dutch elm disease cycle would predict that the disease should follow a compound-disease increase rate. But if one calculates the yearly increase rates using Table 19-3, or if the data are plotted graphically, it is obvious that the rate of increase values are not constant.

A closer look at the data shows three rates of increase. During the minimal-sanitation period, the rates change yearly, but during the 1954–1957 period, the rate averages 0.87. During the maximum-sanitation period, the rates are more constant and average 0.15. During the no-sanitation period, the rate of increase averages 0.43 or 0.88, depending on the base-level population that is used.

The changing rate of increase values for the early period are due in part to the effects of multiple introductions of the disease into the area. Disease increase is made up of a combination of new introductions and secondary

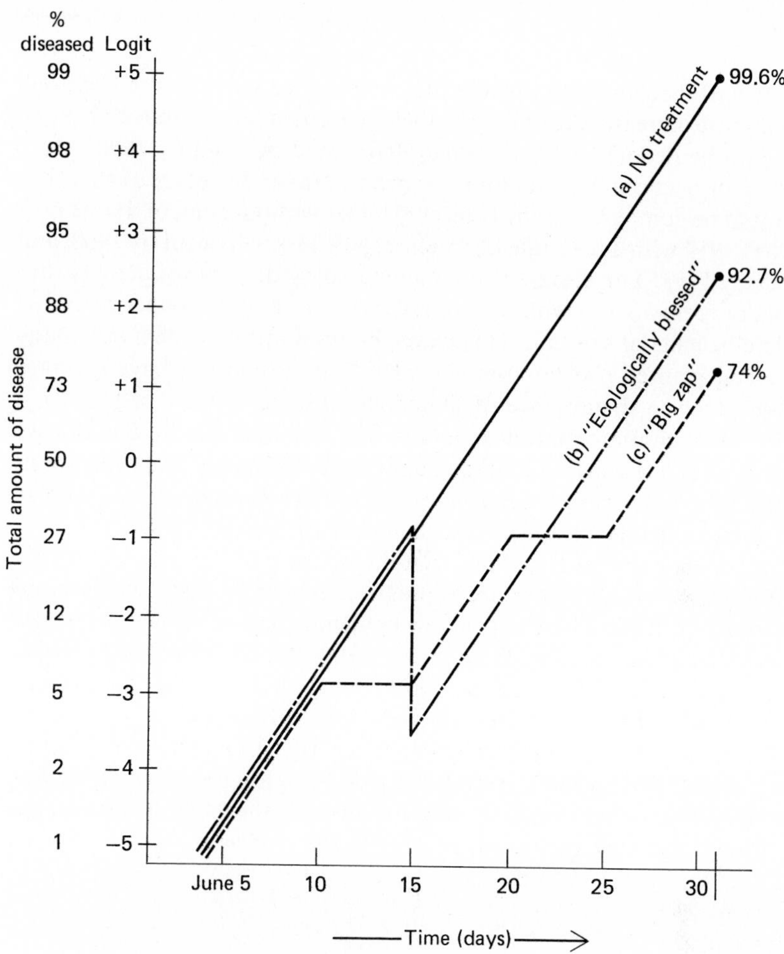

FIGURE 19-3 Red Christmas tree problem graphic plot on a logit scale for percent disease.

spread from established infection centers. When the numbers are small, just a few additional introductions can produce major changes.

A second aspect of the early epidemic is the period of training needed for people diagnosing the disease. If there are very few diseased trees, it is difficult to find them all. As there are more diseased trees, the crew will cover larger areas and find infections that occurred earlier but were not detected. Therefore, the yearly losses to disease may reflect both deflated and inflated numbers. Under these circumstances, the early figures are best evaluated as a group.

Disease control through sanitation was obviously very successful. The disease increase rate remained relatively stable and low. The rate of increase probably reflects the effects of root graft transmission of the disease and the effects of bark beetles from outside the control area moving into the area. With a disease increase rate of 0.15 in 1960, one could predict that 50% of the trees would be dead by 1975 and that 90% would be dead by 1990.

These projections may not look too promising for the future of elms, but they should be compared to the projected losses without control. If one projects from 1957 with a 0.87 rate of increase, 50% loss will occur by 1960, and 90% loss by 1963. The disease is not eliminated by the control activity, but the costs of removal and replacement are spread out over a longer period of time. The benefits of sanitation look even better if one considers that many of the trees, planted shortly after the Civil War, would not have survived until 1990 even in the absence of Dutch elm disease.

The no-sanitation period is marked by an increase in the disease increase rate. If one assumes that the population of elm trees and the amount of beetles were not seriously affected by the sanitation program, an average rate of increase value of 0.43 is calculated for the period 1964–1967.

It is probably better to assume that the removal of the trees during the sanitation period changed the base population of both the disease vector and the nondiseased trees. Therefore, a new epidemic, beginning in 1965, starts with an initial population of 46,462 trees rather than 53,618. The base diseased population is 2597 trees in 1965. Using these figures, an average increase rate of 0.88 is calculated for the 1965–1967 period.

Projections of 90% disease losses, using the 0.43 and 0.88 rates of increase, target 1974 and 1971, respectively. Although accurate yearly figures are not available for the period, it now appears that the 0.88 rate of increase more closely predicted the actual course for the epidemic.

REFERENCES

MANION, P. D., and M. BLUME. 1975. Epidemiology of hypoxylon canker of aspen. Proc. Am. Phytopathol. Soc. *2*:101.

MERRILL, W. 1967. Analyses of some epidemics of forest tree diseases. Phytopathology *57*:822.

MERRILL, W. 1967. The oak wilt epidemics of Pennsylvania and West Virginia: analysis. Phytopathology *57*:1206–1210.

MERRILL, W. 1968. Effect of control programs on development of epidemics of Dutch elm disease. Phytopathology *58*:1060.

MILLER, H. C., S. B. SILVERBORG, and R. J. CAMPANA. 1969. Dutch elm disease: relation of spread and intensification to control by sanitation in Syracuse, New York. Plant Dis. Rep. *53*:551–555.

VAN DER PLANK, J. E. 1963. Plant diseases: epidemics and control. Academic Press, Inc., New York. 349 pp.

WEIDENSAUL, T. C., and F. A. WOOD. 1974. Analysis of a maple canker epidemic in Pennsylvania. Phytopathology *64*:1024–1027.

20

DISEASE CONTROL THROUGH GENETIC RESISTANCE

- Methods of disease control
- Considerations relative to a resistance breeding program with trees
- Methods of inheritance of resistance
- Examples of resistance breeding strategies

Throughout this book, we have tried to develop an appreciation of the importance of tree diseases. It may be frustrating to you—at least it is to me—that once one appreciates the role of disease and even learns much about the biology of the interaction, not very much can be done. Most control practices are totally impractical on a large scale. We professional forest pathologists, therefore, spend most of our applied time developing theories of why something died, and putting down scattered problems with techniques that are too expensive and not completely effective.

METHODS OF DISEASE CONTROL

I like to make the analogy between problem solving in forest pathology and problem solving of international conflicts. Conflicts between nations can be resolved in three ways. One is war, which may solve the problem but also destroy the reason for solving it. We can apply direct control procedures to a disease in localized spots, but do we really accomplish anything significant? Direct control involves tremendous amounts of money. In the end, is it worth the cost of both dollars and environmental impact? This is a question currently being asked of all uses of pesticides.

A second way of resolving a conflict is isolation, just not communicating at all. This works and nations survive. But is the potential progress of both inhibited by the lack of interaction? I equate quarantines with this type of activity. One has to continuously keep building the barriers and avoiding confrontation or a conflict (disease) will flare up.

The third type of solution involves getting together and resolving the differences. Both cultures will be affected by the interaction such that neither will be as it was before. The interaction process does not stop with one meeting but must continue indefinitely. Once embarked on such a course, one must maintain interaction to resolve minor misunderstandings that will develop. This type of approach I equate with breeding for disease resistance and silvicultural stand manipulations.

This chapter will emphasize disease resistance breeding. Chapter 23 will look at the total picture of silvicultural stand manipulations in intensive forest management.

There have been times in history when each of the three approaches to control was indeed the most appropriate. With increasing population and changes in values, the third type of solution emerges as the most appropriate. We are presently in a stage where increasing pressure on our tree resources is forcing us to move in the direction that agriculture started to move in about the turn of the century. We must develop resistance or tolerance in our trees to the biotic and physical pressures of our environment. And we must learn to manage both the diseases and the trees through silvicultural practices.

Forest pathologists do not universally agree that disease-resistance breeding is a viable approach to disease control. The skeptics point to the many problems and possible pitfalls. They also say that it will take generations of pathologists to accomplish anything significant. As we shall see, there are possible problems, but they are not absolute barriers. We shall also see that the time period is not necessarily measured in generations of pathologists.

CONSIDERATIONS RELATIVE TO A RESISTANCE BREEDING PROGRAM WITH TREES

With that as an introduction, let me now develop the topic of breeding for disease resistance in trees with three general considerations.

First, there is no simple, inexpensive method of developing a disease-resistant plant, so whatever program is started must indeed warrant the cost. High value and widely used ornamentals, or intensively managed fiber or timber crop trees, are the only ones to consider.

Second, once committed to a breeding program, there is no ending point. Resistance breeding does not stop once a resistant variety is released. It is important to monitor the performance of the new variety as it is used in the field. Unknown or unpredicted problems may flare up and destroy everything you have accomplished.

Third, one should not develop a resistant line based on just one resistant plant. Genetic diversity of natural populations provides a certain amount of elasticity to the population for survival in a variable environment. A broad-based search for a number of resistant plants with varying types of resistance is an important first step.

METHODS OF INHERITANCE OF RESISTANCE

You may already be aware of the various types of inheritance from introductory biology or genetics. Resistance is just like any other character. It can be controlled by a single gene and will be inherited much like the Mendelian characters. It can be controlled by many interacting genes and be inherited much like height growth in human beings. It can also be controlled by cytoplasmic factors and will therefore demonstrate a greater contribution from the female parent.

It is important to determine how the resistance is inherited, because the tree breeder utilizes different techniques to obtain genetic improvement with each of these methods of inheritance.

Many of our resistant varieties of agricultural crops are based on single-gene resistance. This type of resistance is ideal for incorporation into a hybrid population.

Hybrid plant varieties are usually produced through forced interfertilization of two independently inbred plant lines. One line may be bred for growth form or quality characters, the other may be for disease resistance or other characters. The hybrid, produced by crossing the two lines, is a very uniform population with both resistance and growth form.

If one of the parents carried a homozygous-dominant resistance gene, all the hybrid plants will carry at least one dominant gene for resistance. Although the hybrid population is then totally resistant, a second generation of plants produced by crossing the hybrids usually results in a 3:1 segregation for resistance.

Production of a resistant variety based on quantitatively inherited factors is somewhat more difficult. One difficulty arises from the combined effect of single-gene and quantitive inheritance. Quantitatively inherited genetic improvement is difficult to accurately determine if single genes for resistance are maintained in the population. The single genes may mask the quantitative genes for resistance. The plant breeder must utilize a number of test crosses to verify the absence of single genes.

Genetic improvement of a plant species using quantitative inheritance is usually less spectacular than with single-gene inheritance. Quantitative inheritance of resistance will produce varying improvements in the population, but never total resistance as seen with single-gene inheritance.

Cytoplasmic inheritance is utilized effectively in the production of hybrid varieties if seed is collected only from one plant line. The female parent contributes the largest volume of cytoplasm to the embryo and therefore the greatest probability of cytoplasmic-inherited characters.

Cytoplasmically inherited characteristics have not been utilized in tree breeding but are commonly used in hybrid corn breeding. Cytoplasmically inherited male sterility makes it possible to avoid self-fertilization. In the past, the tassels had to be cut to avoid self-fertilization.

A corn leaf blight epidemic in 1970 is discussed in Chapter 1. This disease, caused by *Helminthosporium maydis,* was extremely damaging to varieties that contained a specific cytoplasmically inherited male sterility factor. To control the disease, a different cytoplasmically inherited male sterility gene was utilized.

EXAMPLES OF RESISTANCE BREEDING STRATEGIES

Single-Gene Inheritance If selection of resistant individuals and controlled crosses determines that a single gene controls resistance, it is best not to produce a long-term crop such as trees using only that one gene for resistance. Examples in agriculture have demonstrated that pathogens can also change by mutation or genetic recombination. It takes 10 to 15 years for the wheat stem rust fungus (*Puccinia graminis* var. *tritici*) to develop a large population of a new infective race capable of destroying the wheat crop. Therefore, every 10 to 15 years a new variety with different single genes controlling resistance must be available for release to farmers.

The corn leaf blight epidemic of 1970 developed because a single gene was maintained in the population. Genetic variation in the *Helminthosporium* fungus eventually resulted in a pathogen population capable of infecting the corn population. Rapid increase in this virulent pathogen population resulted in a serious disease epidemic.

One method of overcoming this difficulty is to produce a synthetic variety in which you incorporate a number of different genes for resistance in a population of plants. Borlaug (1966) describes such a method, which is summarized in Figure 20-1.

The synthetic-variety approach incorporates single-gene resistance of single-cross, backcross, and double-cross programs into a mass-selection synthetic seed orchard. The plant breeder starts by locating disease-resistant and superior-growth-form parent trees. The first crosses attempt to get both resistance and form in the same plants. Siblings of selected resistant, good-growth-form trees are mated to stabilize the characteristics.

The regional test plantations are set up to evaluate the resistance and growth form of the best plants over a range of environmental conditions. These plantations should be in a number of locations that represent the environmental variability over which the plant materials may eventually be used.

Those trees that look good in the regional test plantation are transplanted to a seed orchard. Open pollination among the various trees in the seed orchard should produce a seed lot with a thorough mixing of genes for growth form and resistance. The synthetic variety thus produced should be genetically improved, yet contain sufficient variability to buffer against rapid buildup of a new race of pathogens or against any other pathogens.

The backcross and double-cross programs are breeding techniques for combining different genetic traits into a new line. Backcrossing an F_1 hybrid to one of the parents is done to enhance the number of genetic characters of that parent in the progeny. If the F_1 hybrids of the single-cross program have a good resistance but a poor form, the breeder will backcross to the form parent. The progeny of the backcross should have more of the good form characteristics and yet retain some of the resistance.

The double-cross program might be used to incorporate more than one set of resistance or form genes in the progeny.

The plan for the development of synthetic resistant varieties demonstrates some of the complexity in developing genetically improved varieties when using single genes for resistance.

A second approach to the use of single genes is to utilize a system where stabilizing, directional, or balancing selection pressure keeps the pathogen change in check. There is a general rule that says that the pathogen most fit to survive has the minimum number of virulence genes. In other words, if there is no need for virulence genes, they will be selected against and reduced to low numbers in the wild population.

The selection-pressure approach works well where a pathogen has two distinct and separate phases of growth and development. One example would be the *Fusarium* wilt of tomato. The fungus has a pathogenic existence during the growing season on tomatoes, and then during the winter it survives saprophytically by growing on dead organic matter. Any excess virulence genes developed on the live tomato are of a negative advantage and are selected against in its saprophytic existence. Therefore, a serious buildup of disease due to virulence in the pathogen does not occur. A single-gene-resistant tomato variety can survive indefinitely as an agricultural variety.

We have no good examples of stabilizing selection in diseases of trees, but one might speculate that stabilizing selection might operate if we could find single genes for resistance in aspen for Hypoxylon canker. The pathogen must survive and compete for 2 to 3 years as a saprophyte on killed tissue prior to sporulation. The capacity for pathogenicity may be of selective disadvantage during this saprophytic sporulation phase. Single genes for resistance, which appear to induce callusing of inoculation wounds, appear to be present in the aspen population, but these are only effective against the least-virulent isolates of the Hypoxylon fungus.

If one can recognize those disease situations where stabilizing selection pressure exists, genetic improvement should be "relatively easy" to develop, and the improved varieties should remain resistant to that one disease. The progeny of single-cross, backcross, or double-cross programs could be used directly as an improved variety.

If screening of F_1 hybrids shows that particular parent trees produce resistant progeny, the genetic improvement can be accomplished through development of a seed orchard. Branches from the resistant tree are grafted to root stock. These grafted plants, planted in seed orchards, will usually flower and produce seed while small trees. Therefore, genetic improvement can be accomplished within a relatively short period of time.

FIGURE 20-1 Plan for the development of fast-growing disease-resistant synthetic varieties of a cross-pollinated forest crop. (From Borlaug, 1966.)

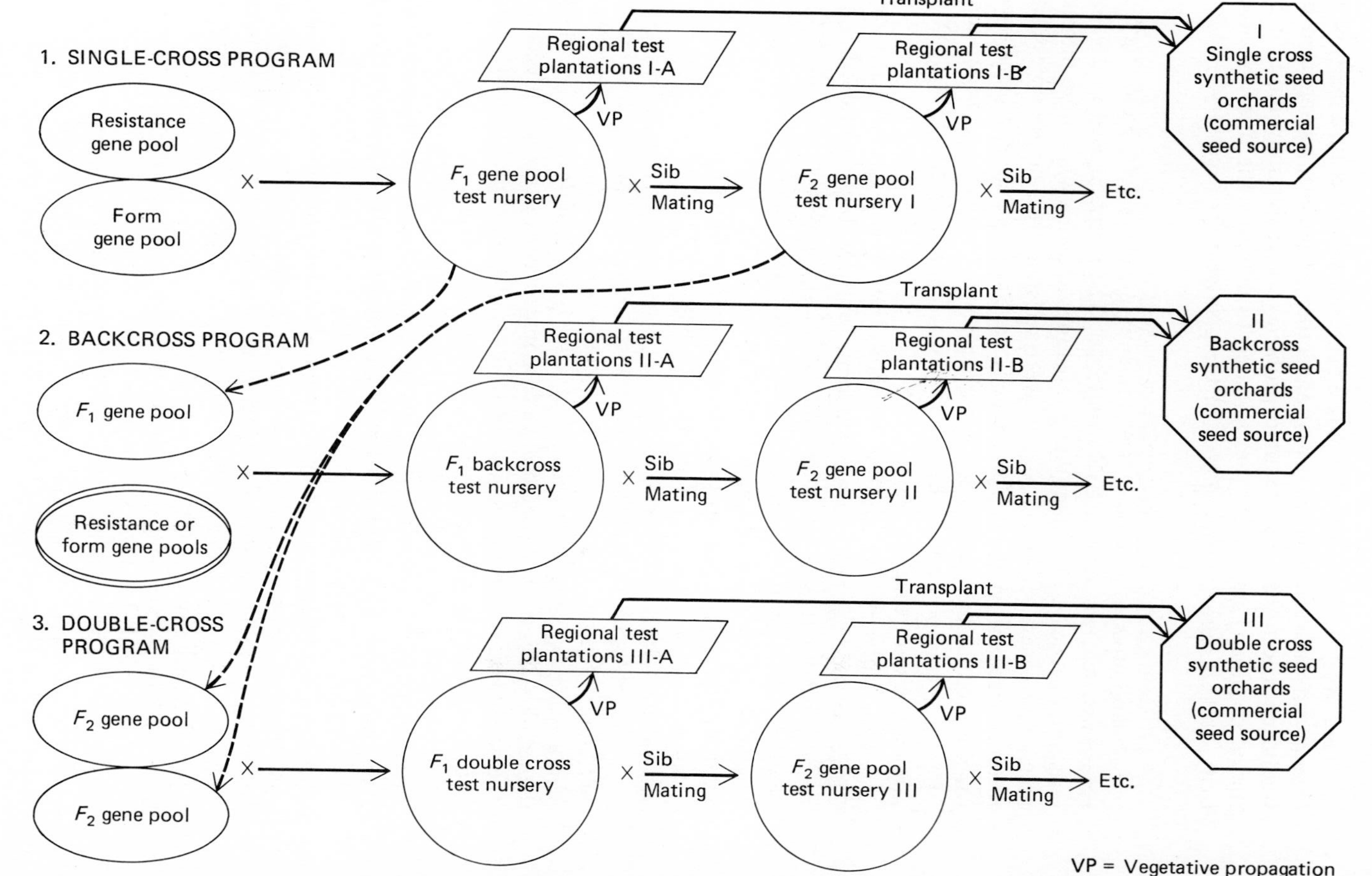
1. SINGLE-CROSS PROGRAM
Resistance gene pool
Form gene pool
×
Regional test plantations I-A
VP
F_1 gene pool test nursery
× Sib Mating
Regional test plantations I-B
VP
F_2 gene pool test nursery I
× Sib Mating
Etc.
Transplant
I Single cross synthetic seed orchards (commercial seed source)
2. BACKCROSS PROGRAM
F_1 gene pool
Resistance or form gene pools
×
Regional test plantations II-A
VP
F_1 backcross test nursery
× Sib Mating
Regional test plantations II-B
VP
F_2 gene pool test nursery II
× Sib Mating
Etc.
Transplant
II Backcross synthetic seed orchards (commercial seed source)
3. DOUBLE-CROSS PROGRAM
F_2 gene pool
F_2 gene pool
×
Regional test plantations III-A
VP
F_1 double cross test nursery
× Sib Mating
Regional test plantations III-B
VP
F_2 gene pool test nursery III
× Sib Mating
Etc.
Transplant
III Double cross synthetic seed orchards (commercial seed source)
VP = Vegetative propagation

One must be cautious with single-gene resistance because the genetic diversity of the variety is narrowed and other pathogens, less affected by stabilizing selections, may become problems. Varieties may have to be shifted or other control activities initiated to maintain the forest productivity, so that periodic surveillance of the plantation is necessary.

Quantitative Inheritance Most phenotypic characters, including disease resistance, are quantitatively inherited. The methods used to develop resistant varieties using quantitative resistance are somewhat more difficult than single genes, because crosses between resistant trees do not necessarily produce resistant seedlings. The progeny have variable levels of resistance, ranging from nonresistant to somewhat resistant. Graphically, the variables of the progeny look much like a normal distribution. Most of the progeny are less resistant than the parents are.

The degree to which the seedlings resemble the parents varies. Therefore, one attempts to select parents with resistance factors that are readily inherited by a good proportion of the progeny. The degree of similarity between parents and progeny is the heritability. A heritability of 1.0 would occur if the progeny were identical to the parents. Most of the quantitative genetic improvement is done with heritabilities less than 0.6. With Hypoxylon canker of aspen, the heritability for reducing canker enlargement was 0.45 or less.

The potential for genetic improvement is not totally expressed in the heritabilty figure. If one selects parents with outstanding resistance characteristics, genetic improvement can be accomplished even if the heritability is low. The potential genetic improvement through breeding is the product of the heritability of the trait and the amount of deviation from the average of the selected parents.

The extra effort in utilizing quantitative selection is well worth the time spent since the resistant (tolerant, not immune) plants do not place a strong selection pressure on the pathogen. With limited use of backcrosses and sib matings, one avoids unconscious narrowing of the gene pool, so that a high degree of heterogeneity can be maintained in the population. The original pathogen plus many unknown future pathogens will have considerable difficulty developing diseases of epidemic proportions. A truly mutual tolerance among pathogen, plant, and man is then developed through the use of quantitative resistance.

Genetic improvement through the use of quantitatively inherited resistance need not take generations of pathologists. In the Hypoxylon resistance program in aspens, the characterization of parents with good heritability took less than 8 years. If genetically improved aspens with resistance to Hypoxylon canker were considered important by industry, one could start a seed orchard, consisting of sucker-reproduced plants, that could be expected to start producing genetically improved seed within another 6 to 8 years.

REFERENCES

ANONYMOUS. 1972. Genetic vulnerability of major crops. National Academy of Sciences, Washington, D.C. 307 pp.

BINGHAM, R. T., R. J. HOFF, and G. I. MCDONALD. 1971. Disease resistance in forest trees. Annu. Rev. Phytopathol. *9*:433–452.

BORLAUG, N. E. 1966. Basic concepts which influence the choice of methods for use in breeding for resistance in cross pollinated and self-pollinated crop plants. *In* Breeding pest resistant trees, ed. H.D. Gerhold et al. Pergamon Press, Oxford, pp. 327–348. 505 pp.

FLOR, H. H. 1971. Current status of the gene-for-gene concept. Annu. Rev. Phytopathol. *9*:275–296.

LEPPIK, E. E. 1970. Gene centers of plants as sources of disease resistance. Annu. Rev. Phytopathol. *8*:323–344.

METTER, L. E., and T. G. GREGG. 1969. Population genetics and evolution. Prentice-Hall, Inc., Englewood Cliffs, N.J. 212 pp.

VALENTINE, F. A., P. D. MANION, and K. E. MOORE. 1976. Genetic control of resistance to Hypoxylon infection and canker development in *Populas tremuloides*. Proc. Twelfth Lakes States For. Tree Improv. Conf. USDA For. Serv. Gen. Tech. Rep. NC-26, pp. 132–146.

VAN DER PLANK, J. E. 1963. Plant diseases: epidemics and control. Academic Press, Inc., New York. 349 pp.

WILLIAMS, P. H. 1975. Genetics of resistance in plants. Genetics (supplement). *79*:409–419.

21

DISEASES OF SEEDLINGS IN THE NURSERY

- Types of seedling diseases and related problems
- Damping-off
- Root rots of nursery-grown seedlings
- Foliage and stem diseases

Management of a tree nursery is a specialized forestry activity with a limited number of participants. Although the vast majority of the readers of this book will never participate in nursery management, many will work with plant materials produced by someone else in the nursery. Some disease management problems associated with forest and urban plantings are directly affected by nursery management practices.

Many of the diseases of seedlings in the nursery are the same diseases we have discussed in Chapters 5, 7, 9, 10, 11, 12, and 16. Nursery problems are also associated with abiotic stress agents, as discussed in Chapters 2, 3, and 4.

Even though most of the basic disease concepts were discussed in earlier chapters, it seems appropriate to bring the threads of the topic together at this point. The resource manager who will never directly participate in nursery management decisions can therefore better understand the origin of some problems and better appreciate the complexity of the nursery manager's problems.

TYPES OF SEEDLING DISEASES AND RELATED PROBLEMS

Intensive, repeated cultivation of tree seedlings in nurseries is an ideal environment for the buildup of diseases that do not generally represent a serious threat to the natural regeneration of trees. In the nursery, foliage disease, canker diseases, and nematode problems intensify rapidly. They may pose an immediate problem to the nursery operator and/or a long-term problem to the forest manager of the outplanted stock. Diseases such as fusiform rust, brown spot needle blight, and Scleroderris canker are accentuated as field problems because of initial infection in the nursery.

Damping-off is a problem unique to germinating seeds and is therefore a consideration of the nursery operator. Damping-off will be discussed as a specific problem of tree seedlings, but it is not unique to trees. Many agricultural crop seeds are treated with fungicides to reduce damping-off.

Trees grown in nurseries have root rot problems which have many similarities to the feeder root rot problems discussed earlier. The damping-off and root rot problems have much in common and we sometimes draw arbitrary lines as to when damping-off stops and root rots begin.

Another consideration for the nusery operator is mycorrhizae. These were discussed earlier, but it is important to recognize here that fumigation for the control of seedling diseases and weeds may have detrimental effects on mycorrhizal development.

Nursery practices are rapidly changing to accommodate to mechanization and maximum production. The practice of growing seedlings in synthetic soil in containers will generate a completely new set of problems. Containerized seedling production is often done in greenhouses with overhead sprinkler systems. These are ideal conditions for the development of foliage pathogens. The seedlings are fertilized heavily, thereby producing an environment unfavorable for mycorrhizae and favorable for some root pathogens. Steamed or fumigated potting mixtures are devoid of a natural microflora of fungi and bacteria. Pathogenic microorganisms can more rapidly colonize a medium in which competition and the antagonism of natural microflora are missing.

DAMPING-OFF

Anyone who has planted seeds and carefully watched for the plants to emerge recognizes that some seeds do not result in plants. Some seeds are not viable, but others are viable yet are killed before the shoot emerges from the soil. Some of the seedlings that emerge seem to collapse and die. These last two phenomena are called damping-off, a problem that may regularly produce losses of 15% or more (Figs. 21-1 to 21-4).

Preemergence damping-off occurs as a result of fungus attack of the radicle prior to seedling emergence from the ground. Postemergence damping-off occurs because of fungus attack at the base of the seedling stem after it emerges from the ground. Both are caused by fungi that invade the succulent stem tissues. Once cells have stopped expanding and their walls become lignified, they become resistant to infection by damping-off fungi.

FIGURE 21-1 Western conifer nursery.

FIGURE 21-2 Many of the missing trees in these rows of Douglas fir were probably lost to damping-off and root rot.

FIGURE 21-3 Damping-off of Austrian pine seedlings in a greenhouse demonstration.

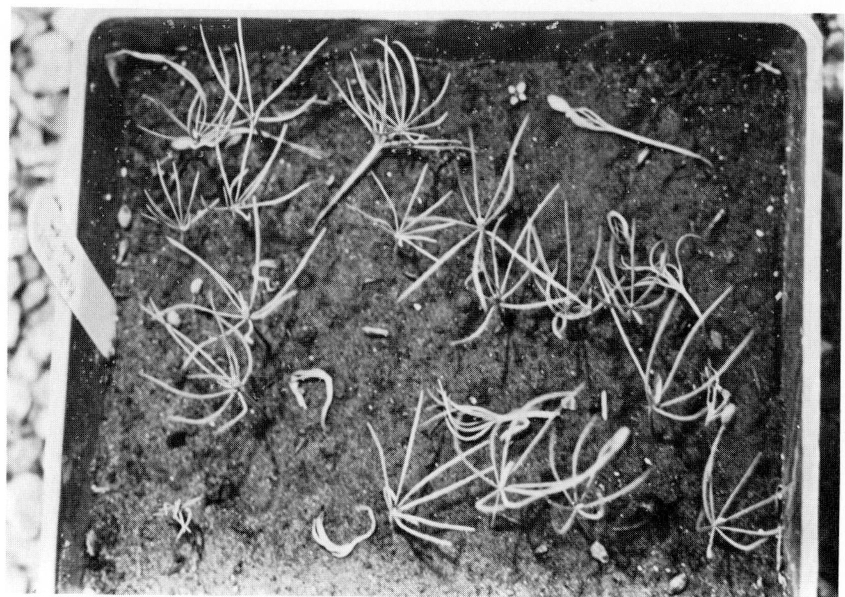

FIGURE 21-4 Closeup of damping-off in Austrian pine.

Phycomycete fungi in the genera *Pythium* and *Phytophthora* are commonly involved in damping-off. Other fungi are *Rhizoctonia* spp., *Fusarium* spp. and *Sclerotium* spp. in the Fungi Imperfecti.

Cool, wet, highly organic, neutral-to-basic soils favor damping-off. Close spacing and continuous cropping also favor damping-off.

Most nursery operators expect a certain amount of damping-off, but keep it to a minimum by avoiding conditions that favor the problem. Soil fumigation with methyl bromide, chloropicrin, or Vorlex are common for weed and root rot control. The fumigation also reduces damping-off. Soil drench with H_2SO_4 to reduce the pH or with fungicides is sometimes used specifically for damping-off.

ROOT ROTS OF NURSERY-GROWN SEEDLINGS

The damping-off problems grade into root rot problems such that a precise line of separation is not possible. One root rot problem of nurseries in the Pacific Northwest is caused by *Fusarium* spp. and other soil-inhabiting fungi that become established soon after the seedlings emerge. If a few days of cool temperature occur during the early development of the seedling, the stage is set and the effects are seen 3 months later, when the seedlings die of root rot.

Another root rot problem of eastern forest tree nurseries is Cylindrocladium root rot caused by *Cylindrocladium scoparium.* This root rot occurs in older seedlings in transplant beds and has been responsible for the abandonment of at least one nursery in the Lake states.

In the South, root rots are the most important hardwood nursery disease. The disease can be caused by many of the soil fungi already listed for damping-off. The first symptoms that the nursery operator recognizes are stunting, top dieback, chlorosis, and premature defoliation. Examination of roots reveals blackened feeder roots. Control is through thorough fumigation.

Charcoal root disease of western forest nurseries is caused by *Macrophomina phaseoli (Sclerotium bataticola),* a fungus that causes root diseases of more than 300 species of plants. This pathogen forms sclerotia (small black resting structures) for survival in the soil in the absence of hosts. These sclerotia make cultural treatments other than thorough fumigation impractical for controlling this disease.

Black root rot is the name applied to Macrophomina root rot in the South. In this region another fungus, *Fusarium oxysporum*, which also forms resting sclerotia, is also involved.

Control of root rot problems is achieved by fumigation of nursery beds prior to seeding. The use of sterilants such as methyl bromide and Vorlex provides a measure of weed and nematode control in addition to root rot control. The sterilants must be used before every crop, in some instances, to prevent major losses. Even with fumigation, damping-off and root rot cause mortality of seedlings.

FOLIAGE AND STEM DISEASES

The close spacing and moist conditions maintained in the nursery are ideal for the rapid spread of foliage and canker pathogens. If allowed to go unchecked, these organisms would seriously reduce or eliminate production.

The category of stem diseases of trees in the nursery would include stem rusts such as white pine blister rust, fusiform rust, western gall rust, and others that infect through foliage and young shoots. It would include Scleroderris canker and other cankers.

Crown gall, described in Chapter 7, is a major concern of hardwood nursery operations. They try to prevent introduction of the bacterium on transplant stock, because once an area is infested, it may need to be taken out of production and grown for 2 years with a resistant crop such as oats or cowpeas. Treatment of plants with *Agrobacterium radiobacter* strain 84 is also used to prevent crown gall (see Chapter 7).

Foliage problems are Lophodermium needle cast, brown spot needle blight, Dothistroma needle blight, Phomopsis blight, and others on conifers. On hardwoods, rusts, anthracnose, and other foliage diseases are problems.

Most stem and foliage diseases are problems of nursery seedlings as well as trees in the field. If the problems are controlled in the nursery, we might expect less of a problem in the field plantings. This is not always the case, particularly where the field planting is done in areas where high levels of natural inoculum occur. Under these conditions, the large artificially maintained population, in the absence of protective fungicides, is rapidly reduced to a much smaller population of naturally resistant individuals. For example, on might question the logic of producing large numbers of seedlings susceptible to fusiform rust, by use of fungicides in the nursery, when these are going to be subjected to natural selection for survival in an environment with high levels of rust. Would it not be better to allow the rust to do some selective thinning in the nursery prior to outplanting?

This is a difficult question to answer because of another possibility—that infected nursery stock may actually contribute to the distribution of highly infective populations of pathogens. Under these circumstances, it is very appropriate to control the pathogens in the nursery.

These and other questions will only be answered by very close ties between the nursery operator, who is producing the stock, and the forest manager, who attempts to maintain the trees for products and environmental benefits over a period of years.

REFERENCES

BLOOMBERG, W. J. 1979. A model of damping-off and root rot of Douglas-fir seedlings caused by *Fusarium oxysporum.* Phytopathology *69*:74–81.

DORWORTH, C. E., H. L. GROSS and D. T. MYREN. 1975. Diseases in Ontario forest tree nurseries, 1966 to 1974. Can. For. Serv. Great Lakes For. Res. Cent. Rep. O-X-230.

HARTLEY, C. 1921. Damping-off in forest nurseries. USDA Bull. 934. 99 pp.

PETERSON, G. W., and R. S. SMITH, JR., tech. coords. 1975. Forest nursery diseases in the United States. Agricultural Handbook 470. USDA Forest Service. 125 pp.

ROWAN, S. J., T. H. FILER, and W. R. PHELPS. 1972. Nursery diseases of southern hardwoods. USDA For. Serv. For. Pest Leafl. 137. 7 pp.

TINUS, R. E., W. I. STEIN, and W. E. BALMER, eds. 1974. Proc. North Am. Containerized For. Tree Seedling Symp. Great Plains Agr. Counc. Publ. 68. 458 pp.

22

PATHOLOGICAL CONSIDERATIONS OF URBAN TREE MANAGEMENT

- Initial considerations
- Goals of tree management for urban environments
- Administration of urban tree management
- Established tree management
- Planning and management of new plantings

Pathological considerations of trees growing in urban and heavily used recreation sites should be a fundamental part of management decisions related to urban forestry. Proper planning and maintenance, for maximum benefit with minimum disruption and cost, cannot be accomplished in the absence of pathological concerns. Yet, in practice, much urban and recreation site management is accomplished without serious planning or thought of the pathological consequences. The pathologist is often called in to diagnose problems, prescribe band-aid or sugar-pill therapy, or write a proper epitaph, in situations where some initial involvement would have avoided or reduced the extent of the problem.

The pathology of urban trees is best understood in the context of the urban tree management plan. Therefore, I will present my concept of what is involved in urban tree management, even though some of my concepts go beyond the realm of pathology.

INITIAL CONSIDERATIONS

Before beginning with the topic of urban tree management, it is appropriate to first recognize both the negative and positive aspects of maintaining a tree population. It should be very obvious that, even though many of us consider trees as an important part of an acceptable environment, they really are not essential.

Why do municipalities appropriate funds to plant and maintain trees? They are expensive investments which may become expensive liabilities when they die and a nuisance to remove when reconstructing roads and buildings. They are also a serious source of property damage during wind or ice storms.

It is hard to determine exactly why trees are planted and maintained. Tradition plays a role. Shade from a hot summer sun is also involved. Today, city planners consider such aspects as softening of harsh man-made structural lines and screening or framing of vistas as contributions of vegetation. Trees generally increase property values. A whole array of other benefits are ascribed to urban trees, but most of them are emotional rather than factually based benefits.

GOALS OF TREE MANAGEMENT FOR URBAN ENVIRONMENTS

The primary goal is to produce an aesthetically appealing community. A second goal is to minimize catastrophic and expensive losses of trees. A third goal is to maximize public support for what you are doing, and the fourth goal is to efficiently utilize a limited resource for the accomplishment of the first three goals.

Urban tree management will be broken down into administration, established tree management, and planning and management of new plantings.

ADMINISTRATION OF URBAN TREE MANAGEMENT

Personnel The administrative personnel in charge of urban tree populations must be holistic enough to recognize the complexity of managing a long-term living resource. Neglect or mismanagement at one time will be felt many years later. One cannot properly manage a tree resource with elected or short-term appointed personnel who are concerned with immediate rather than long-term problems.

Inventory An inventory of location, age, and condition is important for the planning and allocation of resources. Inventories should be periodically updated either by ground survey and/or aerial photographic reconnaissance. It is also important to inventory and categorize sites by soil conditions, including fertility, texture, and moisture-holding capacity. Air pollution may vary and should be recognized. Different uses, such as residential, recreational, industrial, and business, should be recognized because these require different management considerations.

Records Readily accessible inventory and maintenance records should be maintained and utilized, to determine trends and to coordinate management.

ESTABLISHED TREE MANAGEMENT

Mature trees can represent an asset or a liability. They have demonstrated a tolerance for an array of stress-inducing factors associated with the urban environment. Barring a major disease or insect introduction, or a major street or sidewalk renovation, they can generally be expected to continue to survive as a population for a number of additional years.

Elms represent a mature, established tree population. Dutch elm disease, an introduced disease, is a major threat to the population which must be dealt with or the population will be eliminated. The assets of mature, functional trees which are well adjusted to the difficult urban environment should be weighed carefully against the possible benefits of allowing the population to die and then replacing them with other species. Yes, the costs of maintaining elms are high, but so are the costs of replacement. A cost/benefit study by one of my graduate students clearly demonstrated that thorough

Dutch elm disease control is economically sound. It will take many years to eventually replace the elm population. Many of the trees of the new plantings will slowly deteriorate because they are not able to survive in the urban environment, so that investment as well as time will be lost.

The pathological and insect problems of mature trees are rather straightforward and generally predictable. The high value of individual trees often warrants direct control. Most problems, including Dutch elm disease, can be handled with normal tree maintenance. Pruning of cankered, dead, or broken branches eliminates future problems in most instances, and reduces the hazard of decay infection.

Recent surveys of street trees in two upstate New York cities found up to one-third of the large trees with large broken or poorly pruned branch stubs in the upper crown. Good indications of decay, such as conks, punk knots and open decayed wounds, were very common. In some cities, the population of mature trees is obviously not being properly maintained. Such neglect will reduce the potential value derived from these trees, by increasing future problems and reducing the useful life expectancy.

Foliage problems can be kept to a minimum by raking and disposing of leaves. A certain amount of scattered incidence of endemic wilt disease (Verticillium wilt), root rot, decay, and normal old-age deterioration can be expected but will not represent a major expenditure for the urban tree population.

A major problem of the larger trees is associated with pruning for utility wires. Death of large branches in the upper crown of Norway maples was found to be related to the degree of crown disruption for utility wires in a survey of Syracuse, New York, street trees. Severe pruning of some species of mature trees apparently leads to subsequent deterioration of the crown.

The most serious threat to a mature tree population is from decline. The trees are generally growing under predisposed or stress conditions, so that any inciting factor may start the decline syndrome rolling. Declines are discussed in Chapter 18. Therefore, the subject will not be elaborated further, except to emphasize that prevention of losses due to decline must be directed toward avoiding predisposing factors. In a practical sense, it is very difficult to avoid predisposing trees in the urban environment, so that early detection and amelioration of inciting factors is appropriate. If one can avoid massive increases in salt applied to roads, or avoid excavation near trees, or prevent major insect defoliation or any of the other inciting factors, declines can be prevented. If it is impossible to avoid an inciting factor, it is important to help the younger trees recover by pruning and fertilizing. The overmature trees, or trees on highly deficient sites, will not recover, so planning for removal and replacement should begin at the time the inciting factor occurs.

Large, mature trees are often neglected during a sidewalk, curb, street, or utility line construction project. They are in the way during the work phase, and may actually require additional costs to work around. After the construction is completed, they die and need to be removed. Would it not have been better to cut these doomed trees before the project begins, and utilize the money to prune and rejuvenate the more vigorous trees or for the planting of new trees?

There is a point in the development of a tree when it changes from an asset to a liability. In Chapter 14 the concept of pathological rotation was introduced. Pathological rotation in forest trees is the point in time when added increment each year just compensates for the amount of wood lost due to decay fungi. Obviously, when added increment each year is consumed by decay fungi it is appropriate to cut, since the maximum value of the stand has been reached.

I would like to introduce a similar pathological rotation concept for urban trees. Let me call it negative asset rotation. The arborist's method of determining the estimated value of shade trees is based primarily on replacement cost. Obviously, the larger the tree, the larger the replacement cost. Adjustments are made in the replacement cost, depending on species, condition class, and groupings of trees.

Comparison of value increment over time with cost of maintenance or removal indicates that value increment eventually starts to level off while expenses and liabilities are rapidly increasing.

It would seem inappropriate to continue expenditures on maintenance in excess of value increment, and therefore when assets begin to drop (negative asset rotation), one should consider removal of the tree.

PLANNING AND MANAGEMENT OF NEW PLANTINGS

Specifications

One of the first considerations is tree species. The list to select from is very large, as indicated by the list of species planted in the communities of Syracuse, Rochester, and Poughkeepsie, New York, provided in Table 22-1. It is obvious that the species composition of the urban forest is shifting. All three of these cities once had large elm populations. The larger trees today are primarily maples. We see a major shift from native species to exotics in the present plantings.

Considerations for selection of species in actual practice include availability of stock, previous satisfaction, and design specification for shape or size. There seems to be a desire to try anything new. Disease problems are not usually considered, except to discriminate against American elm because of Dutch elm disease.

Because the selection of suitable species is a key to minimum maintenance and maximum satisfaction, let me suggest a list of criteria for

TABLE 22-1 Street Trees of Syracuse, Rochester, and Poughkeepsie, New York (based on Valentine et. al., 1978.)

Older tree population	Younger tree population	
60% of the population	*45% of the population*	
Norway maple	*Malus* spp.	Honeylocust
	Linden	Sugar maple
25% of the population	Green ash	London plane
Silver maple		
Sugar maple	*25% of the population*	
	Norway maple	
10% of the population		
Red maple	*20% of the population*	
White ash	Red maple	*Prunus* spp.
Cottonwood	*Zelkva velcova*	Red oak
Basswood	Silver maple	Cork tree
Linden	Tulip tree	Pin oak
Boxelder	White birch	Ginkgo
Honeylocust	Golden rain tree	
London plane		
	10% of the population	
5% of the population	Hawthorn	Red bud
American elm	White ash	Willow
Horsechestnut	European mountain ash	Basswood
Catalpa	Hackberry	Horsechestnut
Hackberry	American beech	Tree of heaven
Green ash	Boxelder	Black locust
Sycamore	Kentucky coffeetree	Others
Others	Katsura	

consideration before selecting which tree species to plant. One should recognize all the positive and negative attributes of each species under consideration and generate a species-suitability index considering the following criteria:

1. **Retail cost** There is a great variety of production costs among tree species. Low-cost, easily obtainable trees are most suitable for most plantings.

2. **Site requirements** a major asset for an ideal urban tree is tolerance of very poor sites. Tolerance of poor soil aeration makes swamp tree species preferable to upland tree species.

3. **Maintenance requirements** Some trees require more pruning and shaping than others. Water sprouting along the main stem by sycamores, London plane, and American basswood need to be repeatedly removed and therefore are also a consideration.

Growing trees in the open causes lower branches to enlarge quickly. If not pruned early and regularly, Norway maples and other species will develop a low spreading crown which interferes with both sidewalk and street traffic. Some columnar varieties of maple avoid much of this problem. American elm requires minimal shaping during early growth.

4. Susceptibility to breakage Ice and wind storms cause breakage of some tree species more than others. An ice storm in the spring of 1976 in Syracuse caused much breakage of boxelder, Carolina poplar, and silver maple. Broad-spreading Norway maples were also seriously damaged. More upright trees were less damaged. Besides the immediate problem, breakage represents infection sites for decay fungi and future problems.

5. Life expectancy Most people expect a tree to be long-lived. This assumption is correct for sugar maple and oak, but is not necessarily correct for many of the species presently being planted.

As a general rule, long-lived trees such as sugar maple are desirable in parks and recreation areas, but the shorter-lived (30 to 50 years) trees such as honey locust are more appropriate for street plantings. Long-lived trees grow more slowly. Short-lived trees have the advantage of growing quickly, supplying shade to those who planted and paid for them.

As a general rule, urban trees planted today have an average life expectancy of 20 years. If this is correct, we are not planting many shade trees for tomorrow. It might be appropriate to consider planting rapid-growing, long-lived trees such as elms and silver maples and then cutting them before they become too large. These species will develop into shade trees quickly.

6. Major disease problems Susceptibility to a lethal disease-causing organism is a major disadvantage but should not preclude the use of a tree species. Reasonable and effective control measures can offset the negative aspects of a lethal disease.

Actually, a major lethal disease may be an asset for an urban tree. One of the more difficult problems in managing urban maples is deciding when a tree is aesthetically dead. Considerable controversy sometimes arises over cutting trees that a city forester considers functionally dead and the local residents do not. There is usually little doubt when a Dutch-elm-diseased American elm needs to be removed. A major lethal disease also prevents the development of a large overmature, rather hazardous, population of trees. Therefore, from a management standpoint, a major lethal disease should not prohibit the use of a particular species.

7. Minor disease and insect pests All trees have diseases and insect pest problems. Consideration of the visual and psychological impact on people may intensify the negative aspects of a biologically minor problem. The impact of insects and minor diseases is poorly understood, particularly in cities, where the interaction of pests with stressed trees may change the overall effect on trees of insects and minor pathogens.

8. Genetic diversity Domestication of plants, as a general rule, reduces the genetic diversity. Selection for uniformity of desirable traits also unconsciously selects for uniformity of other traits, some of which may be undesirable. Cloning or vegetative propagation is the extreme in uniformity. A desirable trait, uniform columnar growth form in the clone lombardy poplar, is counterbalanced by common susceptibility to environmental stress and weak canker pathogens. The potential for rapid buildup of destructive insects and pathogens is reduced if a population has a degree of genetic diversity. Trees of mixed seed origin or synthetic varieties should be used in preference to clones and narrow-genetic-base varieties. Further elaboration on the genetic considerations was presented in Chapter 20.

9. Tree form and growth characteristics Urban trees are best known for their form and desirable or undesirable growth characteristics. Although size should remain a major factor in species selection, form and other growth characteristics should not be a dominant consideration. How much does exotic form really mean to the general public? Indirect consequences of the other selection criteria are probably more important.

10. Probability of the unexpected The final ingredient of a species suitability index should include a safety factor for the unexpected. There is a real advantage to using native varieties and species that have been widely planted for a number of years or generations. We know what to expect in the way of problems. The emphasis today on new exotics and cultivars should be weighed heavily against the probability of unexpected pests and pathogens.

If the positive and negative aspects of various trees using these 10 criteria are quantified, a ranking of suitable and unsuitable species is obtained. This type of list would be invaluable for planning purposes. At the present time, it is very difficult to quantify or rank species within the minor diseases and insect pest category because of insufficient data. Assessment of the impact of minor pathogens and insects will greatly aid in the eventual establishment of a species-suitability index.

Selection of suitable species is not enough to ensure maximum satisfaction. There are within-species differences for adaptation to variation in climate. This topic, described in Part 1, emphasized the used of local seed sources as best adapted to the climate of an area. With ornamentals, local seed sources may not be available, so a period of 3 to 5 years of screening in a regional or city nursery is suggested. A city can hardly afford not to grow the trees for a few years in its own nursery. Those not adapted to the climate can be easily removed from the population. It is also much easier and less costly to evaluate future insect and pathogen problems in a city nursery.

Maintenance

Maintenance often has a lower priority than planting and removal in the urban tree management budget. This is ironic, since good maintenance often reduces the number of removals and, consequently, the need for planting.

Periodic pruning, to shape trees or remove broken and dead branches, produces immediate visual improvements which the general public can quickly recognize, and also produces long-run benefits such as less decay, fewer cankers, and less future breakage.

Although fertilizing has been demonstrated to produce beneficial effects on tree growth, improved growth may not be what established urban trees need. It may be appropriate to stimulate the growth of young trees to help them recover from an environmental or pest problem. But little documented evidence is available to demonstrate the long-term benefits of urban tree fertilization, so the use of fertilizer should have a low priority in a tight-budget management program.

Pesticides to control insects and diseases are sometimes applied to alleviate entomophobia or pathophobia in people rather than to control problems on trees. They never solve the problem. The unending costs from year to year are rather discouraging.

Careful selection of pesticides and application only where and when absolutely needed should be a general rule in urban tree management.

Pest management is better accomplished through mixing of 5 to 10 species of trees selected on the basis of the suitability criteria presented. By using a mixture, one has a natural buffer against rapid disease increase and catastrophic changes in the population. By working with a reasonably small number of species, it is possible to optimize management activities.

Insect populations are less capable of increasing in a mixed population. In the future, they may be further held at reasonable levels by biological control agents such as parasites, predators, and pathogens. Another tool for insect manipulation may be pheromones, chemicals produced by insects as signals to other insects. An aggregation sex attractant is being experimented with for manipulation of the smaller European elm bark beetle in Dutch elm disease control.

Periodic inventory and assessment should be used to locate trees that are prone to disease or insect problems. These trees represent concentration centers of pest populations. Selective removal of these problem trees should be beneficial to the remaining population.

Rotation

It is appropriate to recognize the useful-life expectancy of trees at the time of planting. Contrary to popular belief, most of the trees planted today will not be around for two or three generations. The stress-inducing pressures of the urban environment greatly reduce the life expectancy of even long-lived trees. There is also a shift toward smaller, shorter-lived tree species.

The resource manager should recognize the useful-life expectancy of trees and anticipate when they should be removed. The negative asset rotation concept, as discussed earlier, is applicable.

By mixing short, medium, and long rotation species, we can avoid major changes in the city landscape from removal of overmature trees. Maximum continuous benefit at minimum costs is assured if planning, maintenance, and forethought are applied to present-day planting of trees in our cities.

REFERENCES

BASSETT, J. R., and W. C. LAWRENCE. 1975. Status of street tree inventories in the U.S. J. Arboric. *1*:48–52.

BAXTER, D. V. 1952. Pathology in forest practice 2nd ed. John Wiley & Sons, Inc., New York. 601 pp.

BOYCE, J. S. 1938. Forest pathology. McGraw-Hill Book Company, Inc.. New York. 600 pp.

SANTAMOUR, F. S., JR. 1971. Trees for city planting: yesterday, today, and tomorrow. Arborists News *36*:25, 27–28.

SANTAMOUR, F. S., JR., H. D. GERHOLD, and S. LITTLE eds. 1976. Better trees for metropolitan landscapes. Proceedings of a symposium. USDA For. Serv. Northeast. For. Exp. Sta.

VALENTINE, F. A., R. D. WESTFALL, and P. D. MANION. 1978. Street tree assessment by a survey sampling procedure. J. Arboric. *4*:49–57.

23

PATHOLOGICAL CONSIDERATIONS OF INTENSIVELY MANAGED FOREST PLANTATIONS

- Historical basis for approaching disease control through the dominant factor
- Plant modification to control diseases
- Pathogen modification to control diseases
- Environmental modification to control diseases
- Conclusions

Before discussing disease considerations in intensively managed plantations, it may be necessary to justify the need for such plantations. If one looks around, it appears that natural forests abound. Why then are intensively managed plantations being considered when there appears to be such surplus?

Actually much of what appears to be forest land is not really available to the timber industry. In many situations, other values placed on the forest far outweigh the commercial value of the timber.

Therefore, one reason for intensively managed plantations comes from the need to produce more wood fiber on fewer and fewer acres

A second advantage of intensive management concerns the need to better control the economic investment of land and trees. The productivity of unmanaged forests is very low if productivity is expressed in terms of growth per acre per year. Low productivity was not a consideration when land was inexpensive and return on investment could be acquired through harvesting the crop of trees that already occupied the site. Forestry has shifted from harvesting to growing of trees, and the economic returns are therefore affected by the rate of fiber production.

Pathological considerations can be more appropriately justified with intensively managed plantations. The economic return is higher; therefore, money can be spent on protecting the crop and the investment. With natural forests it is usually more economically sound to let the diseases run their course and salvage-cut to extract whatever value can be gained from the forest.

Beech bark disease is a prime example of the economic dilemma of natural forests. Even if we had a control for this disease, it could not be economically justified. You may be amazed to learn that a 10-inch (26 cm) beech has an approximate market value to the landowner of $0.50. How much can the forest owner spend to protect this type of investment? He or she will obviously salvage-cut as much timber as possible and use the rest for firewood.

Intensively managed forest plantations will supply the fiber needs of the future. In the South, such plantations already represent a significant proportion of the growing stock. New approaches to harvesting plantations will also become necessary. Equipment that harvests the whole tree, including the tap root, is now being used on a small scale for southern pines. Other approaches to intensive management utilize a modified silage chopper to harvest short-rotation tree crops much like corn. Sycamore and other hardwood species have been utilized for this type of management system on an experimental basis.

Rapid early growth of seedlings has always been a prime objective in establishing plantations, but little could be done to affect early growth. Recent work with mycorrhizal inoculation of seedlings in the nursery suggests that it may be possible to affect the survival ability and initial growth of seedlings in plantations.

These and other intensive management activities are not that distant. We should not move in this direction without recognizing the pathological risks of intensive management. Only through maintaining a watchful eye on the developing plantation can we recognize the problems as they develop. We also have to develop an organized way of looking at disease control problems, because the investment in the plantation will not allow us to blindly apply control procedures.

In Chapter 1, the categories of plant disease were introduced and the methods for recognition for each were presented. The abiotic diseases are recognized primarily on the basis of the random distribution of diseased individuals, lack of consistent relationship with specific biotic agents, and lack of host specificity. The control of abiotic disease involves, first, recognition of the causal agent and, second, adjustment of future management activities through appropriate manipulation of either the plant or the pathogen.

Abiotic diseases are simple-interest diseases and therefore do not generally increase rapidly. Rapid increase in an abiotic disease may occur because of an unusually high level of the disease-inducing agent. Once the abiotic disease is recognized, it is very difficult to significantly affect the amount of future loss due to the disease. Therefore, control is usually directed at reducing losses in the next crop, by changing species or modifying the site.

Decline diseases are slow, progressive deterioration diseases of complex biotic and abiotic origin. As discussed in Chapter 18, declines are best managed by recognition of predisposing and inciting conditions and attempting to avoid or counteract them. Declines are also simple-interest diseases, best controlled through management activities for the next crop.

Biotic plant diseases are characterized by the involvement of a pathogen which is living and capable of reproducing on the host. Reproduction by the pathogen results in nonrandom distribution and intensification of the disease, often at a compound-interest rate. With biotic plant disease, control may be through management of the plant, the pathogen, or the environment. These management activities may be directed toward reducing losses in the present crop, future crops, or both.

I suggest that effective control of biotic-induced diseases be based on the principle that diseases do not generally result from the interaction among the plant, pathogen, and environment with equal importance to each, but that each disease has a dominant side on the disease triangle. The most effective control involves manipulation of the dominant side of the triangle.

HISTORICAL BASIS FOR APPROACHING DISEASE CONTROL THROUGH THE DOMINANT FACTOR

Investigations on the application of forest pathology to silviculture and management in the national forests began with the establishment of the U. S. Forest Service Regional Forest Research Stations in 1907. During the early part of this century, emphasis was placed on cutting practices as they affect disease. In 1916, Meinecke, in his monograph "Forest pathology in forest regulation," emphasized the disease-control benefits of sanitation and hygiene during cutting.

From the beginnings of pathology in Europe, two schools of thought developed. One group, pathogeneticists, of which Robert Hartig was a member, assumed that disease developed because plants inherited a susceptibility to disease. A second group, predispositionists, headed by Sarauer and Ward, assumed that disease developed because environmental factors predisposed plants to disease.

A third, but less-well-defined school of thought, with roots embedded in the overthrow of the spontaneous-generation concept, are concerned with the pathogen in disease. The pathogeneticists and predispositionists acknowledged the role of the pathogens but assumed a general or universal pathogen presence.

If we look at these three groups today, we do not see conflicting concepts of plant disease but out present concept of disease—plant, environment, and pathogen interacting to produce disease.

It is inappropriate to simplify these three schools into common, equally important factors in the development of all diseases. They originated as conflicting concepts because evidence for each could be observed in the real world.

In a practical sense, one of the three is usually the dominant factor in a given disease. The resource manager, attempting to minimize losses due to disease, should recognize the dominant or controlling factor in the diseases of concern and attempt to exploit that factor to the fullest extent.

For economic and environmental reasons, it is most appropriate to integrate disease control into silvicultural practices. These practices will be discussed in relation to their effect on the plant, the pathogen, or the environmental aspects of plant disease.

PLANT MODIFICATION TO CONTROL DISEASES

Genetic variation in susceptibility can be directly utilized in tree improvement breeding programs or can be indirectly exploited in selective cutting practices. Chapter 20 discussed breeding for disease resistance. Genetically improved tree varieties are not yet commonly available, yet the resource manager can partially control diseases through plant modifications now. The probability of improvement of the next stand with selected quality seed trees is much better than with random regeneration from low-quality cull trees with large seed crops. Although large seed crops are often produced on trees that are stressed or deteriorating for various reasons, these trees obviously carry genes for susceptibility to problems and should be avoided. Logging practices of the past have generally removed the best trees and left the less desirable ones to regenerate many of the low-quality second-growth stands we have today. These types of logging practices should obviously be discontinued in favor of more careful selection of superior seed trees.

Fungicides applied to trees to prevent disease is another form of plant modification, but economic and environmental concerns restrict the use of this approach.

PATHOGEN MODIFICATION TO CONTROL DISEASES

Disease control, particularly of introduced diseases, has usually centered on the pathogen. But eradication of the pathogen is generally expensive and almost impossible to accomplish, even in a limited area.

Partial reduction in the pathogen population may or may not reduce the overall impact of the disease. A good example of disease impact reduction with partial pathogen reduction is the selective removal of trees with mistletoe infection in the upper crown. Inoculum produced in the upper crown is more effectively dispersed to intensify the disease than is inoculum in the lower crown.

Pathogens with small wind-disseminated spores are more difficult to control through pathogen manipulation. There is no reason to assume that selective removal of heart-rot- or root-rot-infected trees has any effect on the potential for disease in the surrounding stand. There is also no reason to assume that selective removal of canker-fungi-infected trees or branches has any major impact on diseases spread.

Canker fungi and decay fungi are generally limited by either host or environmental factors. So effective control of these is more appropriately applied to host improvement or environmental manipulation.

Clear-cutting, if done in large blocks, may reduce inoculum of disease agents, but the effects of clear-cutting on the environmental aspect of disease is probably more significant.

Fire is an effective control agent for dwarf mistletoe and brown spot needle blight of longleaf pine. Fire, like clear-cutting, also has environmental effects on disease.

Continuous cropping of the same species on the same site can be expected to increase disease problems. Most tree species develop in forest stands following successional sequences. Trees have evolved, with their pathogens, a balance of dynamically changing successions. The balance is shifted in favor of increasing the pathogen population if a stand is artificially maintained in the same species for more than one generation. These types of problems are well documented in agriculture, but are just beginning to show in forestry. A root rot pathogen such as *Heterobasidion annosum* is more of a problem in plantations than in naturally regenerated stands. Logging and fire control has resulted in extensive regeneration of Douglas fir in the Pacific Northwest. The second-growth Douglas fir is being seriously damaged by *Inonotus weirii,* which produced only minor damage in the original stands.

Continuous cropping is not something that can be eliminated. Economic pressures for specific products and greater yields will force more extensive use of continuous cropping. It is important to recognize and anticipate the problems of this type of cultural practice, and weigh the advantages and disadvantages. If at all possible, one should consider normal succession patterns and adjust crops to more closely approximate the species changes encountered in natural stands.

Stumps produced in thinning or clear-cutting are ideally suited for infection by root rot fungi. Pathogens in the root system will continue to survive for a period, so the stump cut surface is much like a large wound in a living tree. Large numbers of wounds, whether resulting from cutting or from natural disasters, allow a major buildup in fungal pathogens and potential problems for regeneration of the surviving stand. Anything that favors rapid colonization and decomposition of stumps by saprophytic decomposers should be considered control of these diseases through pathogen manipulation.

ENVIRONMENTAL MODIFICATION TO CONTROL DISEASES

Environmental modification, like fertilizing to improve growth or thinning to reduce competiton, enhances the growth of trees and resistance to facultative parasitic microorganisms, but obligate pathogens, particularly rusts, are affected in just the opposite way and may actually increase more rapidly on vigorous trees.

Other types of environmental modification may be directed more at affecting the conditions for dispersal and infection by the pathogen. *Lophodermium* and other foliage diseases of Scotch pine Christmas trees are more serious in plantations where tall weeds grow around the trees and along edges of the plantation adjacent to native forests. These foliage diseases are favored by prolonged moist conditions, which the shading of weeds and large trees provide. Weed control and cutting of large shading trees along the edge of the plantations is an environmental manipulation which affects disease development by affecting dispersal and infection of the pathogen.

The most significant environmental manipulation for control of tree diseases is the matching of species to sites. The selection of species to plant is more often determined by the availability of planting stock or desirability of a particular crop than by the ability of the species to grow well on the site. An even finer matching of trees to site goes beyond species suitability to using local seed sources. If one compares local seed sources with other seed sources of the same species, one usually finds that trees grown from local seed sources generally perform better than most other sources. These types of comparison, called provenance tests, have been made for a number of trees and the results are usually the same. Locally adapted plants are always among the best performers. Environmental stresses on non-locally adapted plants reduce tree vigor and, in many instances, predispose the tree to serious diseases.

CONCLUSIONS

Pathological understanding can contribute to improved intensive forest management. Neglect and misuse of pathological understanding has, in the past, cost a great deal of money. Both time and yield have been lost.

Control of tree diseases should not emphasize local suppression efforts. Control of diseases is more appropriately handled through effective planning and avoiding potential buildups. Just like modern fire control, there must be continuous assessment of potential problems and environmental conditions favoring buildup of pathogens. We must understand disease cycles and recognize where the cycle can be best manipulated.

At any point in time, we in forest pathology are concerned with the problems of the trees of various ages. There is not much we can do for the mature trees. We can prescribe cultural practices for the actively growing crops. But we can have our most significant impact on the future if we apply a little of our wealth of pathological understanding to planning, initiation, and management of the plantations we start today.

REFERENCES

BAXTER, D. V. 1952. Pathology in forest practice. John Wiley & Sons, Inc., New York. 601 pp.

BOYCE, J. S. 1938. Forest pathology. McGraw-Hill Book Company, Inc., New York. 600 pp.

CAMPBELL, W. A., and P. SPAULDING. 1942. Stand improvement of northern hardwoods in relation to diseases in the northeast. USDA For. Serv. Allegheny For. Exp. Sta. Occas. Pap. 5. 25 pp.

DINUS, R. J. 1974. Knowledge about natural ecosystems as a guide to disease control in managed forests. Proc. Am. Phytopathol. Soc. *1*:184–190.

HANSBROUGH, J. R. 1965. Biological control of forest tree diseases. J. Wash. Acad. Sci. *55*:41–44.

HEPTING, G. H. 1961. Forest pathology in forest management in the United States. *In* Recent Advances in Botany. University of Toronto Press, Toronto, pp. 1565–1569.

HEPTING, G. H. 1970. How forest disease and insect research is paying off—the case for forest pathology. J. For. *68*:78–81.

HUBERT, E. E. 1931. An outline of forest pathology. John Wiley & Sons, Inc., New York. 543 pp.

MEINECKE, E. P. 1916. Forest pathology in forest regulation. USDA Bull. *275*:1–62.

INDEX